Essential Quantum Optics

From Quantum Measurements to Black Holes

Covering some of the most exciting trends in quantum optics – quantum entanglement, teleportation, and levitation – this textbook is ideal for advanced undergraduate and graduate students. The book journeys through the vast field of quantum optics following a single theme: light in media. A wide range of subjects are covered, from the force of the quantum vacuum to astrophysics, from quantum measurements to black holes.

Ideas are explained in detail and formulated so that students with little prior knowledge of the subject can follow them. Each chapter ends with several short questions followed by a more detailed homework problem, designed to test the reader and show how the ideas discussed can be applied. Solutions to homework problems are available at www.cambridge.org / 9780521145043.

ULF LEONHARDT is Professor of Theoretical Physics at the University of St Andrews. His research interests include quantum electrodynamics in media and state reconstruction in quantum mechanics. He is one of the inventors of invisibility devices and artificial black holes.

Essential Quantum Optics

From Quantum Measurements to Black Holes

Ulf Leonhardt
University of St. Andrews

CAMBRIDGE
UNIVERSITY PRESS

University Printing House, Cambridge CB2 8BS, United Kingdom

One Liberty Plaza, 20th Floor, New York, NY 10006, USA

477 Williamstown Road, Port Melbourne, VIC 3207, Australia

314-321, 3rd Floor, Plot 3, Splendor Forum, Jasola District Centre, New Delhi - 110025, India

79 Anson Road, #06-04/06, Singapore 079906

Cambridge University Press is part of the University of Cambridge.

It furthers the University's mission by disseminating knowledge in the pursuit of
education, learning and research at the highest international levels of excellence.

www.cambridge.org
Information on this title: www.cambridge.org/9780521145053

© U. Leonhardt 2010

First published 2010

A catalogue record for this publication is available from the British Library

ISBN 978-0-521-86978-2 Hardback
ISBN 978-0-521-14505-3 Paperback

Contents

Acknowledgements

Most of this book was written during a wonderful few months in Singapore. I am very grateful to the National University at Singapore for the privilege of having lived and worked there. I would like to thank Marco Bellini, Akira Furusawa, Awatif Hindi, Zdenek Hradil, Natalia Korolkova, Irina Leonhardt, Alexander Lvovsky, Renaud Parentani, Thomas Philbin, Michael Raymer, Tomáš Tyc, Hidehiro Yonezawa and Chun Xiong for their advice and assistance. My work has been supported by a Royal Society Wolfson Research Merit Award, the University of St Andrews and National University of Singapore.

Many people have contributed to awakening my interest in quantum optics and to begin the research expedition from quantum measurements to black holes which I have tried to describe in this book. As a student in Jena I became fascinated by quantum mechanics listening to Dirk-Gunnar Welsch's lectures on quantum theory. I did not mind the mathematics raining down on the audience, but enjoyed his clear thoughts that I tried to formulate in my own way from the sketchy notes I took. At that time I felt I understood quantum mechanics. The quantum theory of light was already the subject of my Diploma thesis in Jena. I learned the craft of quantum optics from Ludwig Knöll who always understood my questions and was most kind and helpful. I learned the art of quantum optics from Harry Paul in Berlin, during my PhD. I am very grateful for the many conversations that led me to appreciate that quantum mechanics happens in reality and not in Hilbert space, and how astonishing and mysterious it is. Since then I know that I don't understand quantum mechanics. This book is an attempt to convey both the clarity of quantum theory and the sense of profoundly not understanding quantum physics. I was deeply impressed by Harry Paul's style of doing physics in conversations, without focusing on technicalities, and I hope that some of this shines through the technicalities of this book. I also learned to appreciate experiments and to discuss them, to the extent that I joined Michael Raymer's experimental group in Eugene as theorist in the laboratory. From him I learned to respect the dedication and patience it takes to turn ideas into reality. Wolfgang Schleich in Ulm gave me the freedom to give the lecture course that made me fall in love with general relativity. I never learned general relativity from attending lectures, I learned it from giving lectures, fittingly in Ulm, Albert Einstein's place of birth. Since then I have been trying to invent optical applications of the ideas of general relativity, from optical black holes to invisibility. Stig Stenholm was most generous in giving me a haven in Stockholm where I benefited from his broad mind and wide research interests that, I hope, have broadened my horizons. I learned the quantum physics of horizons from Renaud Parentani in many conversations where I enjoyed his wit and clarity, and also from making many errors of my own – according to the classic definition of the expert: an expert is someone who made all possible errors (and learned from them). In this book, I hope to have conveyed some of the lessons learned.

Chapter 1
Introduction

1.1 A note to the reader

Quantum optics has grown from a sub-discipline in atomic, molecular, and optical physics to a broad research area that bridges several branches of physics and that captures the imagination of the public. Quantum information science has put quantum optics into the spotlight of modern physics, as has the physics of ultracold quantum gases with its many spectacular connections to condensed-matter physics. Yet through all these exciting developments quantum optics has maintained a characteristic core of ideas that I try to explain in this slim volume.

Quantum optics focuses on the simplest quantum objects, usually light and few-level atoms, where quantum mechanics appears in its purest form without the complications of more complex systems, often demonstrating in the laboratory the thought experiments that the founders of quantum mechanics dreamed of. Quantum optics has been, and will be for the foreseeable future, quintessential quantum mechanics, the quantum mechanics of simple systems, based on a core of simple yet subtle ideas and experiments.

One of the strengths of quantum optics is the close connection between theory and experiment. Although this book necessarily is theoretical, many of the theoretical ideas I describe are guided by experiments or, in turn, have inspired experiments themselves. Another strength of quantum optics is that it is done by individuals or small teams. One single person or a small group can build and perform an entire experiment. One theorist can do all the calculations

1

for an important problem, often with just pencil and paper. As John A. Wheeler said, "It is nice to know that the computer understands the problem, but I want to understand it, too."

This book does not attempt to cover the entire field of quantum optics. The book is focused and selective in the material it applies and expounds, for three good reasons. One of these clearly is that students or other readers do not need to know much in advance to understand this book. You should have experience in working with quantum mechanics at the level of British senior honours students or American junior postgraduate students, and you should know the basics of classical electromagnetism – that is all. Everything else follows and is deduced and explained without any need of further reference. But you should have an inquisitive mind and be able to do mathematics. I use mathematics as a tool, but I have tried to derive the main results with as little technical effort as possible. Most of the eight chapters of this book make a one semester course, the senior honours course on quantum optics that I have taught at St Andrews.

Another reason for being focused is that understanding is more valuable than knowledge. It is better to study one subject in depth than many on the surface. Understanding one subject well gives you confidence to tackle many more and to anticipate their workings by analogy and using your imagination. As Albert Einstein put it, "Imagination is more important than knowledge. For knowledge is limited to all we now know and understand, while imagination embraces the entire world, and all there ever will be to know and understand."

The third reason is the subject this book focuses on: light in media. Media are transparent materials like glass, water or air, but space itself may be regarded a medium for light, in particular the curved space of gravity. Quantum electrodynamics in media has been the backbone of quantum optics for the last 50 years and is likely to remain so for the next. Another vital ingredient of quantum optics is the theory of irreversible quantum processes that has been highly developed in the theory of the laser or in models of the quantum measurement process. A historically important inspiration came from astronomy in the 1950s when quantum fluctuations were used to measure the size of stars, whereupon quantum fluctuations of light became a research subject in its own right. Thinking about quantum fluctuations eventually led to experimental tests of quantum nonlocality – testing the nature of reality itself – and applications in quantum cryptography. These days, when cosmologists no longer are "often in error but seldom in doubt" (Lev D. Landau) quantum optics is beginning to play a serious role in cosmology, as the observed fluctuations of the cosmic microwave background show. Ideas from astrophysics also inspired applications of quantum optics in laboratory analogues of the event horizon. Quantum optics returns to one of its roots. The quantum features of light in media even have technological implications, they appear as vacuum forces in micro- and nano machines. As I will describe, the subject of light in media contains all these themes, from quantum measurements to black holes.[1]

So although this book focuses on the "light side" of quantum optics, you will be amazed to see how many topics such a modest subject contains. Here is an alphabet of samples: Aspect's experiment, Bell's Theorem, Casimir Forces, Dielectric Media, Einstein's Relativity, Fluctuations, Gibbs Ensembles, Hawking Radiation, Irreversibility, Joint Measurements of Position and Momentum, Katzen (in German Cats, here Schrödinger Cats), Lindblad's Theorem, Models of the Measurement Process, Non-Classical Light, Optical Homodyne Tomography, Polarization Correlations, Quantum Communication, Reversible Dynamics, Squeezed Light, Teleportation, Unruh Effect, Vacuum Noise, Wave–Particle Dualism, x and y Coordinates, and Zero Point Energy. I hope the story of this journey through quantum optics is told with sufficient clarity and occasional amusement for the reader. I also hope to have captured not only the sights and insights on the way, but also the sense of excitement and adventure.

1.2 Quantum theory

Let us recall the basic axioms of quantum theory, and let us try to motivate them. This book is of course not the place for a comprehensive development of the theory. We assume that the reader is already familiar with the basic formalism of quantum mechanics. However, because some of the ideas touched on in this book illustrate fundamental issues of quantum physics, we would find it appropriate to turn "back to the roots of quantum mechanics" in a brief and certainly incomplete survey. Moreover, not all readers may have mastered quantum statistics – the formalism of density matrices – and so it is worthwhile to explain this theory here, with apologies to those who know it already. Let us first sketch, in a couple of lines, *one* possible way of motivating the principal ideas of quantum theory.

1.2.1 Axioms

"At the heart of quantum mechanics lies the *superposition principle*" – to quote from the first chapter of Dirac's classic treatise (Dirac, 1984);

---

[1] This book evolved from my monograph *Measuring the Quantum State of Light* (Leonhardt, 1997a). Hopefully it turned into a fully fledged textbook of Quantum Optics.

"... any two or more states may be superposed to give a new state" (Schleich et al., 1991). We denote the state of a perfectly prepared quantum object by $|\psi\rangle$. Then, according to this principle, the complex superposition $c_1|\psi_1\rangle + c_2|\psi_2\rangle$ of two states $|\psi_1\rangle$ and $|\psi_2\rangle$ is a possible state as well. In other words, perfectly prepared states, called *pure states*, are vectors in a complex space. The superposition principle alone does not make physical predictions, it only prepares the ground for quantum mechanics. Nevertheless, the principle is highly nontrivial and can hardly be derived or taken for granted. In the history of quantum mechanics the superposition principle was motivated by the wavelike interference of material particles. Note, however, that this simple principle experienced a dramatic generalization such that we cannot consider its historical origin as a physical motivation any more.

Let us now turn to more physical assumptions. When we observe a physical quantity of an ensemble of equally prepared states, we obtain certain measurement values a (real numbers) with probabilities p_a. Given a result a, we assume that we would obtain the same result if we repeated the experiment immediately after the first measurement (provided, of course, that the physical object has not been destroyed). This assumption is certainly plausible. As a consequence, the object must have jumped into a state $|a\rangle$, called an *eigenstate*, which gives the measurement result with certainty, an event called the *collapse of the state vector*. Or, if we prefer to assign states only to ensembles of objects, a measurement produces a statistical ensemble of states $|a\rangle$ with probability p_a. According to the superposition principle we can expand the state vector $|\psi\rangle$ before the measurement in terms of the eigenstates $|a\rangle$, written as $|\psi\rangle = \sum_a \langle a|\psi\rangle |a\rangle$, with some complex numbers denoted by the symbol $\langle a|\psi\rangle$. What is the probability for the transition from $|\psi\rangle$ to a particular $|a\rangle$? Clearly, the larger the $\langle a|\psi\rangle$ component is (compared to all other components) the larger should be p_a. However, this component is a complex number in general. So the simplest possible expression for the transition probability is the ratio

$$p_a = \frac{|\langle a|\psi\rangle|^2}{\langle\psi|\psi\rangle}. \tag{1.1}$$

Here $\langle\psi|\psi\rangle$ abbreviates simply the sum of all $|\langle a|\psi\rangle|^2$ values. It is a special case of the more general symbol

$$\langle\psi'|\psi\rangle = \sum_a \langle\psi'|a\rangle\langle a|\psi\rangle \tag{1.2}$$

with the convention

$$\langle\psi|a\rangle = \langle a|\psi\rangle^*. \tag{1.3}$$

The mathematical construction (1.2) of the symbol $\langle\psi'|\psi\rangle$ fulfils all requirements of a scalar product in a vector space. However, at this stage the scalar product depends critically on a particular set of eigenstates $|a\rangle$ or, in other words, on a particular experiment. Let us assume that all possible sets of physical eigenstates form the same scalar product so that no experimental setting is favoured or discriminated against in principle. This assumption seems to be natural yet it is nontrivial. If we accept this, then the symbol $\langle\psi'|\psi\rangle$ describes *the* scalar product in the linear state space. We can employ Dirac's convenient bracket formalism, and in particular we can understand the $\langle a|\psi\rangle$ components as orthogonal projections of the $|\psi\rangle$ vector onto the eigenstates $|a\rangle$.

Formula (1.1) is the key axiom of quantum mechanics. It makes a quantitative prediction about an event in physical reality (the occurrence of the measurement result a), and it contains implicitly the superposition principle for describing quantum states. The historical origin of this fundamental principle is Born's probability interpretation of the modulus square of the Schrödinger wave function.

Now we are in a position to reproduce the basic formalism of quantum mechanics. Because the probability p_a does not depend on the normalization of the state vector $|\psi\rangle$, we may simplify formula (1.1) by considering only normalized states, that is we set

$$\langle\psi|\psi\rangle = 1. \tag{1.4}$$

Because the eigenstates produce the measurement result a with certainty, they must be *orthonormal*

$$\langle a|a'\rangle = \delta_{a'a}. \tag{1.5}$$

Furthermore, the system of eigenvectors must be *complete*

$$\sum_a |a\rangle\langle a| = 1 \tag{1.6}$$

if we assume that any observation gives at least one of the values a so that $\sum_a p_a = \langle\psi|\sum_a |a\rangle\langle a||\psi\rangle$ equals unity for all states $|\psi\rangle$. The average $\langle A\rangle$ of the measurement values a is given by

$$\langle A\rangle = \sum_a a\, p_a = \langle\psi|\hat{A}|\psi\rangle \tag{1.7}$$

where we have introduced the *operator*

$$\hat{A} \equiv \sum_a a\, |a\rangle\langle a| \qquad \cdot \tag{1.8}$$

with *eigenvalues* a and *eigenvectors* $|a\rangle$. (The structure (1.8) explains the term *eigenvectors* for the measurement produced states $|a\rangle$.) The operator $\hat{A}$ is *Hermitian*,

$$\hat{A}^\dagger = \hat{A}, \tag{1.9}$$

because the measurement results a are real. Observable quantities thus correspond to Hermitian operators; their eigenstates are the states assumed after measurement and the eigenvalues are the measurement results.

Suppose that after the measurement of the observable quantity $\hat{A}$ the object is in the eigenstate $|a\rangle$. What happens when we subsequently perform another measurement of the observable $\hat{B}$? Clearly, the quantum object will now assume one of the eigenstates of $\hat{B}$. But the two sets of eigenstates, $\{|a\rangle\}$ and $\{|b\rangle\}$, may differ. As the overlap $|\langle a|b\rangle|^2$ gives the probability of observing b after a, the measurement result b is statistically uncertain if the overlap is not perfect, $|\langle a|b\rangle|^2 < 1$. In this case, the two operators $\hat{A}$ and $\hat{B}$ do not commute,

$$\hat{A}\hat{B} \neq \hat{B}\hat{A}, \tag{1.10}$$

because otherwise they would share the same system of eigenstates. The degree of commutation is characterized by the *commutator*

$$[\hat{A}, \hat{B}] \equiv \hat{A}\hat{B} - \hat{B}\hat{A}, \tag{1.11}$$

an anti-Hermitian operator $\hat{C}$ for Hermitian $\hat{A}$ and $\hat{B}$, with $\hat{C}^\dagger = -\hat{C}$. Incompatible observables correspond to non-commuting operators. They cause mutual statistical uncertainty of measurement results, an uncertainty that can be quantified in *uncertainty relations*, for example in relation (5.61) that we use later on.

We must mention another fundamental axiom of quantum mechanics concerning the composition of physical objects. If one system consists of, say, two subsystems, then the theory should allow us to experiment on each of the subsystems independently. We would obtain two real measurement values (a_1, a_2) and if we had repeated the same experiment immediately after the first measurement we would read the same values (a_1, a_2). Furthermore, we would also obtain a_1 if we had performed the repeated measurement only on the first subsystem, irrespective of what happens on the other (irrespective of which measurement is performed there) and, of course, vice versa. So it is natural to assume that independent measurements correspond to factorized eigenstates

$$|a_1, a_2\rangle = |a_1\rangle \otimes |a_2\rangle. \tag{1.12}$$

As usual, the symbol $\otimes$ denotes the tensor product. Note, however, that the innocent looking statement (1.12) is capable of peculiar physical effects when it is combined with the superposition principle. The state space of the total system is the tensor product of the subspaces. However, the superposition of two different states $|a_1\rangle \otimes |a_2\rangle$ and $|a_1'\rangle \otimes |a_2'\rangle$ will not factorize in general, producing an *entangled state*. The total system is not a mere composition of its parts, because the subsystems are correlated. This correlation may bridge space and time showing the potential nonlocality of quantum mechanics, as expressed for instance in the Einstein–Podolsky–Rosen paradox (Einstein et al., 1935) and in Bell's inequalities (Bell, 1964, 1987) that we discuss in Chapter 7.

As we have seen, we can reproduce the basic mathematical machinery of quantum mechanics starting from some ideas about states and measurements. These ideas have been distilled and formulated quantitatively in axiom (1.1). However, the basic formalism is far too general to be sufficient for solving specific physical problems. Here we rely on physically motivated guesswork to find the significant physical quantities and their relations within the general framework of quantum mechanics. In particular, we need this physical information to understand the classic quantum effects such as the quantization of the energy (or of other observables).

1.2.2 Quantum statistics

So far we have considered only pure states, presupposing a perfectly controlled preparation of physical objects. It is, however, not difficult to relax this assumption and to extend the concept of quantum states to statistical ensembles of physical states. For this extension we assume that we have at least statistical information about the prepared states: we have an *ensemble* of pure states $|\psi_n\rangle$ with probabilities ρ_n. The prediction $\langle A \rangle$ of any physical quantity must be the average of the expectation values $\langle \psi_n | \hat{A} | \psi_n \rangle$ for the individual states $|\psi_n\rangle$ with respect to the preparation probabilities ρ_n, or

$$\langle A \rangle = \sum_n \rho_n \langle \psi_n | \hat{A} | \psi_n \rangle$$

$$= \sum_a \sum_n \rho_n \langle \psi_n | \hat{A} | a \rangle \langle a | \psi_n \rangle$$

$$= \sum_a \langle a | \sum_n \rho_n | \psi_n \rangle \langle \psi_n | \hat{A} | a \rangle . \tag{1.13}$$

We write the last line in terms of the *trace*

$$\langle A \rangle = \mathrm{tr}\{\hat{\rho} \hat{A}\} \tag{1.14}$$

introducing the *density operator* (Landau and Lifshitz, Vol. III, 1981) sometimes also called the *state operator* (Ballentine, 1990) or, with some laxity, the *density matrix*,

$$\hat{\rho} = \sum_n \rho_n | \psi_n \rangle \langle \psi_n | . \tag{1.15}$$

The trace does not depend on the system of states $\{|a\rangle\}$, as long as this system is complete in the sense of Eq. (1.6), because

$$\sum_a \langle a|\hat{F}|a\rangle = \sum_a \sum_b \langle a|b\rangle \langle b|\hat{F}|a\rangle = \sum_b \sum_a \langle b|\hat{F}|a\rangle \langle a|b\rangle$$

$$= \sum_b \langle b|\hat{F}|b\rangle \,. \tag{1.16}$$

Hence we can use the universal symbol "tr" for the trace. We interpret the density operator (1.15) as the most general description of a quantum state and the formula (1.14) as the general rule of predicting observable quantities. Pure states are of course included in this general concept because their density operators are projectors $|\psi\rangle\langle\psi|$. States that are not pure are called *mixed states*.

We may also use the concept of density operators to generalize our fundamental axiom (1.1) about quantum measurements to broader circumstances. According to formula (1.1) the occurrence of a measurement result a is associated with a "jump" of the pure state $|\psi\rangle$ to the pure eigenstate $|a\rangle$. The probability for this process is given by the scalar product $|\langle a|\psi\rangle|^2$ (assuming the normalization of $|\psi\rangle$). Suppose that we are not completely certain about the state $|\psi_n\rangle$ and about the particular measurement result a but that we can still describe statistically the state as well as the observation. In this case the probability $p(A)$ of the measurement A is given by

$$p(A) = \sum_a \rho_a \sum_n \rho_n |\langle a|\psi_n\rangle|^2 = \mathrm{tr}\{\hat{\rho}_a\, \hat{\rho}\} \tag{1.17}$$

where we have introduced for the eigenstates $|a\rangle$ occurring with probabilities ρ_a the density operator

$$\hat{\rho}_a = \sum_a \rho_a |a\rangle\langle a| \,. \tag{1.18}$$

What can we say about density operators in general? First of all, $\hat{\rho}$ is Hermitian and normalized

$$\mathrm{tr}\{\hat{\rho}\} = \sum_n \rho_n\, \mathrm{tr}\{|\psi_n\rangle\langle\psi_n|\} = \sum_n \rho_n \langle\psi_n|\psi_n\rangle = 1 \tag{1.19}$$

because the individual states and the probability distribution ρ_n are normalized. The density operator is strictly *non-negative*, that is it has only non-negative eigenvalues, because for all $|\psi\rangle$

$$\langle\psi|\hat{\rho}|\psi\rangle = \sum_n \rho_n |\langle\psi|\psi_n\rangle|^2 \geq 0 \,. \tag{1.20}$$

Note that this obvious constraint may be very difficult to handle. Given a mathematically constructed operator $\hat{\rho}$, we cannot decide easily in general whether $\hat{\rho}$ meets the physical criterion (1.20). Representing $\hat{\rho}$ in the eigenbasis, the eigenvalues of $\hat{\rho}$ can be interpreted as probabilities (because they must be normalized and non-negative) for the eigenstates. Consequently, any normalized Hermitian operator can be accepted as describing a quantum state as long as the operator is non-negative. Note that the unravelling of a mixed density operator in terms (1.15) of an ensemble of individual pure states is not unique if these states are not orthogonal to each other. For mixed states there is no unique way of telling whether statistical fluctuations of observed quantities are caused by fluctuations in the state preparation (by our subjective lack of knowledge) or by fluctuations caused by the measurement process (by our fundamental lack of complete control).

How can we discriminate pure from mixed states or, more generally, characterize the purity of a state? One option is the *von Neumann entropy* that describes the thermodynamic entropy (Landau and Lifshitz, Vol. V, 1996)

$$S \equiv -k_B \text{tr}\{\hat{\rho} \ln \hat{\rho}\} \tag{1.21}$$

where k_B denotes the Boltzmann constant. The entropy vanishes for pure states only, exceeds zero for mixed states, and, most importantly, is an extensive quantity for non-entangled subsystems $\hat{\rho}_1 \otimes \hat{\rho}_2$ because in this case

$$S = S_1 + S_2. \tag{1.22}$$

The von Neumann entropy is regarded as the fundamental measure of preparation impurity for quantum states. However, the entropy might be difficult to calculate. Another computationally more convenient option is the *purity* $\text{tr}\{\hat{\rho}^2\}$ or the *purity parameter*

$$P = 1 - \text{tr}\{\hat{\rho}^2\}. \tag{1.23}$$

Using the eigenbasis of the density operator, we see that

$$\text{tr}\{\hat{\rho}^2\} = \sum_n \rho_n^2 \leq \sum_n \rho_n = 1. \tag{1.24}$$

The equality sign holds only for pure states, and the purity $\text{tr}\{\hat{\rho}^2\}$ thus discriminates uniquely between mixed and pure states. Because $1 - \rho_n$ is less than or equal to $-\ln \rho_n$ for $0 < \rho_n \leq 1$, the purity parameter gives a lower bound for the von Neumann entropy,

$$k_B P \leq S. \tag{1.25}$$

What happens if we have a composite system and observe only quantities A_1 referring to one subsystem? In this case we can simplify the general rule (1.14) for predicting A_1, introducing the *reduced density operator*

$$\hat{\rho}_1 = \text{tr}_2\{\hat{\rho}\}, \qquad (1.26)$$

where the trace tr_2 should be calculated with respect to the unobserved degrees of freedom. The expectation value $\langle A_1 \rangle$ is given by

$$\langle A_1 \rangle = \text{tr}_1\{\hat{\rho}_1 \hat{A}_1\}, \qquad (1.27)$$

"tracing" here only in the observed subsystem. The so-constructed reduced operator $\hat{\rho}_1$ obeys all requirements for a physically meaningful density operator – it is normalized, Hermitian, and non-negative because the total density operator meets these criteria. Consequently, we can regard $\hat{\rho}_1$ as describing the quantum state of the reduced system. The parts of a composite system are genuine quantum objects, being in mixed states in general. Note that even if the total system is in a pure state, the reduced system might be statistically mixed. This intriguing feature relies on the entanglement of the subsystems (and hence it can be used as a measure for entanglement, see Section 7.1.3). We cannot observe all aspects of an entangled system by considering the subsystems only. Our lack of knowledge about the partner object causes statistical uncertainty in the state of the subsystem, explaining why the reduced system may be in a mixed state.

1.2.3 Schrödinger and Heisenberg pictures

How to describe the *coherent evolution* of physical quantities in quantum mechanics? By coherent evolution we mean a process that is influenced neither by dissipation nor by quantum measurement; we assume that the physical object is sufficiently well isolated from the environment. Such a process is reversible. In Chapter 6 we discuss processes where physical objects interact with their environment or are monitored; these are irreversible processes. Predictable physical quantities are averages (1.14), they combine the quantum state with the operators of observables. So we can make two principal choices: in one, the *Schrödinger picture*, we assume that the state evolves and the observables are fixed, in the other, the *Heisenberg picture*, the operators change and the state remains frozen. Alternatively, we can consider a compromise between the two pictures, called an *interaction picture* where both the state and the operators evolve.

Let us begin with the Schrödinger picture. Suppose that the quantum state evolves in time t or in any other parameter that characterizes the process. We consider a pure state $|\psi\rangle$; the generalization to mixed states is done in Section 6.1.2 prior to the derivation of the fundamental equation of irreversible processes. In order for pure states to obey the superposition principle, their evolution equation must be a linear differential equation. As the state vector $|\psi\rangle$ should describe the complete physical information about the state, including its fate in coherent evolution processes, the equation must be of first order in the parameter t. Otherwise derivatives of $|\psi\rangle$ with respect to t would be needed to completely define the initial conditions. So we are led to the differential equation

$$i\hbar \frac{d|\psi\rangle}{dt} = \hat{H} |\psi\rangle \tag{1.28}$$

where $\hat{H}$ is a linear operator. We will argue that $\hat{H}$ is related to the energy when t is time. We included a natural constant, $\hbar$, in order to relate the energy scale in quantum mechanics to the macroscopic scale. The constant $\hbar$ turns out to be the Planck constant h divided by 2π. The evolution equation (1.28) should preserve the norm of the quantum state $\langle\psi|\psi\rangle$. As the scalar product $\langle\psi_1|\psi_2\rangle$ between two state vectors $|\psi_1\rangle$ and $|\psi_2\rangle$ can be expressed in terms of the norms of $|\psi_1\rangle$, $|\psi_2\rangle$, $|\psi_1\rangle + |\psi_2\rangle$ and $|\psi_1\rangle + i|\psi_2\rangle$, the scalar product should be preserved, too. We differentiate the scalar product

$$\frac{d\langle\psi_1|\psi_2\rangle}{dt} = \frac{i}{\hbar} \left\langle \psi_1 \left| \left(\hat{H}^\dagger - \hat{H} \right) \right| \psi_2 \right\rangle \tag{1.29}$$

and see that its conservation implies

$$\hat{H}^\dagger = \hat{H}. \tag{1.30}$$

The operator $\hat{H}$ is Hermitian; it may correspond to a physical quantity. Equation (1.28) is known as the *Schrödinger equation*.

The expectation values of physical quantities are the averages $\langle A \rangle = \langle\psi|\hat{A}|\psi\rangle$ for pure states. How do they evolve? We find by differentiation and using the Schrödinger equation (1.28) and the Hermiticity (1.30)

$$\frac{d\langle A \rangle}{dt} = \frac{d\langle\psi|\hat{A}|\psi\rangle}{dt} = \langle\psi|\left(\frac{i}{\hbar} \left[\hat{H}, \hat{A} \right] + \frac{\partial \hat{A}}{\partial t} \right)|\psi\rangle. \tag{1.31}$$

Consequently, we may adopt an alternative point of view, the Heisenberg picture, where the state does not change, but the operators evolve according to the *Heisenberg equation*

$$\frac{d\hat{A}}{dt} = \frac{i}{\hbar} \left[\hat{H}, \hat{A} \right] + \frac{\partial \hat{A}}{\partial t}. \tag{1.32}$$

If the operator $\hat{H}$ does not explicitly depend on time, $\hat{H}$ is a conserved quantity, because it surely commutes with itself. Conservation laws are connected to symmetries, to time invariance in the case when the equation of motion does not change in time. By definition (Landau and

Lifshitz, Vol. I, 1982), the conserved quantity related to time invariance is the *energy*. The operator $\hat{H}$ is also called the *Hamiltonian*. In this way we used the Heisenberg picture to find a useful interpretation of the central physical quantity in the Schrödinger equation.

The two pictures are connected by the *evolution operator* $\hat{U}$ that is defined by the requirements

$$i\hbar \frac{d\hat{U}}{dt} = \hat{H}\,\hat{U}\,, \quad \hat{U}(t_0) = \mathbb{1}\,. \tag{1.33}$$

In the Schrödinger picture, the state $|\psi\rangle$ evolves as

$$|\psi(t)\rangle = \hat{U}\,|\psi(t_0)\rangle\,, \tag{1.34}$$

because this expression solves the Schrödinger equation (1.28) with the initial condition $|\psi(t_0)\rangle$. The density operator (1.15) develops as

$$\hat{\rho}(t) = \hat{U}\,\hat{\rho}(t_0)\,\hat{U}^{\dagger}\,, \tag{1.35}$$

assuming that the probabilities ρ_n for the states $|\psi_n\rangle$ do not change. In the Heisenberg picture, operators evolve as

$$\hat{A} = \hat{U}^{\dagger}\hat{A}\,\hat{U}\,, \tag{1.36}$$

because the so-evolved operator solves the Heisenberg equation (1.32). The evolution laws (1.35) and (1.36) guarantee that any expectation value $\langle A \rangle$ remains the same in either of the two pictures. We will frequently and freely exchange Heisenberg and Schrödinger pictures and take whatever is most convenient. The Heisenberg picture is particularly well suited in quantum field theory.

1.3 On the questions and homework problems

Actively reading a book by closely following the arguments and checking the calculations is absolutely essential; the same applies to following a course. Euclid is said to have replied to King Ptolemy's request for an easier way of learning mathematics that "there is no royal road to geometry"; there is no royal road to physics either. As a study guide, each chapter concludes with a set of short questions and a homework problem. The short questions probe the understanding of the material set out in the specific chapter; the homework problems combine topics of several chapters or show new directions beyond the material of this book, illustrating how its ideas can be applied elsewhere.

1.4 Further reading

Quantum optics is the subject of many textbooks, focusing on different aspects of this research area and developing a wide range of methods. Here is a list of the books currently available: Bachor and Ralph (2004), Barnett and Radmore (2002), Carmichael (1987, 2003, 2007), Cohen Tannoudji, Dupont Roc and Grynberg (1989, 1992), Fox (2006), Gardiner (1991), Gardiner and Zoller (2004), Garrison and Chiao (2008), Gerry and Knight (2004), Haken (1981), Hanamura, Kawabe and Yamanaka (2007), Kenyon (2008), Klauder and Sudarshan (2006), Lambropoulos and Petrosyan (2006), Loudon (2000), Louisell (1973), Mandel and Wolf (1995), Meystre and Sargent (2007), Nussenzveig (1974), Orzag (2007), Paul (1995), Paul and Jex (2004), Peng and Li (1998), Peřina (1991), Peřina, Hradil and Jurco (1994), Schleich (2001), Scully and Zubairy (1997), Shih (2009), Vedral (2005), Vogel and Welsch (2006), Walls and Milburn (2008), Yamamoto and Imamoglu (1999).

As complementary material to this book, I particularly recommend the encyclopedic monograph by Mandel and Wolf (1995) and Paul and Jex's textbook (2004) that explains in words, not formulae, the key experiments of quantum optics and their often-intriguing interpretations.

Chapter 2
Quantum field theory of light

2.1 Light in media

The backbone of theoretical quantum optics is the quantum field theory of light, the quantum theory of the electromagnetic field. Quantum electrodynamics becomes surprisingly complicated if one insists on formulating it in a relativistically and gauge-invariant form, as one surely should do in elementary particle physics. Moreover, in quantum optics, we are concerned with light quanta in materials such as glass, with quantum electromagnetism in media, which is even more complicated. To cut a long story short, here we follow a minimalistic approach where we develop the essentials of quantum electrodynamics with as little technical effort as possible, sailing around the cliffs of relativistic quantum field theories. We do not assume any prior knowledge of classical field theory; we only borrow a few ideas from classical electromagnetism (Jackson, 1998). Complications and problems still remain, but hopefully these are mostly no longer formal difficulties, but rather conceptual problems of the quantum nature of light. As Dr Samuel Johnson said on poetry "We all know what light is; but it is not easy to tell what it is."

2.1.1 Maxwell's equations

What is light? Light is a *quantum object* – we describe its properties by quantum observables, Hermitian operators, and its state by a state vector $|\psi\rangle$ or density matrix $\hat{\rho}$. We use the Heisenberg picture of quantum mechanics where the operators evolve in time, but the quantum

14

state does not change. Light is a *field* – the observables extend over space and time, they are functions of space and time coordinates. Light is an *electromagnetic wave* – the central physical quantities of light are connected to the electromagnetic field strengths, the electric field **E**, the electric displacement **D**, the magnetic field **H** and the magnetic induction **B**. Throughout this book we denote vectors in bold. We know that classical electromagnetic fields obey *Maxwell's equations*

$$\nabla \cdot \mathbf{B} = 0, \quad \nabla \times \mathbf{E} = -\frac{\partial \mathbf{B}}{\partial t}, \quad \nabla \cdot \mathbf{D} = 0, \quad \nabla \times \mathbf{H} = \frac{\partial \mathbf{D}}{\partial t}, \qquad (2.1)$$

supplemented by the *boundary condition* that the fields vanish at infinity. What can we say about quantum electromagnetic fields? Suppose we regard the classical fields **E**, **D**, **B** and **H** as the expectation values of the corresponding quantum observables $\hat{\mathbf{E}}$, $\hat{\mathbf{D}}$, $\hat{\mathbf{B}}$ and $\hat{\mathbf{H}}$, for example **E** as $\langle \psi | \hat{\mathbf{E}} | \psi \rangle$. In other words, the classical fields are the ensemble averages of the quantum fields. Remarkably, from this simple and plausible assumption follows that the quantum field strengths must obey Maxwell's equations (2.1) as well. The reason is the linearity of Maxwell's equations: for example, Faraday's classical law of induction, $\nabla \times \langle \psi | \hat{\mathbf{E}} | \psi \rangle = -\partial \langle \psi | \hat{\mathbf{B}} | \psi \rangle / \partial t$, implies that $\langle \psi | \nabla \times \hat{\mathbf{E}} + \partial \hat{\mathbf{B}} / \partial t | \psi \rangle = 0$. Now, if the expectation value $\langle \psi | \hat{F} | \psi \rangle$ of a Hermitian operator $\hat{F}$ vanishes for all $| \psi \rangle$, it must also vanish for its eigenstates, which implies that all eigenvalues are zero. Therefore, the operator itself vanishes, in our case $\nabla \times \hat{\mathbf{E}} + \partial \hat{\mathbf{B}} / \partial t = 0$. The same logic applies to the other equations (2.1). Consequently, Maxwell's equations also govern electromagnetism in the quantum world.

Optical instruments like mirrors, lenses or beam splitters are often made of *dielectric media* such as glass. In order to understand what happens to photons at beam splitters and in interferometers, we formulate a quantum field theory of light in media. The medium responds to the light by becoming electrically polarized or magnetized. We consider a regime of linear response where the polarizations and magnetizations are proportional to the field strengths. In this case, the fields **E**, **D**, **B** and **H** are connected by linear equations, known as the *constitutive equations* (Jackson, 1998). Since these equations are linear, the quantum fields $\hat{\mathbf{E}}$, $\hat{\mathbf{D}}$, $\hat{\mathbf{B}}$ and $\hat{\mathbf{H}}$ must obey them, too, if the constitutive equations are real (Hermitian in the quantum domain). This is the case for non-absorptive media or, to be more precise, for the region of the electromagnetic spectrum where we can ignore the absorption of the materials we are concerned about. We focus on the simplest case, non-absorptive, non-dispersive and isotropic media that are described by the constitutive equations in SI units (Jackson, 1998)

$$\hat{\mathbf{D}} = \varepsilon_0 \varepsilon \hat{\mathbf{E}}, \quad \hat{\mathbf{B}} = \mu_0 \mu \hat{\mathbf{H}}, \quad \varepsilon_0 \mu_0 = c^{-2}, \qquad (2.2)$$

where ε_0 denotes the permittivity and μ_0 the permeability of the vacuum and c the speed of light in vacuum. In non-absorptive and non-dispersive media the electric permittivities ε and the magnetic permeabilities μ are real and do not depend on time or on frequency. Empty space is a medium as well, where, in the absence of gravity, ε and μ are unity. In inhomogeneous media, such as glass surrounded by air, ε and μ vary in space.

In classical electrodynamics (Jackson, 1998) the fields are often expressed in terms of potentials. For quantum electromagnetism, we use the operator of the *vector potential* $\hat{\mathbf{A}}$. We require that

$$\hat{\mathbf{E}} = -\frac{\partial \hat{\mathbf{A}}}{\partial t}, \quad \hat{\mathbf{B}} = \nabla \times \hat{\mathbf{A}}. \tag{2.3}$$

In this way the first two of Maxwell's equations (2.1) are automatically satisfied. The electromagnetic field is *gauge invariant* (Jackson, 1998); we may add the gradient of any scalar field to $\hat{\mathbf{A}}$ without affecting Maxwell's equations; it is wise, however, to require

$$\nabla \cdot \varepsilon \hat{\mathbf{A}} = 0, \tag{2.4}$$

a condition known as the *Coulomb gauge*. One immediate advantage of the Coulomb gauge is the fact that the third of Maxwell's equations (2.1) is also automatically satisfied: $\nabla \cdot \mathbf{D}$ vanishes, as a consequence of the constitutive equations (2.2) and the representation (2.3). The only remaining non-trivial Maxwell equation is

$$\nabla \times \hat{\mathbf{H}} = \frac{\partial \hat{\mathbf{D}}}{\partial t}, \quad \text{or} \quad \frac{1}{\varepsilon} \nabla \times \frac{1}{\mu} \nabla \times \hat{\mathbf{A}} + \frac{1}{c^2} \frac{\partial^2 \hat{\mathbf{A}}}{\partial t^2} = 0. \tag{2.5}$$

This equation, the *electromagnetic wave equation*, and the Coulomb gauge (2.4) govern both the classical and the quantum vector potential.

2.1.2 Quantum commutator

So far, we have not used anything specifically quantum theoretical for the quantum field theory of light; the laws of electromagnetism are valid both for classical and for quantum fields. The only quantum postulate we shall make is an assumption about the Hamiltonian $\hat{H}$ of light. We use italic for the Hamiltonian $\hat{H}$ in order to distinguish it from the magnetic field $\hat{\mathbf{H}}$ denoted in bold. In the Heisenberg picture, the *Hamiltonian* $\hat{H}$ governs the time evolution of physical quantities $\hat{F}$, described in the Heisenberg equation of motion (1.32). The field operators should not depend explicitly on time, only through their dynamics.

As we consider fields that propagate in space and time, we use the partial time derivative for the time evolution,

$$\frac{\partial \hat{F}}{\partial t} = \frac{i}{\hbar}[\hat{H}, \hat{F}]. \tag{2.6}$$

The Hamiltonian plays a double role; it also describes the total *energy* of the system. We assume that the Hamiltonian of light has the same structure as the classical energy of the electromagnetic field (Jackson, 1998),

$$\hat{H} = \frac{1}{2} \int \left(\hat{\mathbf{E}} \cdot \hat{\mathbf{D}} + \hat{\mathbf{B}} \cdot \hat{\mathbf{H}} \right) dV \tag{2.7}$$

where the volume integration is taken over the entire space. It is easy to see that the Hamiltonian is a conserved quantity as a consequence of Maxwell's equations (2.1) and the constitutive equations (2.2); energy is conserved.

We wish to obtain the laws of electromagnetism as Heisenberg equations of motion with the Hamiltonian (2.7). For this, we express the Hamiltonian in terms of the vector potential and the electric displacement,

$$\hat{H} = \int \left(\frac{\hat{\mathbf{D}}^2}{2\varepsilon_0 \varepsilon} + \frac{\varepsilon_0 c^2}{2\mu} (\nabla \times \hat{\mathbf{A}})^2 \right) dV. \tag{2.8}$$

The crucial quantities in the Heisenberg equations (2.6) for the fields are commutators; and so we need to find out what the commutators between $\hat{\mathbf{A}}$ and $\hat{\mathbf{D}}$ are. The field amplitudes at various points in space, but equal times, cannot interact with each other, they are causally disconnected. In quantum mechanics, non-interacting, independent systems are simultaneously measurable in principle, which implies that their observables commute; and so the commutators between $\hat{\mathbf{A}}(\mathbf{r}, t)$ and $\hat{\mathbf{A}}(\mathbf{r}', t)$ and between $\hat{\mathbf{D}}(\mathbf{r}, t)$ and $\hat{\mathbf{D}}(\mathbf{r}', t)$ must vanish at equal times. On the other hand, $\hat{\mathbf{A}}$ and $\hat{\mathbf{D}}$ are not independent,

$$\hat{\mathbf{D}} = -\varepsilon_0 \varepsilon \frac{\partial \hat{\mathbf{A}}}{\partial t} = \frac{\varepsilon_0 \varepsilon}{i\hbar}[\hat{H}, \hat{\mathbf{A}}]. \tag{2.9}$$

We determine the commutator between $\hat{\mathbf{A}}$ and $\hat{\mathbf{D}}$ from this relationship. Note that the commutator of the vectors $\hat{\mathbf{A}}$ and $\hat{\mathbf{D}}$ is a three-dimensional matrix, the matrix of the commutators between the vector components $\hat{A}_l$ and $\hat{D}_m$. We find from the relationship (2.9) that this matrix is symmetric,

$$[\hat{A}_l, \hat{D}_m] = \frac{\varepsilon_0 \varepsilon}{i\hbar}[\hat{A}_l, [\hat{H}, \hat{A}_m]] = \frac{\varepsilon_0 \varepsilon}{i\hbar}[\hat{A}_m, [\hat{H}, \hat{A}_l]] = [\hat{A}_m, \hat{D}_l]. \tag{2.10}$$

We consider the commutator between the Hamiltonian (2.8) and one of the vector components of $\hat{\mathbf{A}}$. For equal times t, the quantum vector potentials at all positions commute. Therefore, $\hat{\mathbf{A}}$ also commutes with all of its spatial derivatives; and so the commutator between the Hamiltonian (2.8) and $\hat{A}_m$ is reduced to

$$
\begin{aligned}
\left[\hat{H}, \hat{A}_m(\mathbf{r}')\right] &= \int \left[\hat{\mathbf{D}}(\mathbf{r}) \cdot \hat{\mathbf{D}}(\mathbf{r}), \hat{A}_m(\mathbf{r}')\right] \frac{dV}{2\varepsilon_0 \varepsilon} \\
&= \int \left(\hat{\mathbf{D}}(\mathbf{r}) \cdot \left[\hat{\mathbf{D}}(\mathbf{r}), \hat{A}_m(\mathbf{r}')\right] + \left[\hat{\mathbf{D}}(\mathbf{r}), \hat{A}_m(\mathbf{r}')\right] \cdot \hat{\mathbf{D}}(\mathbf{r})\right) \frac{dV}{2\varepsilon_0 \varepsilon} \\
&= \int \left[\hat{\mathbf{D}}(\mathbf{r}), \hat{A}_m(\mathbf{r}')\right] \cdot \hat{\mathbf{D}}(\mathbf{r}) \frac{dV}{2\varepsilon_0 \varepsilon} - \text{H.c.}
\end{aligned} \tag{2.11}
$$

Here and throughout the rest of this book H.c. denotes the Hermitian conjugate of the preceding term: $\hat{F}+\text{H.c.}$ describes $\hat{F}+\hat{F}^\dagger$, and $\hat{F}-\text{H.c.}$ refers to $\hat{F} - \hat{F}^\dagger$. We write in vector notation

$$
\left[\hat{H}, \hat{\mathbf{A}}(\mathbf{r}')\right] = \int \left[\hat{\mathbf{D}}(\mathbf{r}), \hat{\mathbf{A}}(\mathbf{r}')\right] \cdot \hat{\mathbf{D}}(\mathbf{r}) \frac{dV}{2\varepsilon_0 \varepsilon} - \text{H.c.} \tag{2.12}
$$

In a similar way we find

$$
\left[\hat{\mathbf{D}}(\mathbf{r}), \hat{H}\right] = \int \left[\hat{\mathbf{D}}(\mathbf{r}), \hat{\mathbf{A}}(\mathbf{r}')\right] \cdot \nabla' \times \hat{\mathbf{H}}(\mathbf{r}') \frac{dV'}{2} - \text{H.c.} \tag{2.13}
$$

Both the Maxwell equation (2.5) and the relationship (2.9) should follow as Heisenberg equations of motion (2.6) with the commutators (2.12) and (2.13), which suggests that the commutator between $\hat{\mathbf{D}}$ and $\hat{\mathbf{A}}$ is given by a matrix delta function,

$$
\left[\hat{\mathbf{D}}(\mathbf{r}, t), \hat{\mathbf{A}}(\mathbf{r}', t)\right] = i\hbar\, \delta^{\mathrm{T}}(\mathbf{r} - \mathbf{r}'). \tag{2.14}
$$

In this way we found the *fundamental commutation relation* between the quantum vector potential in Coulomb gauge and the operator of the dielectric displacement, a relationship that turns out to serve as the key to the quantum aspects of light in media.

Note there is a subtlety indicated by the superscript T in the delta function (2.14). We know that the divergence of $\hat{\mathbf{D}}$ vanishes, hence $\nabla \cdot \delta^{\mathrm{T}}$ must vanish, too; and so δ^{T} cannot be a pure matrix delta function. On the other hand, it is completely sufficient that δ^{T} acts as a delta function wherever we need it. We applied the commutator (2.14) on fields with zero divergence, such as $\hat{\mathbf{D}}$ in Eq. (2.12), or on curls such as $\nabla \times \hat{\mathbf{H}}$ in Eq. (2.13). Both cases are equivalent, because the divergence of a curl vanishes and a divergence-less vector field can always be written as the curl of another vector field (Jackson, 1998). Therefore, it is completely sufficient that δ^{T} acts as a delta function on curls that are also called *transversal fields*. In the following we show that the matrix components of this *transversal delta function* are given by the Fourier representation

$$\delta_{lm}^{T}(\mathbf{r}) = \frac{1}{(2\pi)^3} \int \left(\delta_{lm} - \frac{k_l k_m}{k^2} \right) e^{i\mathbf{k}\cdot\mathbf{r}} \, d^3k. \qquad (2.15)$$

Suppose that δ^{T} acts as a delta function on arbitrary transversal fields $\nabla \times \mathbf{H}$, which means that the integral of $\delta^{T}(\mathbf{r} - \mathbf{r}') \cdot \nabla' \times \mathbf{H}(\mathbf{r}')$ over space gives $\nabla \times \mathbf{H}(\mathbf{r})$. We see from the matrix symmetry (2.10) of the commutator (2.14) that δ^{T} is symmetric. We apply partial integration and utilize the matrix symmetry to move the curl from $\mathbf{H}$ to δ^{T}. The integral of $\nabla \times \delta^{T}(\mathbf{r} - \mathbf{r}') \cdot \mathbf{H}(\mathbf{r}')$ gives $\nabla \times \mathbf{H}$ for all $\mathbf{H}$ if and only if $\nabla \times \delta^{T}(\mathbf{r})$ is the curl of the matrix delta function $\delta(\mathbf{r})$. Consequently, δ^{T} can only differ from δ by a term of zero curl, by a gradient. Because of the matrix symmetry of δ^{T}, this gradient has the Fourier transform $k_l k_m f(k)$ with some function $f(k)$. Finally, to determine the remaining unknown function, we express $\nabla \cdot \delta^{T}$ in Fourier representation where the Fourier transform of the matrix delta function $\delta(\mathbf{r})$ is the unity matrix δ_{lm}. We see that $\nabla \cdot \delta^{T}$ vanishes if $f(k) = k^{-2}$ and so arrive at the representation (2.15).

In this section, we developed the fundamentals of the quantum field theory of light in media from two simple assumptions – the ensemble averages of the quantum fields are the classical fields, subject to Maxwell's equations (2.1), and the Hamiltonian of light is the electromagnetic energy (2.7). We described light by the operator $\hat{\mathbf{A}}(\mathbf{r}, t)$ of the electromagnetic vector potential in Coulomb gauge (2.4). We found that the quantum vector potential obeys the wave equation (2.5), boundary conditions and the fundamental commutation relation (2.14). Quantum optics, however, rarely uses this form of quantum electromagnetism. Instead, the central model of quantum optics is the light mode; the electromagnetic field is expressed in terms of modes. In the following section we develop the concept of the light mode from quantum electromagnetism in media.

2.2 Light modes

Light is subject to the superposition principle; if two light fields are brought to interference their amplitudes simply add up (if they have opposite amplitudes this wave addition results in destructive interference). The mathematical reason for the superposition principle is the linearity of Maxwell's equations. Since the quantum field of light is also subject to Maxwell's equations, this classical superposition principle remains valid in the quantum world. Note, however, that we cannot simply add two quantum fields $\hat{\mathbf{A}}(\mathbf{r}, t)$, because this would alter the fundamental commutator (2.14). There is a deeper reason: the quantum field $\hat{\mathbf{A}}(\mathbf{r}, t)$ already contains all possible light fields, it describes the world of possibilities for light (and the quantum state $\hat{\rho}$ determines which one of these possibilities is brought into existence). How does classical wave-like interference appear in the quantum field theory of light?

2.2.1 Modes and their scalar product

Consider a complete set of classical waves $\mathbf{A}_k(\mathbf{r}, t)$ that obey the Coulomb gauge (2.4), the Maxwell equation (2.5) and the boundary conditions. For example, the plane waves $\mathcal{A} \exp(\mathrm{i}\mathbf{k} \cdot \mathbf{r} - \mathrm{i}\omega t)$ form a complete set and are solutions of the wave equation (2.5) in uniform media with constant ε and μ. Figure 2.1 shows another example, a Gaussian mode. In general, the $\mathbf{A}_k$ are complex functions of space and time coordinates, but not operators; the $\mathbf{A}_k$ are classical. The complex conjugate $\mathbf{A}_k^*$ is part of the complete set of classical waves, because Maxwell's equations are real. Since the set is complete and satisfies the same linear partial differential equations and the same boundary conditions as the quantum field, we can expand $\hat{\mathbf{A}}$ in terms of $\mathbf{A}_k$ and $\mathbf{A}_k^*$ as

$$\hat{\mathbf{A}}(\mathbf{r}, t) = \sum_k \left(\mathbf{A}_k(\mathbf{r}, t)\, \hat{a}_k + \mathbf{A}_k^*(\mathbf{r}, t)\, \hat{a}_k^\dagger \right). \tag{2.16}$$

The expansion coefficient of the complex conjugate wave $\mathbf{A}_k^*$ is the Hermitian conjugate $\hat{a}_k^\dagger$ of the coefficient $\hat{a}_k$ of $\mathbf{A}_k$, because the quantum vector potential is Hermitian. The coefficients $\hat{a}_k$ and $\hat{a}_k^\dagger$ in the expansion (2.16) are operators; only they carry the quantum properties of light. The functions $\mathbf{A}_k(\mathbf{r}, t)$ are called *modes*, where k denotes the *mode index* used to identify them, for example their wavenumbers and

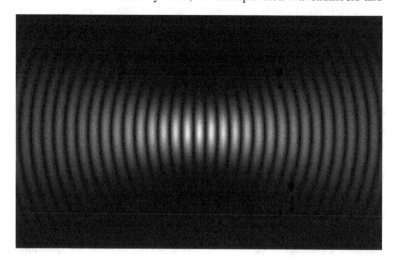

Fig. 2.1. Light mode. The picture illustrates a Gaussian mode (Born and Wolf, 1999), a mode that corresponds to a light beam. The picture shows the real part of one component of the vector potential.

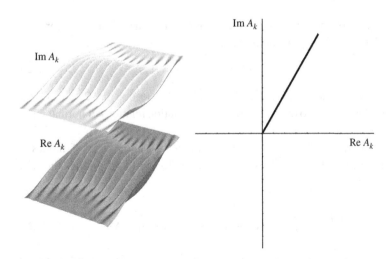

Fig. 2.2. Light modes are complex. The pictures on the left show the real and the imaginary part of the Gaussian mode of Fig. 2.1. The diagram on the right represents, for one space–time point, the mode amplitude A_k in the complex plane.

polarizations. Formula (2.16) describes a *mode expansion* of the electromagnetic field. We discuss an example of such a mode expansion in Section 2.3. Figure 2.2 illustrates the real and imaginary part of a mode.

In principle, we are at liberty to choose any modes $\mathbf{A}_k$, as long as they form a complete set and obey the laws of electromagnetism. However, it is advantageous to formulate certain constraints on the modes that lead to simple commutation rules for the mode operators $\hat{a}_k$ and $\hat{a}_k^\dagger$ and to insights into their physical nature. For this, we introduce a quantity that measures the degree to which two modes differ from each other, the *scalar product* in the function space of the modes

$$\left(\mathbf{A}_1, \mathbf{A}_2\right) \equiv \frac{1}{i\hbar} \int \left(\mathbf{A}_1^* \cdot \mathbf{D}_2 - \mathbf{A}_2 \cdot \mathbf{D}_1^*\right) dV, \quad \mathbf{D} = -\varepsilon_0 \varepsilon \, \frac{\partial \mathbf{A}}{\partial t}. \quad (2.17)$$

The $\hbar$ in front of the scalar product is a matter of convenience. The scalar product has most of the mathematical properties of inner products: *conjugate symmetry*

$$\left(\mathbf{A}_1, \mathbf{A}_2\right) = \left(\mathbf{A}_2, \mathbf{A}_1\right)^* \quad (2.18)$$

and *linearity*, such that for any two complex constants α_1 and α_2

$$\left(\mathbf{A}_0, \alpha_1 \mathbf{A}_1 + \alpha_2 \mathbf{A}_2\right) = \alpha_1 \left(\mathbf{A}_0, \mathbf{A}_1\right) + \alpha_2 \left(\mathbf{A}_0, \mathbf{A}_2\right). \quad (2.19)$$

Here we have adopted the notation commonly used in physics: the conjugate linear argument of the inner product is put in the first position of $(\mathbf{A}_1, \mathbf{A}_2)$. However, the scalar product (2.17) is not positive definite: the product (2.17) of one mode function with itself may become negative or vanish, a property that will become important in distinguishing mode operators and defining the quantum vacuum state. The scalar product (2.17) also shares some of the algebraic features of commutators,

$$\left(\mathbf{A}_1^*, \mathbf{A}_2^*\right) = -\left(\mathbf{A}_1, \mathbf{A}_2\right)^*, \tag{2.20}$$

the analogue of $[\hat{F}^\dagger, \hat{G}^\dagger] = -[\hat{G}, \hat{F}]^\dagger$. In Section 2.2.2 we use the scalar product to derive the Bose commutation relations of normal mode operators.

One defining specific feature of the scalar product of modes is the fact that it is a conserved quantity. To see this, consider the time derivative

$$\frac{d(\mathbf{A}_1, \mathbf{A}_2)}{dt} = \frac{i}{\hbar} \int \left(\mathbf{E}_1^* \cdot \mathbf{D}_2 - \mathbf{A}_1^* \cdot \frac{\partial \mathbf{D}_2}{\partial t} - \mathbf{E}_2 \cdot \mathbf{D}_1^* + \mathbf{A}_2 \cdot \frac{\partial \mathbf{D}_2^*}{\partial t}\right) dV. \tag{2.21}$$

We obtain from Maxwell's equations (2.1) and the relationships (2.3) that

$$\mathbf{A} \cdot \frac{\partial \mathbf{D}}{\partial t} = \mathbf{A} \cdot (\nabla \times \mathbf{H}) = -\nabla \cdot (\mathbf{A} \times \mathbf{H}) + \mathbf{B} \cdot \mathbf{H}. \tag{2.22}$$

According to the constitutive equations (2.2), $\mathbf{E}_1^* \cdot \mathbf{D}_2$ cancels $\mathbf{E}_2 \cdot \mathbf{D}_1^*$ and $\mathbf{B}_1^* \cdot \mathbf{H}_2$ cancels $\mathbf{B}_2 \cdot \mathbf{H}_1^*$; the remaining integral of $\nabla \cdot (\mathbf{A} \times \mathbf{H})$ gives a surface integral of $\mathbf{A} \times \mathbf{H}$ at infinity where $\mathbf{A} \times \mathbf{H}$ vanishes. Consequently,

$$\frac{d(\mathbf{A}_1, \mathbf{A}_2)}{dt} = 0. \tag{2.23}$$

Conservation laws are associated with symmetries. The symmetry behind the conservation of $(\mathbf{A}_1, \mathbf{A}_2)$ is the fact that the modes are complex, but their equations are real. We could multiply a mode by an arbitrary global phase factor $\exp(-i\varphi)$ and would obtain a mode that obeys the same wave equations. Real electromagnetic fields do not have this extra symmetry; we see from the structure of the scalar product (2.17) that the norm $(\mathbf{A}, \mathbf{A})$ vanishes for real $\mathbf{A}$.

2.2.2 Bose commutation relations

Given a mode expansion, we can form appropriate superpositions for arranging their scalar products at a given time; $(\mathbf{A}_k, \mathbf{A}_{k'})$ remain fixed throughout the evolution of the electromagnetic field. In particular, we could orthogonalize the modes, using, for example, a Gram–Schmidt orthogonalization process (Strang, 2005), and they remain orthogonal during their evolution. Suppose the modes obey the *orthonormality conditions*,

$$\left(\mathbf{A}_k, \mathbf{A}_{k'}\right) = \delta_{kk'}, \quad \left(\mathbf{A}_k, \mathbf{A}_{k'}^*\right) = 0. \tag{2.24}$$

In this case we call them *normal modes*. Throughout the rest of the book we will only consider normal modes whenever we mention modes at all; from now on they are synonymous. We take immediate advantage of the orthonormality conditions (2.24) for projecting the mode operators out of the mode expansion (2.16) as

$$\hat{a}_k = (\mathbf{A}_k, \hat{\mathbf{A}}), \quad \hat{a}_k^\dagger = -(\mathbf{A}_k^*, \hat{\mathbf{A}}). \tag{2.25}$$

We use the definition (2.17) of the scalar product and obtain

$$[\hat{a}_k, \hat{a}_{k'}^\dagger] = \frac{1}{\hbar^2} \int \int [\mathbf{A}_k^* \cdot \hat{\mathbf{D}} - \hat{\mathbf{A}} \cdot \mathbf{D}_k^*, \, \mathbf{A}_{k'} \cdot \hat{\mathbf{D}} - \hat{\mathbf{A}} \cdot \mathbf{D}_{k'}] \, dV dV'$$
$$= \frac{1}{i\hbar} \int (\mathbf{A}_k^* \cdot \mathbf{D}_{k'} - \mathbf{A}_{k'} \cdot \mathbf{D}_k^*) \, dV, \tag{2.26}$$

as a consequence of the fundamental commutator (2.14) of the electromagnetic vector potential in Coulomb gauge. Similarly,

$$[\hat{a}_k, \hat{a}_{k'}] = -\frac{1}{i\hbar} \int (\mathbf{A}_k^* \cdot \mathbf{D}_{k'}^* - \mathbf{A}_{k'}^* \cdot \mathbf{D}_k^*) \, dV. \tag{2.27}$$

The integrals (2.26) and (2.27) are the scalar products $(\mathbf{A}_k, \mathbf{A}_{k'})$ and $(\mathbf{A}_k, \mathbf{A}_{k'}^*)$. We thus obtain from the orthonormality conditions (2.24) the *Bose commutation relations*

$$[\hat{a}_k, \hat{a}_{k'}^\dagger] = \delta_{kk'}, \quad [\hat{a}_k, \hat{a}_{k'}] = 0. \tag{2.28}$$

In Chapter 3 we study in detail their consequences, in particular how each pair of mode operators $\hat{a}_k$ and $\hat{a}_k^\dagger$ determines an individual Hilbert space, but one simple conclusion is already evident: the mode operators of different modes commute; it is therefore natural to assume that they represent distinct physical systems. Each mode represents a degree of freedom of the electromagnetic field. The quantum properties of light, the wave amplitudes and their quantum fluctuations, are governed by the degrees of freedom. The total Hilbert space of the degrees of freedom is the tensor product of the Hilbert spaces of all the modes. Applying Occam's razor, *entia non sunt multiplicanda praeter necessitatem*, entities should not be multiplied beyond necessity, this product space, nothing more, nothing less, defines the Hilbert space for the quantum state of light.

2.2.3 Interference

Suppose that $|\psi_1\rangle$ represents one specific quantum state of light and $|\psi_2\rangle$ another. According to the *quantum superposition principle*, a linear combination of $|\psi_1\rangle$ and $|\psi_2\rangle$ also forms a possible quantum state.

Fig. 2.3. Interference. Thomas Young's original drawing of the interference of light at a double slit.

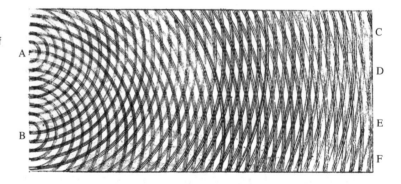

Does quantum interference describe the superposition of electromagnetic waves? For a *wave-like superposition* (shown in Fig. 2.3), the wave amplitude should add up, the expectation value $\langle\psi|\hat{\mathbf{A}}|\psi\rangle$ should be the sum of the amplitudes $\langle\psi_1|\hat{\mathbf{A}}|\psi_1\rangle$ and $\langle\psi_2|\hat{\mathbf{A}}|\psi_2\rangle$. A quantum superposition, however, does not behave like this, unless $|\psi_1\rangle$ and $|\psi_2\rangle$ belong to two different modes. Quantum superposition describes the interference of possibilities, of probability amplitudes, whereas wave superpositions are interferences of waves in reality; these are two rather different concepts. So how does the classical superposition principle emerge? Imagine that one of the modes $\mathbf{A}_k$ is a classical superposition of some different modes $\mathbf{A}'_{k'}$, a linear combination of $\mathbf{A}'_{k'}$. If light is in a quantum state where these modes are excited, the electromagnetic field does appear as a superposition of $\mathbf{A}'_{k'}$. In Chapter 5 we discuss in detail that other modes of the set $\mathbf{A}_k$ should be linear combinations of the $\mathbf{A}'_{k'}$ as well, in order to preserve the orthonormality (2.24) of modes; so

$$\mathbf{A}_k(\mathbf{r},t) = \sum_{k'}\mathbf{A}'_{k'}(\mathbf{r},t)\,B_{k'k} \tag{2.29}$$

with complex constants $B_{k'k}$. Substituting the superposition (2.29) in the mode expansion (2.16) we obtain an expansion in terms of the new modes $\mathbf{A}'_{k'}$ with the associated mode operators

$$\hat{a}'_{k'} = \sum_{k}B_{k'k}\,\hat{a}_k. \tag{2.30}$$

The reflection and transmission of light at media, for example at a beam splitter, generates such superpositions (2.29), because here the $\mathbf{A}_k$ describe the *incoming modes* and the $\mathbf{A}'_{k'}$ the *outgoing modes*. In Chapter 8 we show that event horizons generate another type of superposition: linear combinations of modes $\mathbf{A}_k$ with their complex conjugate partners $\mathbf{A}_k^*$. There we discuss how they lead to the amplification of light and to the spontaneous generation of light quanta at horizons.

The mode functions $\mathbf{A}_k(\mathbf{r}, t)$ describe the classical wave-like aspects of the electromagnetic field and the mode operators $\hat{a}_k$ and $\hat{a}_k^\dagger$ the quantum amplitudes. Given an ensemble of light fields prepared in an identical state and a fixed set of modes, the quantum amplitudes fluctuate. The choice of the mode expansion is a matter of convenience; it is often a good idea to use as few modes as possible to capture the state of a light field. Typically, quantum optics considers such few-mode situations, because here quantum effects appear in their purest form. On the other hand, most natural sources of light excite a broad range of modes. Quantum optics thus benefits from laser sources and the ingenuity of experimentalists to excite few modes of light. In order to observe few-mode quantum physics, the concentration of light in a few modes is not sufficient; it is also essential to be able to select single modes in the detection of light. Here the observer choses the modes. Sometimes this choice is post-selective: although the quantum state of light may comprise a large range of modes, the detection apparatus probes only for specific modes. This *post-selection* of modes is completely sufficient for observing few mode quantum effects, provided the substate of these modes is not entangled with the rest of the quantum field of light.

2.2.4 Monochromatic modes

The mode expansion (2.16) naturally incorporates modes that are wave packets, for example light pulses, modes that contain a range of optical frequencies and hence a range of colours. Consider the special case of *monochromatic modes* that oscillate at single frequencies,

$$\frac{\partial \mathbf{A}_k}{\partial t} = -i\omega_k \mathbf{A}_k, \quad \text{or} \quad \mathbf{A}_k(\mathbf{r}, t) = \mathbf{A}_k(\mathbf{r}) \exp(-i\omega_k t). \tag{2.31}$$

We see from the scalar product (2.17) that monochromatic modes are normalized according to

$$(\mathbf{A}_1, \mathbf{A}_2) = \frac{2\varepsilon_0 \omega_k}{\hbar} \int \mathbf{A}_1^* \cdot \mathbf{A}_2 \, \varepsilon \, dV, \tag{2.32}$$

similar to the normalization of the wave function in non-relativistic quantum mechanics. Since monochromatic modes are *stationary*, each mode conserves energy, and so we expect that the total energy, the Hamiltonian (2.7), is the sum of Hamiltonians of the individual modes. We apply the relationship (2.22) for $\hat{\mathbf{B}} \cdot \hat{\mathbf{H}}$ in the Hamiltonian (2.7) and obtain

$$\hat{H} = \frac{1}{2} \int \left(\hat{\mathbf{E}} \cdot \hat{\mathbf{D}} + \hat{\mathbf{A}} \cdot \frac{\partial \hat{\mathbf{D}}}{\partial t} \right) dV, \tag{2.33}$$

because the integral over the divergence of $\hat{\mathbf{A}} \times \hat{\mathbf{H}}$ vanishes. From the constitutive equations (2.2), the relationship (2.3), the mode expansion (2.16) and the stationarity (2.31) we get

$$
\begin{aligned}
\hat{\mathbf{E}} \cdot \hat{\mathbf{D}} &= -\sum_{kk'} \omega_k \left(\mathbf{A}_k \hat{a}_k - \mathbf{A}_k^* \hat{a}_k^\dagger \right) \varepsilon \, \omega_{k'} \left(\mathbf{A}_{k'} \hat{a}_{k'} - \mathbf{A}_{k'}^* \hat{a}_{k'}^\dagger \right), \\
\hat{\mathbf{A}} \cdot \frac{\partial \hat{\mathbf{D}}}{\partial t} &= -\sum_{kk'} \left(\mathbf{A}_k \hat{a}_k + \mathbf{A}_k^* \hat{a}_k^\dagger \right) \varepsilon \, \omega_{k'}^2 \left(\mathbf{A}_{k'} \hat{a}_{k'} + \mathbf{A}_{k'}^* \hat{a}_{k'}^\dagger \right).
\end{aligned}
\tag{2.34}
$$

We substitute these expressions in the Hamiltonian (2.33) and utilize the orthonormality conditions (2.24) with respect to the scalar product (2.32). In this way we find

$$
\hat{H} = \sum_k \frac{\hbar \omega_k}{2} \left(\hat{a}_k \hat{a}_k^\dagger + \hat{a}_k^\dagger \hat{a}_k \right)
\tag{2.35}
$$

and from the Bose commutation relations (2.28) we obtain, finally,

$$
\hat{H} = \sum_k \hbar \omega_k \left(\hat{a}_k^\dagger \hat{a}_k + \frac{1}{2} \right).
\tag{2.36}
$$

The total energy of the electromagnetic field is the sum of the energies of the stationary modes. Each mode carries an energy of $\hbar \omega_k$ times the equivalent of the modulus squared of the quantum amplitude, $\hat{a}_k^\dagger \hat{a}_k$, plus 1/2. The expectation value of each $\hat{a}_k^\dagger \hat{a}_k$ is larger or equal to zero, because $\langle \psi | \hat{a}_k^\dagger \hat{a}_k | \psi \rangle$ represents the non-negative norm $\langle \varphi | \varphi \rangle$ of the Hilbert space vector $|\varphi\rangle = \hat{a}_k | \psi \rangle$. Consequently, the minimal value of the energy is

$$
E_0 = \sum_k \frac{\hbar \omega_k}{2} = \infty.
\tag{2.37}
$$

Even without any photons present, with light in the vacuum state,[1] the energy of the electromagnetic field is infinite. Clearly, this infinity is an artifact – it would make the electromagnetic field infinitely massive, $E_0 c^2 = \infty$. The electromagnetic field would collapse under the weight of its own gravity, even if no light at all is present. Some unknown mechanism beyond quantum electromagnetism must regularize the infinity of the electromagnetic vacuum energy. Nevertheless, the *zero point energy* (2.37) results in perfectly finite and experimentally confirmed facts, as we discuss in the next section.

---

[1] Throughout this book we always mean by "vacuum" simply "no light" and not an evacuated system.

2.3 Zero-point energy and Casimir force

In 1948 Hendrik B. G. Casimir showed by theoretical calculations (Casimir, 1948) that two metallic plates should exert an attractive force on each other in complete vacuum and without being charged. Remarkably, this force of the quantum vacuum appears in daily life: it relates to the force that causes objects to stick. Casimir's interpretation of the attractive force as a force of vacuum fluctuations was inspired by a cryptic remark of Niels Bohr, as Casimir wrote to Peter Milonni (Milonni, 1994) "I mentioned my results to Niels Bohr during a walk. That is nice, he said, that is something new. I told him that I was puzzled by the extremely simple form of the expressions for the interaction at very large distances and he mumbled something about zero-point energy. That was all, but it put me on a new track...".

2.3.1 An attractive cavity

Consider the electromagnetic field in a cavity made of perfectly conducting, infinitely large plates at distance a, see Fig. 2.4. In the plates the transversal components of the electric field vanish (Jackson, 1998), compensated by the mobile charges in the conducting material. We may thus describe the influence of the plates on the electromagnetic field by these boundary conditions. For reasons of mathematical simplicity, imagine that we surround the plates by an equally perfectly conducting box of finite side length L and then we take the limit $L \to \infty$.

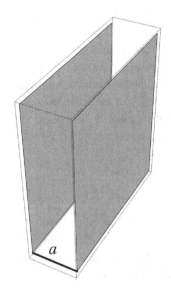
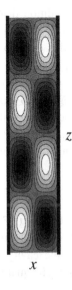

Fig. 2.4. Casimir cavity. The left figure shows two perfectly conducting plates at distance a (surrounded by a fictitious quantization box). The right figure illustrates an electromagnetic mode in the cavity; it shows the y component of the vector potential (2.38) at $y = 0$.

The boundary conditions of this so-called *quantization box* will not matter in the limit of infinite box size, but the box allows us to normalize the electromagnetic modes in a simple way. For the monochromatic modes, the components of the vector potential are

$$
\begin{aligned}
A_x &= \mathcal{A}_x \cos(k_x x)\, \sin(k_y y)\, \sin(k_z z) \exp(-i\omega t), \\
A_y &= \mathcal{A}_y \sin(k_x x)\, \cos(k_y y)\, \sin(k_z z) \exp(-i\omega t), \\
A_z &= \mathcal{A}_z \sin(k_x x)\, \sin(k_y y)\, \cos(k_z z) \exp(-i\omega t),
\end{aligned}
$$

$$
k_x = \frac{l\pi}{a}, \quad k_y = \frac{m\pi}{L}, \quad k_z = \frac{n\pi}{L},
$$
(2.38)

where l, m, n are non-negative integers and $\mathcal{A}_x, \mathcal{A}_y, \mathcal{A}_z$ are real constants. We see that the transversal components $\mathcal{A}_y$ and $\mathcal{A}_z$ vanish at the plates that are located at $x = 0$ and $x = a$, and so do the transversal components of the electric field (2.3), in agreement with our boundary conditions. Similarly, the transversal components of the electric field vanish at the fictitious plates of the quantization box. The modes (2.38) satisfy the Maxwell equation (2.5) in the empty space between the plates if

$$
\omega = c\sqrt{k_x^2 + k_y^2 + k_z^2} = \omega_{lmn}.
$$
(2.39)

The plates thus select a set of standing waves with discrete eigenfrequencies ω_{lmn}. The modes (2.38) obey the Coulomb gauge (2.4) if the vector of the amplitudes $\mathcal{A}$ is orthogonal to the wave vector $\mathbf{k}$. We obtain

$$
\frac{\mathcal{A}_x l}{a} + \frac{\mathcal{A}_y m + \mathcal{A}_z n}{L} = 0.
$$
(2.40)

This transversality condition determines one of the amplitude vector components in terms of the other two, unless $l = 0$ or $m = 0$ or $n = 0$. Apart from this exception, the modes are characterized by a two-dimensional amplitude vector, that is by two polarizations. In the exceptional case of $l = 0$ or $m = 0$ or $n = 0$, the electric field is uniform in the direction orthogonal to a pair of plates, like the static field of a capacitor, and the modes have only one polarization. In either case, the modes are normalized according to the scalar product (2.32). We see that they are orthogonal to each other and find

$$
(\mathbf{A}, \mathbf{A}) = \frac{\varepsilon_0 \omega}{\hbar} \mathcal{A}^2 \frac{L^2 a}{4}.
$$
(2.41)

The total zero-point energy (2.37) of the modes is infinite, but this infinity may depend on the cavity size. Suppose we subtract from the zero-point energy of the cavity modes (2.38) the infinite zero-point energy of empty space. The quantum electromagnetism of empty space is equivalent to the case of an infinitely large cavity where the boundary conditions cease to have an influence. Therefore we concentrate on the quantity

$$U = E_0(a) - E_0(\infty). \tag{2.42}$$

The two polarizations of each mode contribute $\hbar\omega_{lmn}/2$ to the total zero-point energy; so we count them twice,

$$E_0(a) = \hbar \sum_{lmn}{}' \omega_{lmn}, \tag{2.43}$$

except when $l = 0$ or $m = 0$ or $n = 0$. The prime in the summation indicates that we multiply the corresponding term by $1/2$ in this case.

The difference U in the zero-point energies between the cavity and empty space depends on the cavity size. Suppose one of the cavity plates is movable. Depending on the position a, U varies as the potential energy of a mechanical body. Consequently, the zero-point energy generates the mechanical force

$$F = -\frac{\partial U}{\partial a}, \tag{2.44}$$

known as the *Casimir force* (Casimir, 1948). We calculate U in the limit of an infinite quantization box, $L \to \infty$. First, we replace the m, n summation by k_z, k_y integrations where we represent each integer step Δm or Δn as $(L/\pi)\,dk_y$ or $(L/\pi)\,dk_z$, respectively. Then we express the k_z, k_y integral in polar coordinates with $k_y = \kappa \cos\varphi$, $k_z = \kappa \sin\varphi$. The angle φ runs from 0 to $\pi/2$, because $k_y, k_z \geq 0$. The frequencies of the modes do not depend on φ; and so the angular integration gives $\pi/2$. In the limit of empty space, $a \to \infty$, we also replace the l summation by an integration. In order to get a convenient expression for U, we express this integral in terms of the finite cavity distance a as an integration over $l = (a/\pi)k_x$. In this way, we obtain

$$U = \frac{\hbar c L^2}{2\pi} \int_0^\infty \left(\sum_l{}' \sqrt{\kappa^2 + \frac{l^2\pi^2}{a^2}} - \frac{a}{\pi} \int_0^\infty \sqrt{\kappa^2 + k_x^2}\,dk_x \right) \kappa\,d\kappa. \tag{2.45}$$

We change the integration variable from κ to $\xi = l^2 + \kappa^2 a^2/\pi^2$, multiply the zeroth term in the sum by $1/2$ according to our sum convention, and get

$$U = \frac{\pi^2 \hbar c}{4a^3} L^2 \left(\frac{1}{2} I(0) + \sum_{l=1}^\infty I(l) - \int_0^\infty I(l)\,dl \right) \tag{2.46}$$

with

$$I(l) = \int_{l^2}^\infty \sqrt{\xi}\,d\xi. \tag{2.47}$$

The difference U in the zero-point energies of the cavity and empty space thus appears as the difference between an infinite series and an integral. To evaluate this difference, we use the *Euler–Maclaurin sum formula* (Abramowitz and Stegun, 1970),

$$\int_0^\infty I(l)\,dl = \sum_{l=1}^\infty I(l) + \frac{I(0)}{2} + \frac{I'(0)}{12} - \frac{I'''(0)}{720} + \cdots , \qquad (2.48)$$

where the primes denote differentiations with respect to l. We obtain from the definition (2.47) that $I'(l)$ gives $-2l^2$, which implies that the only non-vanishing term in the Euler–Maclaurin series corresponds to $I'''(0) = -4$. Consequently,

$$U = -\frac{\pi^2 \hbar c}{720\, a^3} L^2, \qquad (2.49)$$

which gives the force per area

$$\frac{F}{L^2} = -\frac{\pi^2 \hbar c}{240\, a^4}. \qquad (2.50)$$

The cavity plates attract each other with a force per area that depends only on fundamental constants and on the inverse fourth power of the cavity size. The reason for this power law is the scaling of the zero-point energy: suppose we extend the cavity size by a scaling factor η. The eigenfrequencies (2.39) of the cavity would be reduced by η^{-1}; the zero-point energy would be reduced by the same factor; and so the zero-point energy per area scales with η^{-3}. Therefore, the Casimir force per area depends on a^{-4}.

2.3.2 Reflections

The Casimir effect becomes important on the scale of nano-machines (Chan et al., 2001). The effect has been measured (Lamoreaux, 1997, 1999; Bordag et al., 2001; Lamoreaux, 2005) and the observed results agree very well with the theoretical prediction (2.50). Zero-point energy is real, despite its infinity. In our calculation, we used some slightly dubious tricks; for example, the integral (2.47) is infinite, but we treated it as a finite quantity. All of these dubious features are arte-facts, one can avoid them, apart from one fundamental assumption: the zero-point energy of the electromagnetic field in empty space is constant, with finite energy density, for reasons as yet unknown. In

Appendix A we describe a theory of the Casimir effect that avoids the avoidable artefacts and is more general than the simple calculation presented here, but it is also more technically involved. We also show in Appendix A that Casimir forces occur in inhomogeneous media in general, not only between two cavity plates. They result in a mechanical stress in the material, the stress of the quantum vacuum.

Where does the *vacuum stress* come from? In the quantum mechanical vacuum state, the expectation value of the electromagnetic field is zero, but this is a zero on average. If we measure the electromagnetic field strength of the quantum vacuum we observe that the field strengths fluctuate around zero from measurement to measurement. (We show this in Section 3.2.2; in Section 5.2.2 we describe a device that measures the field strength of a single mode, the balanced homodyne detector.) In a dielectric medium, the electromagnetic field induces dipoles that are proportional to the field strength. Since the response of the medium is linear, it also must respond to the electromagnetic fluctuations of the quantum vacuum. Although the vacuum-induced dipoles have only a fleeting existence, they interact with each other. Dipoles tend to align. The vacuum-induced dipoles respond to each other although they are zero on average, which means that their fluctuations are correlated. For example, in the case of the Casimir cavity shown in Fig. 2.4, the quantum fluctuations of the vacuum induce random surface charges in the cavity plates, but the charges at one plate are correlated to the charges at the other. In turn, because of the dipole–dipole interaction, the plates feel a force, the Casimir force (2.50).

Note that the Casimir force is related to the *van der Waals force* of polarizable atoms or molecules, the force that attracts atoms to each other and to surfaces, and that also appears in the equation of state of a real gas (Landau and Lifshitz, Vol. V, 1996). Here quantum fluctuations induce dipoles that align and attract each other. So the van der Waals force, the force of the quantum vacuum, causes things to stick. To give a simple, Singaporean example, a gecko can hang on a glass surface using only one toe. This extraordinary feat of their extraordinary feet is due to the van der Waals force between the glass and the numerous micro-hairs in the gecko's toes (Autumn et al., 2002), see Fig. 2.5. Both the Casimir and the van der Waals force only need a stimulant, some fluctuations that induce the dipoles. Remarkably, the quantum vacuum provides such fluctuations in the ceaseless probing and pondering of possibilities that is going on in the quantum world.

Fig. 2.5. Gecko feet. (Photo: Paul D. Stewart.)

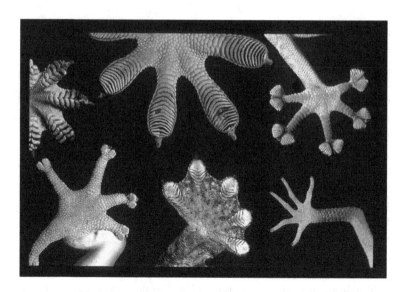

2.4 Questions

2.1 Why do the quantum operators of the electromagnetic field strengths obey Maxwell's classical equations?

2.2 What are dielectric media? What are the constitutive equations?

2.3 What is the electromagnetic vector potential? Why is it a useful quantity in electrodynamics?

2.4 What is a gauge transformation? What is the Coulomb gauge? Suppose the vector potential does not satisfy the Coulomb gauge; how can you transform it in order to impose this gauge?

2.5 What is a Hamiltonian in quantum mechanics? How does the Hamiltonian determine the equations of motion?

2.6 Show, using Maxwell's equations and the constitutive equations, that the Hamiltonian $\hat{H}$ is a conserved quantity, where

$$\hat{H} = \frac{1}{2} \int \left(\hat{\mathbf{E}} \cdot \hat{\mathbf{D}} + \hat{\mathbf{B}} \cdot \hat{\mathbf{H}} \right) dV.$$

2.7 Explain the arguments that lead to the fundamental commutator

$$\left[\hat{\mathbf{D}}(\mathbf{r}, t), \hat{\mathbf{A}}(\mathbf{r}', t) \right] = i\hbar \, \delta^{\mathrm{T}} \left(\mathbf{r} - \mathbf{r}' \right).$$

2.8 What are light modes? What is the physical meaning of the mode functions $\mathbf{A}_k(\mathbf{r}, t)$ and the amplitude operators $\hat{a}$ and $\hat{a}^\dagger$?

2.9 Can you also define light modes in classical optics? What would be the meaning of the mode amplitudes here? Give an example where the mode amplitudes fluctuate.

2.10 What is the meaning of the scalar product between two modes? Why is it a useful concept?

2.11 Deduce the Bose commutation relations from the fundamental field commutator and the orthonormality conditions of the modes.

2.12 Suppose you change the prefactor in the scalar product $(i\hbar)^{-1} \int (\mathbf{A}_1^* \cdot \mathbf{D}_2 - \mathbf{A}_2 \cdot \mathbf{D}_1^*) \, dV$ from $\hbar$ to, say, $\hbar'$. Which commutation relations for the mode operators would you obtain if the modes are normalized according to this modified scalar product?

2.13 What is the difference between quantum interference and the classical interference of electromagnetic waves?

2.14 Derive the expression $(2\varepsilon_0 \omega_k/\hbar) \int \mathbf{A}_1^* \cdot \mathbf{A}_2 \, \varepsilon \, dV$ for the scalar product of two monochromatic modes.

2.15 Derive the expansion of the Hamiltonian in terms of monochromatic modes, $\hat{H} = \sum_k \hbar \omega_k \left(\hat{a}_k^\dagger \hat{a}_k + 1/2 \right)$.

2.16 What is zero-point energy? What is wrong with it? Why is it still a useful concept?

2.17 Show that the modes of the Casimir cavity obey the Maxwell equations and the boundary conditions at perfect conductors.

2.18 When do the Casimir modes occur in two polarizations and when in only one? Why?

2.19 Calculate the scalar product between the Casimir modes.

2.20 In the calculation of the Casimir potential, fill in the steps explained in words. When can you replace sums by integrals? Why?

2.21 Why does the Casimir force between two perfectly conducting plates depend on the inverse fourth power of their distance?

2.22 What is the physical mechanism behind the Casimir force?

2.5 Homework problem

Quantum field theory is not restricted to the fields that traditionally describe interactions, such as the electromagnetic field, but is applicable to matter itself. The quantum-field-theoretical approach to matter is particularly relevant to ultracold quantum gases. Consider a gas of identical bosonic atoms with mass m. The atoms are trapped in the potential $U(\mathbf{r})$ and are interacting with each other by collisions via the potential $V(|\mathbf{r}_{k'} - \mathbf{r}_k|)$. The multi-particle wave function $\psi(\mathbf{r}_1, \ldots, \mathbf{r}_n)$ of the atoms obeys the Schrödinger equation with the Hamiltonian

$$\hat{H} = \sum_k \left(-\frac{\hbar^2 \nabla_k^2}{2m} + U(\mathbf{r}_k) + \sum_{k'>k} V(|\mathbf{r}_{k'} - \mathbf{r}_k|) \right).$$

Let us define the quantum field $\hat{\psi}(\mathbf{r}, t)$ with the Bose commutation relations

$$\left[\hat{\psi}(\mathbf{r}, t), \hat{\psi}^\dagger(\mathbf{r}', t) \right] = \delta(\mathbf{r} - \mathbf{r}'), \quad \left[\hat{\psi}(\mathbf{r}, t), \hat{\psi}(\mathbf{r}', t) \right] = 0.$$

We postulate that $\hat{\psi}$ obeys the field equation

$$i\hbar \frac{\partial \hat{\psi}}{\partial t} = \left(-\frac{\hbar^2 \nabla^2}{2m} + U + \int \hat{\psi}^\dagger(\mathbf{r}')\, V\big(|\mathbf{r} - \mathbf{r}'|\big)\, \hat{\psi}(\mathbf{r}')\, d\mathbf{r}' \right) \hat{\psi}.$$

You will show that the quantum field $\hat{\psi}$ is equivalent to the Schrödinger wave function $\psi(\mathbf{r}_1, \ldots, \mathbf{r}_n)$ of the atoms and work out how the field theoretical approach leads to physically useful approximations.

(a) Define the state $|\psi\rangle$ of the quantum field as

$$|\psi\rangle = \int \psi(\mathbf{r}_1, \ldots, \mathbf{r}_n)\, \hat{\psi}^\dagger(\mathbf{r}_1) \cdot \ldots \cdot \hat{\psi}^\dagger(\mathbf{r}_n)|0\rangle\, dV_1 \cdot \ldots \cdot dV_n$$

where $|0\rangle$ denotes the vacuum state with the property

$$\hat{\psi}|0\rangle = 0.$$

The wave function $\psi(\mathbf{r}_1, \ldots, \mathbf{r}_n)$ evolves according to the Schrödinger equation with Hamiltonian $\hat{H}$ and the operator $\hat{\psi}$ obeys the field equation. Show that $|\psi\rangle$ does not change in time.

(b) The result of (a) proves that the Schrödinger wave function ψ of the atoms can be represented by the quantum state $|\psi\rangle$ of the matter field $\hat{\psi}$. The number of creation operators in the definition of $|\psi\rangle$ corresponds to the number of atoms in the gas. Treating the quantum gas as a quantum field is a convenient form of book-keeping. As an additional bonus, this theory automatically encodes the bosonic nature of the atoms. To prove this, show that the quantum state $|\psi\rangle$ remains the same if you exchange two particles in the wave function $\psi_1(\mathbf{r}_1, \ldots, \mathbf{r}_n)$. This property is the defining feature of bosons.

(c) The atoms in ultracold quantum gases mainly experience contact interaction where the potential V can be approximated by a delta function such that

$$i\hbar \frac{\partial \hat{\psi}}{\partial t} = \left(-\frac{\hbar^2 \nabla^2}{2m} + U + g\, \hat{\psi}^\dagger \hat{\psi} \right) \hat{\psi}, \quad g = \frac{4\pi \hbar^2 a}{m}$$

where a denotes the scattering length. The electromagnetic field is represented as a series of modes. Why is this not possible for the matter field $\hat{\psi}$?

(d) In Bose–Einstein condensation, the lion's share of the atoms are condensed in a state with the macroscopic wave function $\psi_0(\mathbf{r}, t)$ that satisfies the classical field equation

$$i\hbar \frac{\partial \psi_0}{\partial t} = \left(-\frac{\hbar^2 \nabla^2}{2m} + U + g\, |\psi_0|^2 \right) \psi_0.$$

The quantum field only gives a small correction $\hat{\varphi}$ to the classical matter wave amplitude ψ_0,

$$\hat{\psi} = \psi_0 + \hat{\varphi}.$$

Show that $\hat{\varphi}$ approximately obeys the equation

$$i\hbar \frac{\partial \hat{\varphi}}{\partial t} = \left(-\frac{\hbar^2 \nabla^2}{2m} + U + 2g \, |\psi_0|^2 \right) \hat{\varphi} + g \, \psi_0^2 \, \hat{\varphi}^\dagger.$$

(e) Why can you represent $\hat{\varphi}$ in terms of a mode expansion?
(f) We write $\hat{\varphi}$ as

$$\hat{\varphi} = \sum_k \left(u_k \hat{a}_k + v_k^* \hat{a}_k^\dagger \right).$$

Why is $u_k \neq v_k$ in general?
(g) Suppose that

$$i\hbar \frac{\partial}{\partial t} \begin{pmatrix} u \\ -v \end{pmatrix} = \left(-\frac{\hbar^2 \nabla^2}{2m} + U + 2g \, |\psi_0|^2 \right) \begin{pmatrix} u \\ v \end{pmatrix} + g \begin{pmatrix} \psi_0^2 v \\ \psi_0^{*2} u \end{pmatrix}.$$

Show that $\hat{\varphi}$ given by the mode expansion (f) satisfies the field equation derived in part (d).
(h) We define the scalar product of two modes $w_k = (u_k, v_k)$ as

$$(w_1, w_2) = \int \left(u_1^* u_2 - v_1^* v_2 \right) dV.$$

Show that (w_1, w_2) is a conserved quantity.
(i) For orthonormal modes we have

$$\hat{a}_k = \int \left(u_k^* \hat{\varphi} - v_k^* \hat{\varphi}^\dagger \right) dV.$$

Show that the $\hat{a}_k$ operators obey the usual Bose commutation relations

$$[\hat{a}_k, \hat{a}_{k'}^\dagger] = \delta_{kk'}, \quad [\hat{a}_k, \hat{a}_{k'}] = 0.$$

2.6 Further reading

Quantum field theory is widely used in particle physics and condensed matter physics. Classic textbooks on the relativistic quantum field theory of particle physics are the books by Bjorken and Drell (1965), Itzykson and Zuber (2006), and Weinberg (2000a, 2000b, 2000c). The classic book on second quantization for book-keeping in condensed matter physics is Negele and Orland's text (1998). Fetter and Walecka (1971) is a very readable book on quantum field theory and Green's functions in condensed matter physics; Fradkin (1991) and the recent books by Altland and Simon (2006) and Tsvelik (2007) are also recommended.

Our approach to the quantum field theory of light in media is inspired by the quantum field theory in curved space (Birrell and Davies, 1982), because dielectric media act like space–time geometries (Leonhardt and Philbin, 2009). Glauber and Lewenstein (1991) developed another, equivalent theory for light in non-dispersive media. The quantum field theory of light in dispersive and dissipative media is summarized by Knöll, Scheel and Welsch (2001), Lukš and Peřinova (2002), and Buhmann and Welsch (2007). The intriguing physics of the quantum vacuum is the subject of Milonni's book (1994) that, however, appeared before the first precise measurements of the Casimir effect (Lamoreaux, 1997). The literature and theory of the Casimir effect is described by Lamoreaux (1999; 2005), Bordag, Mohideen and Mostepanenko (2001) and in Milton's book (2001).

Chapter 3
Simple quantum states of light

3.1 The electromagnetic oscillator

In Chapter 2 we arrived at a relatively simple description of light in the quantum world: light consists of a superposition of modes. The mode functions $\mathbf{A}_k(\mathbf{r}, t)$ contain the classical, wave-like aspects of light, the fact that light is an electromagnetic wave; the $\mathbf{A}_k(\mathbf{r}, t)$ obey the Coulomb gauge (2.4), the wave equation (2.5) and the boundary conditions. The quantum degrees of freedom are the mode amplitudes $\hat{a}_k$; their expectation values and fluctuations depend on the quantum state of light. For a single mode, the electromagnetic field is proportional to the mode amplitude; if the amplitude fluctuates the field strengths at all points in space and time fluctuate in unison. In this chapter we discuss examples of single-mode states – some simple quantum states of a single mode. We determine the state of the quantum vacuum, the state of pure particle-like photons, the state of thermal light, and the state that describes the coherent wave-like light produced by lasers. We discuss minimum uncertainty states and the squeezing of quantum noise. We begin by defining the central physical quantities of a single mode of light.

The wave-like, classical properties of a light mode are in the mode function; all we need to know for deducing the quantum physics of the mode is the Bose commutation relation for the mode operators

$$[\hat{a}, \hat{a}^{\dagger}] = 1. \tag{3.1}$$

In the following we introduce the key operators of the single mode. The *photon number* operator $\hat{n}$ is given by the quantum counterpart of a classical modulus squared amplitude

$$\hat{n} \equiv \hat{a}^{\dagger}\hat{a}. \tag{3.2}$$

We see in Section 3.2 that $\hat{n}$ accounts for the photons of the mode. We introduce the *phase shifting* operator

$$\hat{U}(\theta) \equiv \exp(-i\theta\hat{n}). \tag{3.3}$$

As the name suggests, the phase shifting operator provides the amplitude $\hat{a}$ with a phase shift θ when acting on $\hat{a}$

$$\hat{U}^{\dagger}(\theta)\,\hat{a}\,\hat{U}(\theta) = \hat{a}\exp(-i\theta). \tag{3.4}$$

This property is easily seen by calculating the derivative of $\hat{U}^{\dagger}\,\hat{a}\,\hat{U}$ with respect to θ

$$\frac{d}{d\theta}\,\hat{U}^{\dagger}\,\hat{a}\,\hat{U} = i\,[\hat{n},\hat{U}^{\dagger}\,\hat{a}\,\hat{U}] = \hat{U}^{\dagger}\,i\,[\hat{n},\hat{a}]\hat{U} = -i\,\hat{U}^{\dagger}\,[\hat{a},\hat{a}^{\dagger}]\hat{a}\,\hat{U}$$
$$= -i\,\hat{U}^{\dagger}\,\hat{a}\,\hat{U}. \tag{3.5}$$

Because the right-hand side of Eq. (3.4) obeys the same differential equation with the initial operator $\hat{a}$ for $\theta = 0$, both sides must be equal. There is another way of looking at formula (3.4). When the observer wishes to change the phase of the mode,

$$\hat{\mathbf{A}}_{k}(\mathbf{r},t) = \mathbf{A}_{k}(\mathbf{r},t)\,\exp(-i\theta)\,\hat{a}, \tag{3.6}$$

this phase is picked up by the quantum state in the Schrödinger picture. Here operators are not changed, but the state is altered,

$$\hat{\rho}(\theta) = \hat{U}\hat{\rho}\,\hat{U}^{\dagger}. \tag{3.7}$$

Finally, we introduce a pair of operators, $\hat{q}$ and $\hat{p}$, called the *quadratures*, see Fig. 3.1. They appear as the "real" and the "imaginary" part, respectively, of the "complex" amplitude $\hat{a}$ multiplied by $\sqrt{2}$

$$\hat{q} = \frac{1}{\sqrt{2}}\,(\hat{a}^{\dagger} + \hat{a}), \quad \hat{p} = \frac{i}{\sqrt{2}}\,(\hat{a}^{\dagger} - \hat{a}) \tag{3.8}$$

so that

$$\hat{a} = \frac{1}{\sqrt{2}}\,(\hat{q} + i\hat{p}). \tag{3.9}$$

For monochromatic modes $\hat{q}$ and $\hat{p}$ correspond to the in-phase and the out-of-phase component of the field amplitude (with respect to a phase reference). The in-phase component oscillates with $\exp(-i\omega_{k}t)$ and the out-of-phase component at the phase $\pi/2$ later. It is easy to see from the basic bosonic commutation relation (3.1) that $\hat{q}$ and $\hat{p}$ are canonically conjugate variables

$$[\hat{q},\hat{p}] = i. \tag{3.10}$$

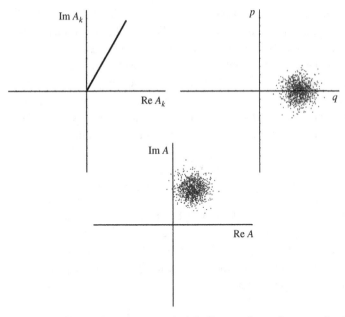

Fig. 3.1. Modes and quadratures. The left diagram shows the normalized complex mode amplitude A_k at one space–time point. The quadratures q and p of the right picture represent the degrees of freedom for each individual light mode; they depend on the quantum state and may fluctuate. The quadratures constitute the complex amplitude (3.9) and act like the position and momentum of a harmonic oscillator. The mode function A_k varies across space and time, whereas the "vibrational amplitude" (3.9) is uniform for each mode. The field strength A shown below is the complex product of the mode function A_k with the amplitude (3.9).

The quadratures $\hat{q}$ and $\hat{p}$ can be regarded as the position and the momentum in physical units where $\hbar = 1$. What do they describe? We express the photon number operator $\hat{n}$ in terms of the quadratures $\hat{q}$ and $\hat{p}$ and obtain, using the canonical commutation relation (3.10),

$$\hat{E} \equiv \hat{n} + \frac{1}{2} = \frac{\hat{q}^2}{2} + \frac{\hat{p}^2}{2}. \tag{3.11}$$

The right-hand side of this equation stands for the energy of a harmonic oscillator with unity mass and frequency; the single mode thus appears as an oscillator with position $\hat{q}$ and momentum $\hat{p}$, the *electromagnetic oscillator*. The photon number plus 1/2 gives the energy of the electromagnetic oscillator. The additional 1/2 is the zero-point energy of the single mode. For a single mode, the field strengths at all points in space in time oscillate in unison.

The quadratures $\hat{q}$ and $\hat{p}$ are the components of the "complex" vibrational amplitude $\hat{a}$ of the electromagnetic oscillator; they have nothing to do with the position and the momentum of a photon concepts that are problematic in any case (Mandel and Wolf, 1995). Nevertheless, the canonical commutation relation (3.10) entitles us to treat $\hat{q}$ and $\hat{p}$ as perfect examples of position and momentum – like quantities. We will see later in this book that this analogy is one of the key points why quantum optics allows some fundamental Gedanken experiments of quantum physics to be carried out – not literally but certainly in the spirit of their inventors. We note that phase shifting rotates the quadratures

$$\begin{aligned}
\hat{q}_\theta &\equiv \hat{U}^\dagger(\theta)\hat{q}\,\hat{U}(\theta) = \hat{q}\,\cos\theta + \hat{p}\,\sin\theta, \\
\hat{p}_\theta &\equiv \hat{U}^\dagger(\theta)\hat{p}\,\hat{U}(\theta) = -\hat{q}\,\sin\theta + \hat{p}\,\cos\theta,
\end{aligned}$$ (3.12)

as is easily verified using definition (3.8) and the phase shifting property (3.4) of the annihilation operator $\hat{a}$. We see that we can go from a position representation to a momentum representation via a phase shift θ of $\pi/2$. After these definitions and simple calculations we are prepared to introduce some single-mode states.

3.2 Single-mode states

In this section we discuss several states of the electromagnetic oscillator that have a number of useful applications (in a purely mathematical or in a truly physical sense). We begin with the quadrature states, then turn to the Fock states and the state of thermal light. Finally we consider coherent states as the most important realistic states of light, describing coherent laser light. All states are introduced as eigenstates of prominent observables such as the quadratures, the photon number, and the annihilation operator, or as their statistical mixtures.

3.2.1 Quadrature states

Let us call the eigenstates $|q\rangle$ and $|p\rangle$ of the quadratures $\hat{q}$ and $\hat{p}$ *quadrature states*, satisfying

$$\hat{q}\,|q\rangle = q\,|q\rangle, \quad \hat{p}\,|p\rangle = p\,|p\rangle.$$ (3.13)

Because the quadratures obey the canonical commutation relation (3.10) their spectrum must be unbounded and continuous (Cohen Tannoudji et al., 1977), as we would expect for position and momentum. They are orthogonal

$$\langle q\,|\,q'\rangle = \delta(q - q'), \quad \langle p\,|\,p'\rangle = \delta(p - p')$$ (3.14)

and complete

$$\int_{-\infty}^{+\infty} |q\rangle\langle q|\,dq = \int_{-\infty}^{+\infty} |p\rangle\langle p|\,dp = 1.$$ (3.15)

As is well known, position and momentum states are mutually related to each other by Fourier transformation

$$|p\rangle = \frac{1}{\sqrt{2\pi}} \int_{-\infty}^{+\infty} \exp(+iqp) |q\rangle \, dq,$$

$$|q\rangle = \frac{1}{\sqrt{2\pi}} \int_{-\infty}^{+\infty} \exp(-iqp) |p\rangle \, dp.$$

(3.16)

However, the quadrature states are not truly normalizable, and so they cannot be generated experimentally (at least in a strict sense). Nevertheless, they will appear in many mathematical tricks. For instance, they are needed to introduce the *quadrature wave functions*

$$\psi(q) = \langle q \mid \psi \rangle, \quad \widetilde{\psi}(p) = \langle p \mid \psi \rangle.$$

(3.17)

In contrast to the quadrature states, the quadrature wave functions have a physical meaning. Their moduli squared account for the quadrature probability distributions $|\psi(q)|^2$ and $|\widetilde{\psi}(p)|^2$ of the pure state $|\psi\rangle$, which can be measured using homodyne detection, as we consider in detail in Section 5.2.2.

3.2.2 Fock states

Let us introduce *Fock states*, or $|n\rangle$, as the eigenstates of the photon number operator $\hat{n}$

$$\hat{n}|n\rangle = n|n\rangle.$$

(3.18)

Fock states are named after Vladimir A. Fock (Владимир Александрович Фок) and are widely used in quantum field theory. As eigenstates of the number operator $\hat{n}$, Fock states have a perfectly fixed photon number. Single-photon Fock states $|1\rangle$ and photon pairs can be made using parametric amplification, see Section 7.1.3, while higher number states of light are still very difficult to generate with present technology. Fock states have been produced by the interaction of single atoms with microwave radiation stored in extremely high quality cavities (Varcoe et al., 2000).

Let us study the Fock states in some detail. First, we see that if $|n\rangle$ is an eigenstate of $\hat{n}$, then $\hat{a}|n\rangle$ must be an eigenstate as well, with the eigenvalue $n - 1$, because

$$\hat{n}\hat{a}|n\rangle = \hat{a}^{\dagger}\hat{a}^2|n\rangle = \left(\hat{a}\hat{a}^{\dagger}\hat{a} - \hat{a}\right)|n\rangle = (n-1)\hat{a}|n\rangle.$$

(3.19)

In a similar way, we easily show that $\hat{a}^{\dagger}|n\rangle$ is an eigenstate of $\hat{n}$ with the eigenvalue $n + 1$. So we derive the fundamental relations

$$\hat{a}|n\rangle = \sqrt{n}|n - 1\rangle,$$

(3.20)

$$\hat{a}^{\dagger}|n\rangle = \sqrt{n + 1}|n + 1\rangle.$$

(3.21)

The prefactors have been obtained using the fact that $\langle n | \hat{a}^\dagger \hat{a} | n \rangle$ must equal the eigenvalue n. Because of these relations, $\hat{a}$ is called the *annihilation operator* (it takes one photon out of $|n - 1\rangle$) and $\hat{a}^\dagger$ is called the *creation operator*. The annihilation operator or the creation operator lowers or raises the photon number in integer steps. What would happen if we had a Fock state with non-integer eigenvalue n? A sufficiently large number of lowerings would certainly produce a Fock state with a photon number less than zero. On the other hand, we know that the average $\langle \psi | \hat{a}^\dagger \hat{a} | \psi \rangle$ surely is non-negative, because it represents the modulus squared of the Hilbert space vector $\hat{a} | \psi \rangle$. This bound leads to a contradiction, because for eigenstates of $\hat{n}$ the average $\langle \hat{n} \rangle$ should equal the eigenvalue n. Consequently, no fractional photons exist, at least if the photon number is fixed precisely, that is for photon number eigenstates.

What happens if the photon number is integer – if we reach zero after lowering n in integer steps? Two options satisfy

$$\hat{a}^\dagger \hat{a} | 0 \rangle = 0. \tag{3.22}$$

One is to require that

$$\hat{a} | 0 \rangle = 0, \tag{3.23}$$

the other that

$$\hat{a} | 0 \rangle \neq 0, \quad \text{but} \quad \hat{a}^\dagger (\hat{a} | 0 \rangle) = 0. \tag{3.24}$$

Let us study the first option (3.23) first. Using the quadrature decomposition (3.9) of the annihilation operator and Schrödinger's famous formula $\hat{p} = -i \, \partial / \partial q$ in the q representation, we obtain a differential equation for the wave function $\psi_0(q)$ of the state $|0\rangle$

$$\hat{a} \, \psi_0(q) = \frac{1}{\sqrt{2}} \left(q + \frac{\partial}{\partial q} \right) \psi_0(q) = 0. \tag{3.25}$$

The solution of this equation is

$$\psi_0(q) = \pi^{-1/4} \exp \left(-\frac{q^2}{2} \right) \tag{3.26}$$

(normalized to yield $\int_{-\infty}^{+\infty} | \psi_0(q) |^2 \, dq = 1$). In the momentum representation we obtain the same formula for $\widetilde{\psi}_0(p)$

$$\widetilde{\psi}_0(p) = \pi^{-1/4} \exp \left(-\frac{p^2}{2} \right). \tag{3.27}$$

In this way we have shown that a well-behaved state with precisely zero photons exists, called the *vacuum state*. So even if the light mode is

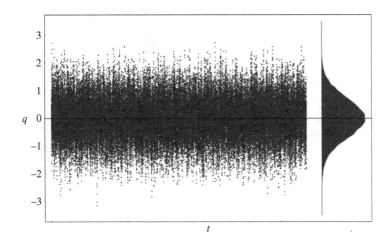

Fig. 3.2. Vacuum noise. The position quadrature q of light in the vacuum state (in darkness) was measured using homodyne detection, see Section 5.2.2. The histogram on the right side closely follows the theoretical prediction, $|\psi(q)|^2$ with the wave function (3.26). (Data: Akira Furusawa and Hidehiro Yonezawa.)

completely empty, a physically meaningful state that might cause physical effects is still associated with this "emptiness". Figure 3.2 shows a plot of the quadrature probability distribution $|\psi_0(q)|^2$ for a vacuum that has been measured using homodyne detection. (Homodyne detection is described in Section 5.2.2.) The Gaussian curve illustrates beautifully that even in a complete vacuum the quadratures are still restlessly fluctuating, and so do the field strengths. Of course, the quadratures must fluctuate; if both position and momentum quadratures were fixed, Heisenberg's uncertainty principle would be violated. The fluctuation energy of the vacuum state gives rise to the vacuum term $1/2$ in the energy (3.11) of the electromagnetic oscillator and to the Casimir force studied in Section 2.3 and Appendix A.

The excited Fock states are the solutions of the relation (3.21) for an initial vacuum

$$|n\rangle = \frac{\hat{a}^{\dagger n}}{\sqrt{n!}} |0\rangle. \tag{3.28}$$

We obtain a formula for their wave functions by expressing the relation (3.21) for $n = m - 1$ in the Schrödinger representation

$$\hat{a}^{\dagger} \psi_{m-1}(q) = \frac{1}{\sqrt{2}} \left(q - \frac{\partial}{\partial q} \right) \psi_{m-1}(q) = \sqrt{m}\, \psi_m(q). \tag{3.29}$$

This formula is satisfied by

$$\psi_n(q) = \frac{H_n(q)}{\sqrt{2^n\, n!}\, \sqrt{\pi}} \exp\left(-\frac{q^2}{2} \right). \tag{3.30}$$

Here the H_n denote the Hermite polynomials, and we have used their relation 10.13(14) of (Erdélyi et al., Vol. II, 1953). Figure 3.3 shows plots of some Fock wave functions. They appear as standing

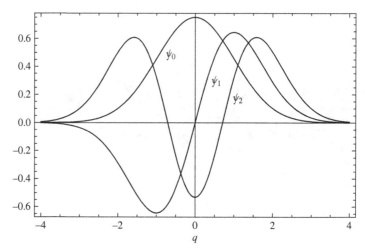

Fig. 3.3. Fock states. The picture shows the quadrature wave functions (3.30) of some Fock states. The wave functions are even, $\psi_n(-q) = \psi_n(q)$, for even numbers n and odd, $\psi_n(-q) = -\psi_n(q)$, for odd numbers. They oscillate in the classically allowed region between the turning points of a classical harmonic oscillator with energy $n + 1/2$. Outside this region, that is, in the classically forbidden zone, the wave functions decay exponentially.

Schrödinger waves for quadrature values ranging between the classical turning points of the electromagnetic oscillator. Consequently, the quadrature distributions, or squared wave functions, are quite broad. They illustrate that, because the Fock states are particle-like, they have noisy quadrature amplitudes and exhibit few features of a classical, stable wave.

Let us return to the second possibility (3.24) for a vacuum state with a wave function $\varphi_0(q)$. It means that the function

$$\varphi_{-1}(q) \equiv \hat{a}\, \varphi_0(q) = \frac{1}{\sqrt{2}} \left(q + \frac{\partial}{\partial q} \right) \varphi_0(q) \tag{3.31}$$

satisfies

$$\hat{a}^\dagger\, \varphi_{-1}(q) = \frac{1}{\sqrt{2}} \left(q - \frac{\partial}{\partial q} \right) \varphi_{-1}(q) = 0. \tag{3.32}$$

However, the solution of this equation

$$\varphi_{-1}(q) = C \exp\left(+\frac{q^2}{2} \right) \tag{3.33}$$

is not normalizable. Hence $\varphi_0(q)$, or the inhomogeneous solution of the differential equation (3.31) is not normalizable as well. It is called the *irregular wave function* of the vacuum state and is given by the expression

$$\varphi_0(q) = C \sqrt{\frac{\pi}{2}} \exp\left(-\frac{q^2}{2} \right) \mathrm{erfi}(q) \tag{3.34}$$

with erfi being the imaginary error function

$$\mathrm{erfi}(z) = \frac{2}{\sqrt{\pi}} \int_0^z \exp(t^2)\, dt. \tag{3.35}$$

Because the irregular vacuum is not normalizable, it must be rejected
as a physically meaningful state. However, one may encounter useful
physical applications of irregular wave functions, see Appendix B.

We have shown that a unique vacuum state exists for any single
mode of light, and hence for the entire electromagnetic field. All Fock
states are given as excitations (3.28) of the vacuum. Because the stationary Schrödinger equation is of second order in q there are only two
fundamental solutions – regular and irregular wave functions. The irregular ones are discarded as physical states. So the Fock states must be
complete,

$$\sum_{n=0}^{\infty} |n\rangle\langle n| = 1, \tag{3.36}$$

that is they span the whole Hilbert space of the electromagnetic
oscillator. Additionally, Fock states are orthonormal

$$\langle n|n'\rangle = \delta_{nn'} \tag{3.37}$$

because they are eigenstates of the Hermitian operator $\hat{n}$. The Fock
states form the most convenient and most frequently used orthonormal
Hilbert space basis in quantum optics, called the *Fock basis*.

3.2.3 Thermal states

Most natural light is thermal radiation: sunlight, for example, or the
light of most lamps. Thermal radiation is a state of the electromagnetic
field in thermal equilibrium (or part of the field in thermal equilibrium,
for example the part radiating out of a thermal source). Note that light
on its own cannot reach an equilibrium state, because it is not interacting
with itself, but when light is brought into contact with material media
or generated by thermal sources it may thermalize, as we discuss in
Chapter 6 and in particular in Section 6.2.2. In interacting with a hot
material the electromagnetic field exchanges photons. So the energy and
the photon number of thermal light fluctuates.

Let us derive the quantum state of thermal light without assuming
that the reader is familiar with quantum statistical mechanics. In thermal
equilibrium light is in a stationary state that does not evolve in time. The
most general stationary state is a statistical mixture of the eigenstates of
the Hamiltonian $\hat{H}$, a mixture of Fock states,

$$\hat{\rho} = \sum_{n} \rho_n |n\rangle\langle n|. \tag{3.38}$$

In addition to being stationary, thermal light is in a state of maximal
disorder, maximal entropy S for a given energy E that is set by the

ceaseless exchange of photons with the thermal environment, the hot material. Another natural constraint on ρ_n is the conservation of the total probability: the probabilities ρ_n must sum up to unity. It is mathematically convenient to describe the maximization of the entropy in the presence of constraints using Lagrange multipliers. We simply subtract the constraints from the entropy (1.21) with variable prefactors,

$$S = -k_B \sum_n \rho_n \ln \rho_n - a \left(\sum_n \rho_n - 1 \right) - b \left(\sum_n \rho_n E_n - E \right), \qquad (3.39)$$

where E denotes the average energy of the mode. If we optimize the entropy (3.39) for the parameters a and b we demand that the derivatives of S with respect to a and b vanish. In this case, both constraints on the ρ_n follow. On the other hand, when the constraints are satisfied the modified entropy (3.39) agrees with the original (1.21). So the extremum of S with respect to all the ρ_n and the parameters a and b, the Lagrange multipliers, solves the optimization problem with constraints. We obtain by differentiation

$$0 = \frac{\partial S}{\partial \rho_n} = -k_B \left(\ln \rho_n + 1 \right) - a - b E_n \qquad (3.40)$$

that has the solution

$$\rho_n = \frac{1}{Z} \exp \left(-\frac{E_n}{k_B T} \right) \qquad (3.41)$$

where T denotes $1/b$ and Z abbreviates $\exp(1 + a/k_B)$. As we will see in a moment, T is the *temperature* while Z is the *partition function* or *statistical sum*, because

$$Z = \sum_n \exp \left(-\frac{E_n}{k_B T} \right) \qquad (3.42)$$

in order to satisfy the conservation of the total probability. When thermal equilibrium is reached the entropy assumes the value

$$S = -k_B \sum_n \rho_n \ln \rho_n = k_B \ln Z + \frac{E}{T} \qquad (3.43)$$

where E denotes the average energy

$$E = \sum_n \rho_n E_n. \qquad (3.44)$$

In thermodynamics, E is the internal energy. Suppose that T varies. We obtain from the definition (3.42) of the statistical sum that

$k_B T^2 dZ = ZE\, dT$. Hence we get for the derivative of the entropy (3.43) the expression

$$\frac{dS}{dE} = \frac{1}{T}.$$ (3.45)

This formula is the definition of the thermodynamic temperature (Landau and Lifshitz, Vol. V, 1996), which justifies our terminology. We thus obtained *Boltzmann's formula* (3.41) for the statistical distribution of states with energies E_n in thermal equilibrium with temperature T.

So far our analysis has been rather general, we have not used the specific physics of the electromagnetic oscillator, but derived the essentials of the quantum statistical mechanics of a *canonical Gibbs ensemble*. We can combine our results (3.38), (3.41) and (3.42) in the compact expressions

$$\hat{\rho} = \frac{1}{Z} \exp\left(-\frac{\hat{H}}{k_B T}\right), \quad Z = \mathrm{tr}\left\{\exp\left(-\frac{\hat{H}}{k_B T}\right)\right\}.$$ (3.46)

For the specific case of a single light mode we use the parameter

$$\beta = \frac{\hbar\omega}{k_B T}.$$ (3.47)

In statistical physics, the inverse temperature $1/(k_B T)$ is typically denoted by β, but here we have included $\hbar\omega$ in β for convenience. For the electromagnetic oscillator of the light mode, the statistical sum is the geometric series

$$Z = \sum_{n=0}^{\infty} e^{-n\beta} = \frac{1}{1 - e^{-\beta}}.$$ (3.48)

Therefore, the thermal state of light is described by the density matrix

$$\hat{\rho} = \left(1 - e^{-\beta}\right) \sum_{n=0}^{\infty} e^{-n\beta} |n\rangle\langle n|.$$ (3.49)

For the average photon number, we get the *Planck spectrum* of the harmonic oscillator in thermal equilibrium,

$$\bar{n} = \frac{1}{Z} \sum_{n=0}^{\infty} e^{-n\beta} n = -\frac{1}{Z}\frac{\partial Z}{\partial \beta} = \frac{1}{e^{\beta} - 1}.$$ (3.50)

When the entire electromagnetic field is in a thermal state, we obtain the total energy density $\varrho(\omega)$ per volume and frequency by summing over the individual energies $\hbar\omega\,\bar{n}$ of all the modes with the same frequency ω and dividing by the total volume. In empty space, we arrive at Planck's radiation formula that, historically, opened one of the windows to the quantum world (Jammer, 1989),

$$\varrho(\omega) = \left(\frac{\hbar\omega^3}{\pi^2 c^3}\right) \frac{1}{\exp\left(\hbar\omega/(k_B T)\right) - 1}. \tag{3.51}$$

3.2.4 Coherent states

Laser light differs from thermal light – it is a coherent electromagnetic wave, the closest possible analogue to a classical electromagnetic wave, whereas thermal light represents a statistical mixture of photons. Here we discuss quantum states that describe ideal laser light, *coherent states*. Since laser light has a well-defined amplitude, we define coherent states as the eigenstates of the amplitude operator, the annihilation operator $\hat{a}$,

$$\hat{a}\,|\alpha\rangle = \alpha\,|\alpha\rangle. \tag{3.52}$$

Coherent states are also named *Glauber states* after Roy J. Glauber who introduced them into quantum optics inspired by Erwin Schrödinger's coherent wave packets of the harmonic oscillator (Schrödinger, 1926). Because the annihilation operator $\hat{a}$ is not Hermitian, the eigenvalues of $\hat{a}$ are complex. They correspond to the complex wave amplitudes in classical optics, with magnitudes $|\alpha|$ and phases $\arg\alpha$. Note, however, that the phase of laser light drifts in time, due to spontaneous emission in the active material of the laser. So, strictly speaking, the quantum state of continuous laser light is a phase-randomized ensemble of coherent states. On the other hand, in typical quantum optical experiments all light used originates from a master laser. The phases of all light beams involved fluctuate in unison with the master laser, they are constant relative to the master light. Since the master laser sets a global reference for the entire experiment, we would never notice the fluctuating phase and can regard coherent laser light as being in the coherent state (3.52) with fixed phase.

Coherent states come as close as quantum mechanics allows to wave-like states of the electromagnetic oscillator. As the wave aspects of light are commonly regarded as classical, coherent states are often called classical states. Furthermore, fields in statistical mixtures of coherent states are classical as well, whereas any state that cannot be understood as an ensemble of coherent states is called *non-classical*. The experimental generation and application of non-classical light fields is one of the top issues of modern quantum optics. Despite much progress made, producing non-classical states is still challenging, because they are easily destroyed, reduced to classical light, by any kind of losses.

Let us return to coherent states and study their properties. First we note that vacuum is a coherent state as well, because it satisfies Eq. (3.52) for $\alpha \doteq 0$; vacuum is a zero-amplitude coherent state. Without much mathematical effort we see directly from the definition (3.52) that the mean energy of a coherent state is simply

$$\langle \hat{E} \rangle = \langle \alpha | \, \hat{a}^\dagger \hat{a} + \frac{1}{2} \, | \alpha \rangle = |\alpha|^2 + \frac{1}{2}, \tag{3.53}$$

or the sum of the classical wave intensity $|\alpha|^2$ and the vacuum energy $1/2$. We also see easily from the definition (3.52) that a phase shift by the angle θ simply shifts the phase $\arg \alpha$ of the coherent state amplitude

$$\hat{U}(\theta) \, | \alpha \rangle = | \alpha \exp(-i\theta) \rangle. \tag{3.54}$$

This result is what we would expect for wave-like states.

To study coherent states more carefully, we introduce the *displacement operator*

$$\hat{D}(\alpha) = \exp \left(\alpha \, \hat{a}^\dagger - \alpha^* \hat{a} \right). \tag{3.55}$$

Because $i(\alpha \, \hat{a}^\dagger - \alpha^* \hat{a})$ is Hermitian, $\hat{D}$ must be unitary. We show in Section 7.1.4 how to implement the displacement operator in a practical device. The operator (3.55) displaces the amplitude $\hat{a}$ by the complex number α

$$\hat{D}^\dagger(\alpha) \, \hat{a} \, \hat{D}(\alpha) = \hat{a} + \alpha. \tag{3.56}$$

To prove this statement we imagine the displacement α as being decomposed into infinitesimal steps $\delta \alpha$ and we obtain in first order of $\delta \alpha$

$$\hat{D}^\dagger(\delta\alpha) \, \hat{a} \, \hat{D}(\delta\alpha) = \hat{a} + \left[\hat{a}, \hat{a}^\dagger \, \delta\alpha - \hat{a} \, \delta\alpha^* \right] = \hat{a} + \delta\alpha. \tag{3.57}$$

Because the total displacement operator $\hat{D}(\alpha)$ with $\alpha = \sum \delta\alpha$ is the product $\prod \hat{D}(\delta\alpha)$ of the infinitesimal displacements $\hat{D}(\delta\alpha)$, we can apply the infinitesimal steps (3.57) as often as we need to show that $\hat{D}^\dagger(\alpha) \, \hat{a} \, \hat{D}(\alpha)$ equals indeed $\hat{a} + \sum \delta\alpha$, which proves Eq. (3.56). (Readers who are familiar with Lie groups notice that we have used the properties of the Lie algebra to study the group elements. Quite generally, coherent states are intimately linked to Lie groups; see Perelomov (1986)). What has the displacement operator to do with coherent states? Let us apply a negative displacement to $|\alpha\rangle$. We see from the basic property (3.56) of the displacement operator that

$$\hat{a} \, \hat{D}(-\alpha) \, | \alpha \rangle = \hat{D}(-\alpha) \, \hat{D}^\dagger(-\alpha) \, \hat{a} \, \hat{D}(-\alpha) \, | \alpha \rangle$$
$$= \hat{D}(-\alpha) \, (\hat{a} - \alpha) \, | \alpha \rangle, \tag{3.58}$$

which must equal zero because of the definition (3.52) of coherent states. This implies that $\hat{D}(-\alpha)|\alpha\rangle$ is *the* vacuum state $|0\rangle$. Consequently, a coherent state $|\alpha\rangle$ is a *displaced vacuum*

$$|\alpha\rangle = \hat{D}(\alpha)|0\rangle. \tag{3.59}$$

Of course, this does not mean that coherent states are physically similar to vacuum states. They have only some quantum noise properties in common. (Displacing the vacuum may appear as a rather inappropriate description of generating high quality laser light.)

To study the relation between coherent states and the vacuum in more detail, we calculate the quadrature wave functions $\psi_\alpha(q)$ and $\tilde{\psi}_\alpha(p)$. We decompose the complex amplitude α into real and imaginary parts

$$\alpha = \frac{1}{\sqrt{2}}(q_0 + ip_0) \tag{3.60}$$

and represent the displacement operator in terms of the quadratures $\hat{q}$ and $\hat{p}$

$$\hat{D} = \exp\left(ip_0\hat{q} - iq_0\hat{p}\right). \tag{3.61}$$

We take advantage of the *Baker–Hausdorff formula*

$$\exp\left(\hat{F} + \hat{G}\right) = \exp\left(-\frac{1}{2}\left[\hat{F}, \hat{G}\right]\right)\exp\left(\hat{F}\right)\exp\left(\hat{G}\right)$$

$$= \exp\left(+\frac{1}{2}\left[\hat{F}, \hat{G}\right]\right)\exp\left(\hat{G}\right)\exp\left(\hat{F}\right) \tag{3.62}$$

for any two operators with the property that the commutator $\left[\hat{F}, \hat{G}\right]$ commutes with both $\hat{F}$ and $\hat{G}$. This fundamental operator relation is proved for instance in (Gardiner, 1991). Here we use the Baker–Hausdorff formula to split $\hat{D}$ into three parts

$$\hat{D}(\alpha) = \exp\left(+\frac{ip_0q_0}{2}\right)\exp\left(-iq_0\hat{p}\right)\exp\left(ip_0\hat{q}\right)$$

$$= \exp\left(-\frac{ip_0q_0}{2}\right)\exp\left(ip_0\hat{q}\right)\exp\left(-iq_0\hat{p}\right). \tag{3.63}$$

In the position representation the momentum operator $\hat{p}$ equals $-i\partial/\partial q$ and the exponential $\exp\left(-q_0\,\partial/\partial q\right)$ is a translation operator

$$\exp\left(-q_0\frac{\partial}{\partial q}\right)\psi(q) = \psi(q - q_0), \tag{3.64}$$

as is easily verified by differentiating both sides with respect to q_0: both sides of Eq. (3.64) satisfy the same differential equation in q_0 and they also share the same initial condition $\psi(q)$ for $q_0 = 0$. In this way, we see that the displacement operator acts in three steps on a position wave function: first it displaces the wave function, then it multiplies it with $\exp(ip_0\hat{q})$, that is, with $\exp(ip_0q)$ in the position representation, and finally, the displacement operator attaches the phase factor $\exp(-ip_0q_0/2)$ to the wave function. Because coherent states are displaced vacua the position wave function is simply

$$\psi_\alpha(q) = \psi_0(q - q_0) \exp\left(+ip_0q - \frac{ip_0q_0}{2}\right) \tag{3.65}$$

$$= \pi^{-1/4} \exp\left(-\frac{(q - q_0)^2}{2} + ip_0q - \frac{ip_0q_0}{2}\right) \tag{3.66}$$

with ψ_0 being the vacuum wave function. In a similar way we obtain the momentum wave function

$$\tilde{\psi}_\alpha(p) = \pi^{-1/4} \exp\left(-\frac{(p - p_0)^2}{2} - iq_0p + \frac{ip_0q_0}{2}\right). \tag{3.67}$$

Equations (3.66) and (3.67) show that the quadrature probability distributions $|\psi_\alpha(q)|^2$ and $|\tilde{\psi}_\alpha(p)|^2$ of coherent states are Gaussian with the same width as the Gaussian curve for vacuum. They are shifted only by the real amplitudes q_0 and p_0; we would obtain a completely similar picture for the coherent states, see Fig. 3.4, as for the vacuum in Fig. 3.2. In this sense coherent states are similar to the vacuum. Only the vacuum fluctuations contaminate the quadrature amplitudes, illustrating that coherent states are wave-like – they have just as much amplitude noise as is unavoidable. This is an important reason why high-quality laser light is such a fantastic tool for experimentation.

So much for the wave features of coherent states; let us now study the particle aspects. For this purpose we seek the Fock representation of $\hat{D}(\alpha)|0\rangle$. We express $\hat{D}$ in terms of $\hat{a}$ and $\hat{a}^\dagger$, Eq. (3.55), and use

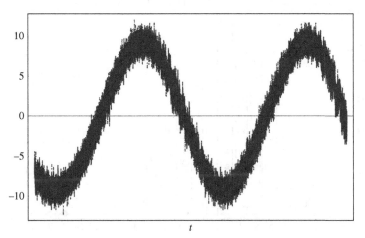

Fig. 3.4. Coherent state. The quadrature q_θ of light in a coherent state was measured using homodyne detection, see Section 5.2.2, while the phase θ was varied. The coherent state has a well-defined amplitude apart from vacuum fluctuations. (Data: Akira Furusawa and Hidehiro Yonezawa.)

the Baker–Hausdorff formula (3.62) again to split the displacement operator into three parts

$$\hat{D} = \exp\left(-\frac{1}{2}|\alpha|^2\right) \exp\left(\alpha\hat{a}^\dagger\right) \exp\left(-\alpha^*\hat{a}\right). \tag{3.68}$$

Because the annihilation operator annihilates the vacuum, all powers of $\hat{a}$ contained in the Taylor series expansion of the exponential $\exp(-\alpha^*\hat{a})$ give zero when applied to $|0\rangle$, with the only exception of the zeroth-order term 1. So, the exponential $\exp\left(-\alpha^*\hat{a}\right)$ does not affect the vacuum state at all. We expand the other exponential $\exp(\alpha\hat{a}^\dagger)$ in the splitting (3.68) in the Taylor series and use formula (3.28) for the Fock states $|n\rangle$. We obtain

$$|\alpha\rangle = \exp\left(-\frac{1}{2}|\alpha|^2\right) \sum_{n=0}^{\infty} \frac{\alpha^n}{\sqrt{n!}} |n\rangle. \tag{3.69}$$

The Fock representation (3.69) shows that a coherent state has *Poissonian* photon statistics

$$p_n = \left|\langle n|\alpha\rangle\right|^2 = \frac{|\alpha|^{2n}}{n!} \exp\left(-|\alpha|^2\right), \tag{3.70}$$

see Fig. 3.5. Counting the photons of a coherent state means making repeated measurements on a statistical ensemble of identically prepared fields. The probability for counting n photons is given by the Poissonian (3.70) and the average photon number is $|\alpha|^2$, as we have seen before. Classical particles obey the same statistical law when they are taken at random from a pool with $|\alpha|^2$ on average (Tijms, 2007). For example, a Poissonian describes the raisins in a cake. Suppose they are randomly distributed with $|\alpha|^2$ on average per unit volume. Under these conditions, the probability of finding n raisins per unit volume follows the

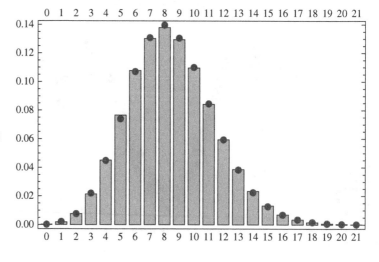

Fig. 3.5. Photon number distribution of light in a coherent state. The distribution was reconstructed from experimental homodyne data. We see that the reconstructed histogram (dots) is approximately Poissonian (bar chart). (Data: Akira Furusawa and Hidehiro Yonezawa.)

Poisson formula (3.70). Therefore we may say that, when the photons of a coherent state are counted, they behave like randomly distributed classical particles. This classical randomness seems not too surprising, because coherent states are wave-like.

Let us finally derive some mathematical properties of coherent states that turn out to be quite useful. First, we note that coherent states are not exactly orthogonal to each other, because they are not eigenstates of a Hermitian operator. Instead, they are approximately orthogonal when their amplitudes differ sufficiently. In fact, we obtain from the Fock representation (3.69)

$$\langle \alpha' | \alpha \rangle = \exp\left(-\frac{|\alpha|^2}{2} - \frac{|\alpha'|^2}{2}\right) \sum_{n=0}^{\infty} \frac{(\alpha'^* \alpha)^n}{n!}$$

$$= \exp\left(-\frac{|\alpha|^2}{2} - \frac{|\alpha'|^2}{2} + \alpha'^* \alpha\right) \tag{3.71}$$

and consequently

$$\left|\langle \alpha' | \alpha \rangle\right|^2 = \exp\left(-|\alpha - \alpha'|^2\right). \tag{3.72}$$

The Gaussian in Eq. (3.72) rapidly approaches zero when the amplitudes α and α' differ significantly more than the quadrature noise level of the vacuum. We also note that coherent states are complete

$$\int_{-\infty}^{+\infty} \int_{-\infty}^{+\infty} |\alpha\rangle\langle\alpha| \frac{dq_0 \, dp_0}{2\pi} = 1. \tag{3.73}$$

That is, we may express physical quantities in a coherent-state basis. (Coherent states are even overcomplete because fewer than all of them form a basis already (Cahill, 1965; Bacry et al., 1975). This property is, by the way, a side effect of their lack of strict orthogonality.) The proof of the completeness relation (3.73) is a matter of substituting the Fock representation for $|\alpha\rangle$ and performing the necessary integrations using polar coordinates in the complex plane.

3.3 Uncertainty and squeezing

In the preceding section we introduced Fock states, thermal states and coherent states as some physically meaningful states of a single light mode. We have seen that the quadrature amplitudes of these states fluctuate according to certain probability distributions. Coherent states are distinguished for having only as much statistical uncertainty in their quadrature amplitudes as the vacuum. Is this the quantum mechanical optimum, or, to put it differently, what are the *minimum uncertainty states*? Wolfgang Pauli settled this question in a brilliant few-line proof published in his *Handbuch der Physik* article (Pauli, 1933). Let us first denote the average complex amplitude of possible candidate states $|\psi\rangle$ by α

$$\langle \psi | \hat{a} | \psi \rangle = \alpha = \frac{1}{\sqrt{2}} (q_0 + i p_0). \tag{3.74}$$

We remove the amplitude α from consideration by applying the displacement operator to obtain a new state,

$$|\varphi\rangle = \hat{D}(-\alpha) |\psi\rangle \tag{3.75}$$

that contains the same amount of quadrature noise as the state $|\psi\rangle$. The variances of q and p are given by

$$\begin{aligned}
\Delta^2 q &= \langle \psi | (\hat{q} - q_0)^2 | \psi \rangle = \langle \varphi | \hat{q}^2 | \varphi \rangle, \\
\Delta^2 p &= \langle \psi | (\hat{p} - p_0)^2 | \psi \rangle = \langle \varphi | \hat{p}^2 | \varphi \rangle.
\end{aligned} \tag{3.76}$$

Now we use Pauli's argument (Pauli, 1933) and state that for the position wave function $\varphi(q)$ of $|\varphi\rangle$ the quantity

$$\delta \equiv \left| \frac{q}{2 \Delta^2 q} \varphi + \frac{\partial \varphi}{\partial q} \right|^2 \tag{3.77}$$

must necessarily be greater or at least equal to zero. On the other hand,

$$\begin{aligned}
\delta &= \frac{1}{4} \left(\frac{q}{\Delta^2 q} \right)^2 \varphi^* \varphi + \frac{q}{2 \Delta^2 q} \left(\varphi \frac{\partial \varphi^*}{\partial q} + \varphi^* \frac{\partial \varphi}{\partial q} \right) + \frac{\partial \varphi^*}{\partial q} \frac{\partial \varphi}{\partial q} \\
&= \frac{1}{4} \left(\frac{q}{\Delta^2 q} \right)^2 \varphi^* \varphi + \frac{1}{2} \frac{\partial}{\partial q} \left(\frac{q}{\Delta^2 q} \varphi^* \varphi \right) - \frac{1}{2} \frac{\varphi^* \varphi}{\Delta^2 q} + \frac{\partial \varphi^*}{\partial q} \frac{\partial \varphi}{\partial q} \\
&= \frac{1}{4 (\Delta^2 q)^2} \left(q^2 - 2 \Delta^2 q \right) \varphi^* \varphi + \frac{\partial \varphi^*}{\partial q} \frac{\partial \varphi}{\partial q} + \frac{1}{2} \frac{\partial}{\partial q} \left(\frac{q}{\Delta^2 q} \varphi^* \varphi \right). \tag{3.78}
\end{aligned}$$

We integrate the last line, drop the total differential $\partial \left(\varphi^* \varphi \, q / \Delta^2 q \right) / \partial q$ and use $\hat{p} = -i \, \partial / \partial q$ for the $(\partial \varphi^* / \partial q)(\partial \varphi / \partial q)$ term to see that

$$\int_{-\infty}^{+\infty} \delta \, dq = -\frac{1}{4 \Delta^2 q} + \Delta^2 p \geq 0 \tag{3.79}$$

which implies for $\Delta q = \sqrt{\Delta^2 q}$ and $\Delta p = \sqrt{\Delta^2 p}$ the relation

$$\Delta q \, \Delta p \geq \frac{1}{2}. \tag{3.80}$$

This formula is nothing other than Heisenberg's uncertainty relation (with $\hbar$ being scaled to unity). Because Pauli used the integration of δ in his derivation, he produced essentially a local version of Heisenberg's uncertainty principle. The great advantage of this mathematical trick is that it reveals the minimum uncertainty states in the twinkling of an eye. In fact, because $\delta \geq 0$, the equality sign in relation (3.80) holds only if

$$\frac{1}{2}\frac{q}{\Delta^2 q}\varphi + \frac{\partial\varphi}{\partial q} = 0. \tag{3.81}$$

The normalized solution of this differential equation is

$$\varphi(q) = \left(2\pi\Delta^2 q\right)^{-1/4} \exp\left(-\frac{q^2}{4\Delta^2 q}\right). \tag{3.82}$$

So, apart from a displacement, the minimum-uncertainty states have Gaussian wave functions like the coherent states. However, the variance $\Delta^2 q$ should not necessarily equal $1/2$, as is the case for coherent states. In other words, both variances $\Delta^2 q$ and $\Delta^2 p$ are not required to be equal to minimize Heisenberg's uncertainty relation (3.80). The statistical uncertainty of the position quadrature q may be *squeezed* below the vacuum level $1/2$ at the cost, however, of enhancing the uncertainty in the canonically conjugate quadrature p and vice versa.

Let us study this squeezing effect more carefully. We parameterize the deviation of the variances from their vacuum values by a real number ζ called the *squeezing parameter*

$$\Delta^2 q = \frac{1}{2}e^{-2\zeta}, \quad \Delta^2 p = \frac{1}{2}e^{+2\zeta}. \tag{3.83}$$

Obviously, the product of Δq and Δp equals the minimal value $1/2$. How can we squeeze the vacuum? Mathematically, we could just scale the position wave function ψ_0 for the vacuum

$$\varphi(q) = e^{\zeta/2}\,\psi_0(e^\zeta\,q). \tag{3.84}$$

The prefactor $\exp(\zeta/2)$ in this scaling serves for maintaining the normalization of $\varphi(q)$. The momentum wave function $\tilde\varphi(p)$ is the Fourier transformed position wave function. Consequently,

$$\tilde\varphi(p) = e^{-\zeta/2}\,\tilde\psi_0(e^{-\zeta}\,p). \tag{3.85}$$

This formula implies that the momentum wave function is streched when the position wave function is squeezed and vice versa. We differentiate φ in Eq. (3.84) with respect to the squeezing parameter ζ and obtain

$$\frac{\partial\varphi}{\partial\zeta} = \frac{1}{2}\left(q\frac{\partial}{\partial q} + \frac{\partial}{\partial q}q\right)\varphi = \frac{1}{2}(i\hat{q}\hat{p} + i\hat{p}\hat{q})\varphi. \tag{3.86}$$

Since $i\hat{q}\hat{p} + i\hat{p}\hat{q}$ equals $\hat{a}^2 - \hat{a}^{\dagger 2}$ we can express the solution of this differential equation in terms of the unitary *squeezing operator*

$$\hat{S}(\zeta) \equiv \exp\left(\frac{\zeta}{2}\left(\hat{a}^2 - \hat{a}^{\dagger 2}\right)\right) \tag{3.87}$$

and obtain for the *squeezed vacuum state*

$$|\varphi\rangle = \hat{S}(\zeta)|0\rangle. \tag{3.88}$$

According to Pauli's proof, all minimum uncertainty states are displaced Gaussian states, that is, they have displaced rescaled vacuum wave functions. Consequently, all minimum uncertainty states are *displaced squeezed vacua*

$$|\psi\rangle = \hat{D}(\alpha)\,\hat{S}(\zeta)|0\rangle \tag{3.89}$$

having position wave functions of

$$\psi(q) = \pi^{-1/4}\,e^{\zeta/2}\,\exp\left(-e^{2\zeta}\frac{(q - q_0)^2}{2} + ip_0 q - \frac{ip_0 q_0}{2}\right). \tag{3.90}$$

The quadratures of the squeezed vacuum are illustrated in Figs. 3.6 and 3.7.

We have found not only a convenient mathematical notation for the squeezed states, but also one possible physical process for generating squeezed light experimentally. We simply interpret the squeezing operator $\hat{S}(\zeta)$ as an evolution operator (1.33) describing the result of the interaction

$$\hat{H}_{int} = \chi\left(b^*\hat{a}^2 - b\,\hat{a}^{\dagger 2}\right). \tag{3.91}$$

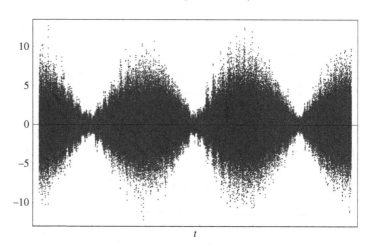

Fig. 3.6. Quadrature noise of a squeezed vacuum. The quadrature q_θ was measured using homodyne detection, see Section 5.2.2, while the phase θ was gradually varied. The quadrature noise is reduced for the squeezed component, but significantly enhanced for the anti-squeezed quadrature. (Data: Akira Furusawa and Hidehiro Yonezawa.)

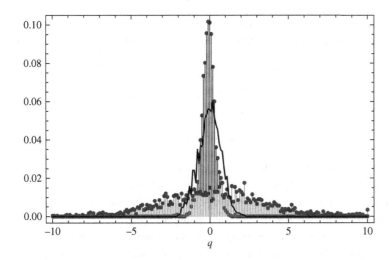

Fig. 3.7. Quadrature histograms of a squeezed vacuum. The histograms were taken from the data of Fig. 3.6. The squeezed (dark grey) and the anti squeezed (light grey) components are shown in comparison with the quadrature distribution of a vacuum (black). (Data: Akira Furusawa and Hidehiro Yonezawa.)

The squeezing parameter ζ contains the product of the amplitude b, the coupling constant χ and the interaction time. Processes described by nonlinear Hamiltonians such as $\hat{H}_{int}$ in Eq. (3.91) belong to the area of *nonlinear optics* (Shen, 1984; Boyd, 1992). In particular, the squeezing interaction (3.91) is realized by the degenerate parametric amplification of the light mode that we describe in detail in Section 7.1. A crystal such as KTP (potassium titanyl phosphate) is pumped by another laser beam with amplitude b and twice the frequency of the mode of interest. The B photons of the pump beam are converted into pairs of A photons with a probability that depends on the coupling constant χ. There is another picture for understanding this process: the crystal acts like an electromagnetic swing; the pump modulates the oscillation of the A mode at twice its frequency. This is precisely the condition for a playground swing, instinctively known to every child. The child modulates the effective length of the swing at twice its oscillation frequency. In this way, tiny wobbles are amplified; the swing starts to oscillate. Such a classical swing relies on initial fluctuations that are in-phase with respect to the parametric pump. A quantum swing like the degenerate parametric amplifier experiences at least the vacuum fluctuations from the very beginning. Vacuum fluctuations that are in-phase with respect to the pump are amplified, whereas out-of-phase fluctuations get deamplified or, in other words, squeezed.

A squeezed vacuum requires a pump for generation, and, hence, when produced it carries energy. To quantify the amount of squeezing energy we note that the squeezing operator changes the quadratures

$$\hat{S}^\dagger(\zeta)\,\hat{q}\,\hat{S}(\zeta) = \hat{q}\,e^{-\zeta}, \quad \hat{S}^\dagger(\zeta)\,\hat{p}\,\hat{S}(\zeta) = \hat{p}\,e^{+\zeta} \qquad (3.92)$$

because it scales the eigenfunctions of $\hat{q}$ and $\hat{p}$ accordingly. Substituting for $\hat{a}$ the quadrature decomposition (3.9) we see that

$$\hat{S}^\dagger(\zeta)\,\hat{a}\,\hat{S}(\zeta) = \hat{a}\cosh\zeta - \hat{a}^\dagger\sinh\zeta. \tag{3.93}$$

We use this formula to express the energy (3.11) of a squeezed state (3.89) and obtain

$$\langle\psi|\,\hat{E}\,|\psi\rangle = |\alpha|^2 + \frac{1}{2} + \sinh^2\zeta. \tag{3.94}$$

Three terms contribute to the energy. The first accounts for the coherent energy given by the modulus squared of the coherent amplitude α, the second is the vacuum energy $1/2$, and the last term quantifies the fluctuation energy of squeezed states. Originally, the contribution to this squeezing energy comes from the pump used to generate the squeezed light. It is stored in the enhanced fluctuations of the stretched component. Because both the squeezed and the stretched quadratures contribute to $\hat{E} = (\hat{q}^2 + \hat{p}^2)/2$, even a squeezed vacuum carries energy.

Let us calculate the energy distribution, that is, the photon number statistics of a squeezed vacuum

$$p_n = \left|\langle n|\,\hat{S}(\zeta)\,|0\rangle\right|^2. \tag{3.95}$$

We express the scalar product in the position representation

$$\langle n|\,\hat{S}(\zeta)\,|0\rangle = \int_{-\infty}^{+\infty}\psi_n(q)\,e^{\zeta/2}\,\psi_0(e^\zeta\,q)\,dq. \tag{3.96}$$

A squeezed vacuum as well as the vacuum state is perfectly symmetric if we change the sign of the quadrature amplitude q, because it has an even wave function $\psi(q) = \psi(-q)$. The wave functions $\psi_n(q)$ for the Fock states are even for even photon numbers and odd if n is odd. Consequently, the integral (3.96) vanishes for odd photon numbers and we obtain

$$p_{2m+1} = 0 \quad (m = 0, 1, 2, \ldots). \tag{3.97}$$

A squeezed vacuum contains only photon pairs. We may regard this fact as a simple consequence of the mirror symmetry of squeezing. A deeper physical reason for this remarkable property is that a squeezed vacuum may be generated in a parametric process described by the quadratic Hamiltonian (3.91). Loosely speaking, photons are created in pairs: each pump photon is converted into two signal photons of half the pump frequency. The probability for finding a photon pair is

$$p_{2m} = \binom{2m}{m}\frac{1}{\cosh\zeta}\left(\frac{1}{2}\tanh\zeta\right)^{2m} \quad (m = 0, 1, 2, \ldots) \tag{3.98}$$

expressed in terms of the binomial coefficient

$$\binom{n}{k} = \frac{n!}{k!\,(n-k)!}.\qquad\qquad(3.99)$$

Here Eq. 2.20.3.3. of Prudnikov et al. (Vol. II, 1992), has been used to perform the necessary integration (3.96). We can explain formula (3.98) in terms of the statistics of classical particles, similar to our interpretation of the Poissonian photon distribution of coherent states. Formula (3.98) appears like a probability distribution of independently produced particle pairs. Photons are generated independently from one another with a probability proportional to $(\tanh \zeta)/2$, that is, proportional to half of the squeezing parameter ζ for weak pumping. For stronger pumping the generation process becomes saturated. We observe pairs of $2m$ independently produced photons. All photons appear as distinguishable classical particles, but the detector cannot discriminate between them. It detects any m pairs of $2m$ particles, giving rise to a statistical enhancement described by the binomial coefficient (3.99) in formula (3.98). Note that although this explanation is consistent with the pair statistics (3.98), it loses its meaning when the wave features of light (for instance, interference effects) become important. Figure 3.8 shows an experimental plot of the photon statistics of a squeezed vacuum

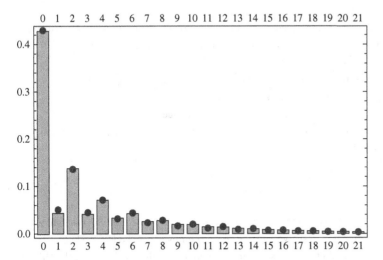

Fig. 3.8. Photon number distribution of a squeezed vacuum. Photons are produced in pairs. However, due to detection inefficiencies sometimes only one photon of a pair is observed. As a result, odd photon numbers occur with non-zero probability. The histogram (dots) was reconstructed from the measured quadratures of Fig. 3.6 and fitted to theory (bar chart). The theory is based on formula (3.98) and the effect of detection losses modelled by a beam splitter (explained in Section 6.2.2). (Data: Akira Furusawa and Hidehiro Yonezawa.)

measured using a method described in Section 5.2.3 and Appendix B. We see clearly near-zero probability for odd photon numbers and a decreasing probability for an increasing number of pairs. The small non-zero probability for odd numbers is caused by detection inefficiencies, where only one photon of a photon pair is detected. Because higher number pairs are produced by higher-order processes, they are less likely to be observed than lower number pairs. Vacuum has always the lion's share in the photon number distribution of a squeezed vacuum.

Finally, we note that squeezed states are non-classical states. Because they are pure states and different from coherent states, they cannot be described in terms of statistical mixtures of coherent states. Apart from this trivial formal statement, the quadrature noise reduction below the vacuum level and the photon pairing of squeezed vacuums illustrate beautifully that these states have indeed distinguished quantum properties.

3.4 Questions

3.1 What is the photon-number operator?

3.2 What is the phase-shifting operator? What is its effect on the mode operators?

3.3 What are the quadrature operators? What is their physical interpretation?

3.4 What are quadrature wave functions? Why are they useful?

3.5 What are photons?

3.6 Why are $\hat{a}$ and $\hat{a}^\dagger$ called annihilation and creation operators?

3.7 Why are photons quantized?

3.8 Why is the vacuum state of a mode unique?

3.9 Show that the wave function of the vacuum state is normalized to unity.

3.10 Where does the following formula come from?

$$|n\rangle = \frac{\hat{a}^{\dagger n}}{\sqrt{n!}} |0\rangle$$

3.11 Use this formula and the Gaussian expression for the vacuum wave function to show that $\psi_n(q)$ is even for even n and odd for odd n.

3.12 Why do the Fock states form a complete orthogonal basis of the Hilbert space for one mode?

3.13 What are the general requirements on a quantum state in thermal equilibrium?

3.14 Deduce the thermal state

$$\hat{\rho} = \frac{1}{Z} \exp\left(-\frac{\hat{H}}{k_B T}\right), \quad Z = \mathrm{tr}\left\{\exp\left(-\frac{\hat{H}}{k_B T}\right)\right\}.$$

3.15 Why is T the temperature?

3.16 For a single mode of thermal light, why is

$$\bar{n} = \frac{1}{e^{\beta} - 1}, \quad \beta = \frac{\hbar\omega}{k_B T} ?$$

3.17 How many photons are in a single mode of bright sunlight? Consider light in the optical range of the spectrum with wavelengths around 500 nm and use that the surface of the Sun has a temperature of about 6000 K. Why is sunlight bright?

3.18 What are coherent states?

3.19 State several reasons why the coherent states are wave-like.

3.20 Why are coherent states displaced vacuum states?

3.21 What does the displacement operator do to position wave functions? How does it act on momentum wave functions?

3.22 Show that the photons of a coherent state are Poisson-distributed.

3.23 What does the Poisson distribution describe?

3.24 Show that the Poisson distribution $\exp(-\bar{n})\,\bar{n}^n/n!$ is normalized to unity, that the mean value $\langle n \rangle$ is $\bar{n}$ and that the variance $\langle n^2 \rangle$ is equal to the mean.

3.25 What are minimum uncertainty states?

3.26 Why are the squeezed states the minimum uncertainty states for position and momentum?

3.27 How does the squeezing operator act on the quadrature wave functions?

3.28 Which physical mechanism to generate squeezed light does the squeezing operator suggest?

3.29 State several reasons why the squeezed vacuum carries energy.

3.30 Show that the energy of a squeezed state is $|\alpha|^2 + 1/2 + \sinh^2 \zeta$. What is the meaning of each term in this expression?

3.31 Give interpretations for the photon statistics of the squeezed vacuum.

3.5 Homework problem

In 1900 Max Planck wrote down a formula that accurately described the observed spectral energy density of thermal light. Planck used his unrivalled expertise in the foundations of thermodynamics to guess his formula. He then tried to derive it from statistical physics, but could only do it by postulating a strange hypothesis – the quantization of energy. Planck's radiation law thus opened one of the first glimpses into quantum mechanics. In this problem you will derive Planck's radiation law from the present quantum theory of light.

(a) Consider the radiation field in a quadratic "quantization" box of length L. The spectrum of modes is given by

$$\omega^2 = \frac{c^2\pi^2}{L^2}\left(l^2 + m^2 + n^2\right), \quad l, m, n = 0, 1, 2, \ldots .$$

Consider the energy per given frequency ω written as ϱL^3 where ϱ denotes the spectral energy density,

$$\varrho L^3 = {\sum_{lmn}}' 2\hbar\,\omega_{lmn}\,\bar{n}, \quad \bar{n} = \left(\exp\left(\frac{\hbar\omega}{k_B T}\right) - 1\right)^{-1}$$

where $\omega = \omega_{lmn}$. The factor 2 appears due to the two polarizations of light, except when $l = 0$ or $m = 0$ or $n = 0$, which is indicated by a prime as in the calculation of the Casimir force in Section 2.3.1. Deduce Planck's law

$$\varrho = \frac{\hbar\omega^3}{\pi^2 c^3}\,\bar{n}$$

in the limit of $L \to \infty$ (replacing the summation by integrations).

(b) Consider the energy flux σ through an infinitesimal area. Note that the flux depends on the direction of light waves propagating through the area element. Derive the Stefan–Boltzmann law

$$\sigma = \frac{\pi^2}{60\,c^2\hbar^3}\,(k_B T)^4.$$

You may use the integral

$$\int_0^\infty \frac{x^3\,dx}{e^x - 1} = \frac{\pi^4}{15}.$$

3.6 Further reading

Most textbooks on quantum optics describe the single-mode quantum states discussed here. The reader is referred to the book list of Chapter 1. Various aspects of generalized coherent states are considered in Perelomov's book (1986). The classic review article on squeezed states is the paper by Loudon and Knight (1987). Other forms of squeezing different from the quadrature squeezing discussed here do exist as well, in particular polarization squeezing. For a review see Korolkova (2007).

Chapter 4
Quasiprobability distributions

4.1 Wigner representation

In classical optics (Born and Wolf, 1999), the state of the electromagnetic oscillator is perfectly described by the statistics of the classical amplitude α. The amplitude may be completely fixed (then the field is coherent), or α may fluctuate (then the field is partially coherent or incoherent). In classical optics as well as in classical mechanics, we can characterize the statistics of the complex amplitude α or, equivalently, the statistics of the components position q and momentum p introducing a phase space distribution $W(q, p)$. (As explained in Section 3.1, the real and the imaginary part of the complex amplitude α can be regarded as the position and the momentum of the electromagnetic oscillator.) The distribution $W(q, p)$ quantifies the probability of finding a particular pair of q and p values in their simultaneous measurement. Knowing the phase space probability distribution, all statistical quantities of the electromagnetic oscillator can be predicted by calculation. In this sense the phase space distribution describes the state in classical physics.

All this is much more subtle in quantum mechanics. First of all, Heisenberg's uncertainty principle prevents us from observing position momentum simultaneously *and* precisely. So it seems there is no point in thinking about quantum phase space. But wait! In quantum mechanics we cannot directly observe quantum states either. Nevertheless, we are legitimately entitled to use the concept of states as if they were existing entities (whatever they are). We use their properties to predict the statistics of observations. Why not use a quantum phase space distribution $W(q, p)$ solely to calculate observable quantities in a classical-like fashion? Clearly, the concept of quantum phase space must contain a certain flaw. The probability distribution $W(q,p)$ could become negative, for instance, or ill behaved. Also the classical-like fashion of making statistical predictions may seem to be classical-like at the first glance but not at the second. For these very reasons we should call $W(q,p)$ a *quasiprobability distribution*. Furthermore, there are certainly infinitely many ways of making up quasiprobability distributions (simply because there is no way of defining them properly). Which one shall we choose? Is there a royal road to quantum phase space?

4.1.1 Wigner's formula

Bertrand and Bertrand (1987) had the brilliant idea of defining the quasiprobability distribution $W(q,p)$ by postulating its properties. Just one postulate turns out to be sufficient (Leonhardt, 1997a). Let us assume that the distribution $W(q,p)$ *behaves* like a joint probability distribution for q and p without ever mentioning any simultaneous observation of position and momentum. What can we say about classical probability distributions? The *marginal distributions* or, in other words, the reduced distributions $\int_{-\infty}^{+\infty} W(q,p)\,\mathrm{d}p$ or $\int_{-\infty}^{+\infty} W(q,p)\,\mathrm{d}q$ must give the position or the momentum distribution, respectively. Additionally, if we perform a phase shift θ all complex amplitudes are shifted in phase, meaning that the components q and p rotate in the two-dimensional phase space (q,p). A classical probability distribution for position and momentum values would rotate accordingly. In view of this fact, we postulate that the position probability distribution $\mathrm{pr}(q,\theta)$ after an arbitrary phase shift θ should equal the projection illustrated in Fig. 4.1 or, expressed in a formula,

$$\mathrm{pr}(q,\theta) \equiv \langle q| \, \hat{U}(\theta) \, \hat{\rho} \, \hat{U}^{\dagger}(\theta) \, |q\rangle$$
$$= \int_{-\infty}^{+\infty} W(q\cos\theta - p\sin\theta, q\sin\theta + p\cos\theta)\,\mathrm{d}p. \qquad (4.1)$$

This single formula connects the quasiprobability distribution $W(q,p)$ to quantum mechanics. The same formula ties $W(q,p)$ to observable quantities. And, even more remarkably, the formula links quantum states to observations. Considering special cases of formula (4.1) we see that the marginal distributions of $W(q,p)$ produce the correct position and momentum probabilities, respectively. For $\theta = 0$ we obtain

$$\int_{-\infty}^{+\infty} W(q,p)\,\mathrm{d}p = \langle q| \, \hat{\rho} \, |q\rangle \qquad (4.2)$$

and for $\theta = \pi/2$

$$\langle p| \, \hat{\rho} \, |p\rangle = \int_{-\infty}^{+\infty} W(-q,p)\,\mathrm{d}q = \int_{-\infty}^{+\infty} W(q,p)\,\mathrm{d}q. \qquad (4.3)$$

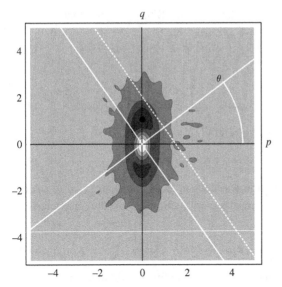

Fig. 4.1. Wigner function. The Wigner function is a quasiprobability distribution for position and momentum. Although $W(q, p)$ cannot be observed as a probability distribution, the marginal distributions of the Wigner function are the observable quadrature distributions, the correct probability distributions for rotated quadratures (linear combinations of position and momentum). The quadratures are projections – "quantum shadows" – seen under the angle θ. The picture illustrates formula (4.1) that defines the Wigner function uniquely. It shows the experimentally reconstructed Wigner function of a Schrödinger kitten state (4.50). (Data: Akira Furusawa and Hidehiro Yonezawa.)

(To justify Eq. (4.3) we note that $\hat{U}^\dagger(\pi/2)|q\rangle$ is a momentum eigenstate $|p = q\rangle$ with eigenvalue q according to Eq. (3.12).) Integrals such as (4.1) are called *Radon transformations* (Radon, 1917), and they are thoroughly studied in the mathematical theory of tomographic imaging (Herman, 1980; Natterer, 1986). The inversion of the Radon transformation (Radon, 1917) plays a distinguished role in tomography. In quantum state tomography the *inverse Radon transformation* turns out to be the mathematical key for quantum state reconstruction, see Section 5.2.3.

Why is postulate (4.1) sufficient? To understand the reason we introduce the Fourier transformed distribution $\widetilde{W}(u, v)$ called the *characteristic function*

$$\widetilde{W}(u, v) \equiv \int_{-\infty}^{+\infty} \int_{-\infty}^{+\infty} W(q, p) \exp(-iuq - ivp) \, dq \, dp \qquad (4.4)$$

and the Fourier-transformed position probability distribution

$$\widetilde{pr}(\xi, \theta) \equiv \int_{-\infty}^{+\infty} pr(q, \theta) \exp(-i\xi q) \, dq. \qquad (4.5)$$

On the other hand, our postulate (4.1) for $W(q,p)$ implies that

$$\widetilde{\mathrm{pr}}(\xi,\theta) = \int_{-\infty}^{+\infty}\int_{-\infty}^{+\infty} W(q',p')\exp(-\mathrm{i}\xi q)\,\mathrm{d}q\,\mathrm{d}p, \qquad (4.6)$$

with the abbreviations

$$q' = q\cos\theta - p\sin\theta \quad\text{and}\quad p' = q\sin\theta + p\cos\theta. \qquad (4.7)$$

In terms of q' and p', q is given by

$$q = q'\cos\theta + p'\sin\theta. \qquad (4.8)$$

We change the integration variables in Eq. (4.5) from (q,p) to (q',p') and obtain, according to the very definition of the characteristic function (4.4),

$$\widetilde{\mathrm{pr}}(\xi,\theta) = \widetilde{W}(\xi\cos\theta, \xi\sin\theta). \qquad (4.9)$$

The Fourier transformed position probability distribution is the characteristic function in polar coordinates. In this way the two functions are closely connected.

So far, we have used only the second line of the fundamental postulate (4.1) or, so to say, the classical nature of the quasiprobability distribution $W(q,p)$. The quantum features come into play when we substitute the first line, that is, the definition of $\mathrm{pr}(q,\theta)$ in the Fourier transformation (4.5). We obtain explicitly

$$\widetilde{\mathrm{pr}}(\xi,\theta) = \int_{-\infty}^{+\infty} \langle q|\,\hat{U}(\theta)\,\hat{\rho}\,\hat{U}^{\dagger}(\theta)\,|q\rangle\exp(-\mathrm{i}\xi q)\,\mathrm{d}q$$

$$= \int_{-\infty}^{+\infty} \langle q|\,\hat{U}(\theta)\,\hat{\rho}\,\hat{U}^{\dagger}(\theta)\exp(-\mathrm{i}\xi\hat{q})\,|q\rangle\,\mathrm{d}q$$

$$= \mathrm{tr}\{\hat{U}(\theta)\,\hat{\rho}\,\hat{U}^{\dagger}(\theta)\exp(-\mathrm{i}\xi\hat{q})\}$$

$$= \mathrm{tr}\{\hat{\rho}\,\hat{U}^{\dagger}(\theta)\exp(-\mathrm{i}\xi\hat{q})\,\hat{U}(\theta)\}. \qquad (4.10)$$

We use the rotation formula (3.12) for the quadrature operators to get

$$\hat{U}^{\dagger}(\theta)\exp(-\mathrm{i}\xi\hat{q})\,\hat{U}(\theta) = \exp(-\mathrm{i}\hat{q}\xi\cos\theta - \mathrm{i}\hat{p}\xi\sin\theta). \qquad (4.11)$$

This is the *Weyl operator* $\exp(-\mathrm{i}u\hat{q} - \mathrm{i}v\hat{p})$ in polar coordinates, the displacement operator (3.61) with $q_0 = v$ and $p_0 = -u$ (Weyl, 1950). The Fourier transform $\widetilde{\mathrm{pr}}(\xi,\theta)$ gives the characteristic function in polar coordinates, as we have learned from Eq. (4.9). Consequently,

$$\widetilde{W}(u,v) = \mathrm{tr}\{\hat{\rho}\,\exp(-\mathrm{i}u\hat{q} - \mathrm{i}v\hat{p})\}. \qquad (4.12)$$

The characteristic function is the "quantum Fourier transform" of the density operator. Because the characteristic function $\widetilde{W}(u,v)$ is the Fourier transform of $W(q,p)$ by definition, the quasiprobability distribution $W(q, p)$ should be very closely related to the density operator. Indeed, both are one-to-one representations of the quantum state, as we will show in the next section. Let us calculate the trace in Eq. (4.12) in the position representation. We use the Baker–Hausdorff formula (3.62) to re-express the Weyl operator

$$\exp(-iu\hat{q} - iv\hat{p}) = \exp\left(i\frac{uv}{2}\right)\exp(-iu\hat{q})\exp(-iv\hat{p}). \qquad (4.13)$$

The operator $\exp\left(-iv\hat{p}\right)$ shifts the position eigenstates $|q\rangle$ by v to produce $|q + v\rangle$ because of the relation (3.16) between the position and momentum eigenstates. Consequently,

$$\widetilde{W}(u, v) = \int_{-\infty}^{+\infty} \langle q| \hat{\rho} \exp(-iu\hat{q} - iv\hat{p}) |q\rangle \, dq$$

$$= \exp\left(i\frac{uv}{2}\right) \int_{-\infty}^{+\infty} \langle q| \hat{\rho} \exp(-iuq) |q + v\rangle \, dq. \qquad (4.14)$$

We replace q by $x - v/2$, and obtain the compact formula

$$\widetilde{W}(u, v) = \int_{-\infty}^{+\infty} \exp(-iux)\left\langle x - \frac{v}{2}\right| \hat{\rho} \left|x + \frac{v}{2}\right\rangle \, dx. \qquad (4.15)$$

To derive an explicit expression for the quasiprobability distribution $W(q,p)$, we simply invert the Fourier transformation in definition (4.4) and get by virtue of formula (4.15) for the characteristic function

$$W(q,p) = \frac{1}{(2\pi)^2} \int_{-\infty}^{+\infty} \int_{-\infty}^{+\infty} \widetilde{W}(u, v) \exp(iuq + ivp) \, du \, dv$$

$$= \frac{1}{(2\pi)^2} \int_{-\infty}^{+\infty} \int_{-\infty}^{+\infty} \int_{-\infty}^{+\infty} \left\langle q' - \frac{v}{2}\right| \hat{\rho} \left|q' + \frac{v}{2}\right\rangle$$

$$\times \exp\left(-iuq' + iuq + ivp\right) \, dq' \, du \, dv$$

$$= \frac{1}{2\pi} \int_{-\infty}^{+\infty} \int_{-\infty}^{+\infty} \left\langle q' - \frac{v}{2}\right| \hat{\rho} \left|q' + \frac{v}{2}\right\rangle \exp(ivp)$$

$$\times \delta(q' - q) \, dv \, dq'. \qquad (4.16)$$

Setting $v = x$ we obtain, finally,

$$W(q,p) = \frac{1}{2\pi} \int_{-\infty}^{+\infty} \exp(ipx)\left\langle q - \frac{x}{2}\right| \hat{\rho} \left|q + \frac{x}{2}\right\rangle \, dx. \qquad (4.17)$$

This is Wigner's legendary formula for a classical-like phase space distribution in quantum mechanics called the *Wigner function*

(Wigner, 1932). We may also find its form in momentum representation along similar lines as in our derivation of Wigner's formula (4.17). We obtain for the characteristic function

$$\widetilde{W}(u,v) = \int_{-\infty}^{+\infty} \exp(-ivy) \left\langle y + \frac{u}{2} | \hat{\rho} | y - \frac{u}{2} \right\rangle dy \qquad (4.18)$$

and, by Fourier transformation, for the Wigner function

$$W(q,p) = \frac{1}{2\pi} \int_{-\infty}^{+\infty} \exp(iqy) \left\langle p + \frac{y}{2} | \hat{\rho} | p - \frac{y}{2} \right\rangle dy. \qquad (4.19)$$

For a pure state described by the quadrature wave functions $\psi(q)$ or $\widetilde{\psi}(p)$ according to definition (3.17), the Wigner function is given by the expressions

$$W(q,p) = \frac{1}{2\pi} \int_{-\infty}^{+\infty} \exp(ipx)\, \psi^* \left(q + \frac{x}{2}\right) \psi \left(q - \frac{x}{2}\right) dx$$

$$= \frac{1}{2\pi} \int_{-\infty}^{+\infty} \exp(iqy)\, \widetilde{\psi}^* \left(p - \frac{y}{2}\right) \widetilde{\psi} \left(p + \frac{y}{2}\right) dy. \qquad (4.20)$$

Formula (4.17) appeared for the first time in Wigner's paper "On the Quantum Correction for Thermodynamic Equilibrium" (Wigner, 1932). It was "chosen from all possible expressions, because it seems to be the simplest".[1]

4.1.2 Basic properties

Wigner's representation of quantum mechanics has found many applications in several areas of quantum physics. It was used whenever quantum corrections to classical laws were of interest, as in Wigner's paper (Wigner, 1932). Here we will focus on only the basic properties of the Wigner function. First, we note that the Wigner function is real for Hermitian operators $\hat{\rho}$,

$$W^*(q,p) = W(q,p). \qquad (4.21)$$

This property is verified by considering the complex conjugate of Wigner's formula (4.17) and replacing x by $-x$. Second, the trace of an operator F is the integral of its Wigner function,

$$\int_{-\infty}^{+\infty} \int_{-\infty}^{+\infty} W_F(q,p)\, dq\, dp = \mathrm{tr}\{\hat{F}\}. \qquad (4.22)$$

---

[1] A footnote in Wigner's paper, however, says that "this expression was found by L. Szilard and [Eugene P. Wigner] some years ago for another purpose...".

In order to define the Wigner function of an arbitrary operator $\hat{F}$ we simply replace the density matrix in Wigner's formula (4.17) by $\hat{F}$. Property (4.22) implies that the Wigner function of a quantum state is normalized to unity, because $\text{tr}\{\hat{\rho}\} = 1$. Additionally, according to the definition (4.4) of the characteristic function,

$$\tilde{W}(0,0) = 1. \tag{4.23}$$

We also see that the Wigner function of the reduced state $\text{tr}_B\{\hat{\rho}_{AB}\}$ of a two-mode system is the integral of the total Wigner function over the phase space of the second subsystem. So far, the Wigner function shows features of a proper probability distribution.

A remarkable property of the Wigner representation is the *overlap formula*

$$\text{tr}\{\hat{F}_1\hat{F}_2\} = 2\pi \int_{-\infty}^{+\infty}\int_{-\infty}^{+\infty} W_1(q,p)\,W_2(q,p)\,dq\,dp \tag{4.24}$$

for the Wigner functions W_1 and W_2 of two arbitrary operators $\hat{F}_1$ and $\hat{F}_2$. The proof of the overlap formula (4.24) is a straightforward calculation using Wigner's expression (4.17) for the right-hand side

$$\frac{1}{2\pi}\int_{-\infty}^{+\infty}\int_{-\infty}^{+\infty}\int_{-\infty}^{+\infty}\int_{-\infty}^{+\infty} \exp\left(ip(x_1+x_2)\right)\left\langle q - \frac{x_1}{2}\middle|\hat{F}_1\middle|q + \frac{x_1}{2}\right\rangle$$

$$\times \left\langle q - \frac{x_2}{2}\middle|\hat{F}_2\middle|q + \frac{x_2}{2}\right\rangle dx_1\,dx_2\,dq\,dp$$

$$= \int_{-\infty}^{+\infty}\int_{-\infty}^{+\infty}\left\langle q - \frac{x}{2}\middle|\hat{F}_1\middle|q + \frac{x}{2}\right\rangle\left\langle q + \frac{x}{2}\middle|\hat{F}_2\middle|q - \frac{x}{2}\right\rangle dq\,dx$$

$$= \int_{-\infty}^{+\infty}\int_{-\infty}^{+\infty}\left\langle q'\middle|\hat{F}_1\middle|q''\right\rangle\left\langle q''\middle|\hat{F}_2\middle|q'\right\rangle dq'\,dq''$$

$$= \int_{-\infty}^{+\infty}\left\langle q'\middle|\hat{F}_1\hat{F}_2\middle|q'\right\rangle dq' = \text{tr}\{\hat{F}_1\hat{F}_2\}. \tag{4.25}$$

Why is the overlap formula remarkable? First, we can use it for calculating expectation values

$$\text{tr}\{\hat{\rho}\hat{F}\} = \int_{-\infty}^{+\infty}\int_{-\infty}^{+\infty} W(q,p)\,2\pi\,W_F(q,p)\,dq\,dp. \tag{4.26}$$

(We have simply replaced $\hat{F}_1$ by $\hat{\rho}$ and $\hat{F}_2$ by $\hat{F}$.) This equation would be the rule for predicting expectations in classical statistical physics, too. The Wigner function $W(q,p)$ plays the role of a classical phase space density, whereas $2\pi\,W_F(q,p)$ appears as the physical quantity that is averaged with respect to $W(q,p)$. This is exactly the classical-like way of calculating quantum mechanical expectation values we were seeking.

We can understand formula (4.26) in another way by seeing $2\pi W_F(q,p)$ as a filter function. Consequently, all that quantum mechanics allows us to predict are filtered projections of the Wigner function, "quantum shadows".

Another simple consequence of the overlap formula (4.24) is the expression

$$|\langle\psi_1|\psi_2\rangle|^2 = 2\pi \int_{-\infty}^{+\infty}\int_{-\infty}^{+\infty} W_1(q,p)\,W_2(q,p)\,dq\,dp \qquad (4.27)$$

for the transition probability between the pure states $|\psi_1\rangle$ and $|\psi_2\rangle$. However, this quantity vanishes if the states $|\psi_1\rangle$ and $|\psi_2\rangle$ are orthogonal. The overlap (4.27) of two positive functions W_1 and W_2 cannot be zero, unless the functions are disjoint, which already indicates that Wigner functions cannot be positive in general. In fact, states having Gaussian wave functions are the only pure states with non-negative Wigner functions (Hudson, 1974; Tatarskii, 1983; Lütkenhaus and Barnett, 1995). In this way, the overlap formula (4.24) reveals both the similarities and the differences between a classical probability distribution and the Wigner function. Quantum interference implies that the classical-like Wigner function cannot be regarded as a probability distribution, but as a quasiprobability distribution only. This property is one way in which the unavoidable flaw in the concept of quantum phase space may appear. Negative regions in the Wigner function of a given state can be seen as signatures of non-classical behaviour (Lütkenhaus and Barnett, 1995).

Using the overlap formula (4.24) we are also able to quantify the purity of a quantum state. In fact, identifying both $\hat{F}_1$ and $\hat{F}_2$ with $\hat{\rho}$ we obtain

$$\mathrm{tr}\{\hat{\rho}^2\} = 2\pi \int_{-\infty}^{+\infty}\int_{-\infty}^{+\infty} W(q,p)^2\,dq\,dp. \qquad (4.28)$$

The *purity* $\mathrm{tr}\{\hat{\rho}^2\}$ ranges between zero and unity and equals exactly unity if and only if the state is pure, when $\hat{\rho} = |\psi\rangle\langle\psi|$. According to relation (1.25) the entropy S is bounded by

$$S \equiv -k_B\mathrm{tr}\{\hat{\rho}\ln\hat{\rho}\} \geq k_B\left(1 - 2\pi \int_{-\infty}^{+\infty}\int_{-\infty}^{+\infty} W(q,p)^2\,dq\,dp\right). \qquad (4.29)$$

We see that the overlap of the Wigner function with itself provides a convenient way of expressing statistical purity in quantum mechanics.

Finally, we can use the overlap formula (4.24) to represent the density matrix elements in a given basis in terms of the Wigner function

$$\langle a'|\hat{\rho}|a\rangle = \mathrm{tr}\{\hat{\rho}\,|a\rangle\langle a'|\} = 2\pi \int_{-\infty}^{+\infty}\int_{-\infty}^{+\infty} W(q,p)\,W_{a'a}(q,p)\,dq\,dp. \qquad (4.30)$$

Here $W_{a'a}(q,p)$ denotes the Wigner representation of the projector $|a\rangle\langle a'|$ obtained by replacing $\hat{\rho}$ by $|a\rangle\langle a'|$ in Wigner's formula (4.17). This property shows that the Wigner function is indeed a one-to-one representation of the quantum state.

We may turn the tables and ask, is any normalized real function $W(q,p)$ always a Wigner function, that is, does it correspond to a state? Obviously it does not, because the integral of the squared function must be less than or equal to $(2\pi)^{-1}$ according to the purity relation (4.28). Another quantum constraint is imposed on any realistic Wigner function: the values of it may range between only $\pm\pi^{-1}$, that is

$$\left|W(q,p)\right| \le \frac{1}{\pi}. \tag{4.31}$$

To prove this inequality we consider a pure state $\hat{\rho} = |\psi\rangle\langle\psi|$ first. We use the Schwarz inequality for two square-integrable complex-valued functions f and g,

$$\left|\int f^*(x)\,g(x)\,dx\right|^2 \le \left(\int |f(x)|^2\,dx\right)\left(\int |g(x)|^2\,dx\right), \tag{4.32}$$

to estimate the Wigner function given in terms of Wigner's formula (4.17). We obtain

$$\left|W(q,p)\right|^2 \le \frac{1}{(2\pi)^2}\int_{-\infty}^{+\infty}\left|\langle q-\frac{x}{2}|\psi\rangle\right|^2 dx\int_{-\infty}^{+\infty}\left|\langle q+\frac{x}{2}|\psi\rangle\right|^2 dx = \frac{1}{\pi^2}, \tag{4.33}$$

because the state vector $|\psi\rangle$ is normalized. In case of a statistical mixture the density matrix can be represented as a sum of pure states $|\psi_a\rangle\langle\psi_a|$ weighted by their probabilities ρ_a according to the definition (1.15) of the density operator. Consequently, the Wigner function for a mixed state is a weighted sum of pure Wigner functions as well. By estimating the individual pure Wigner functions by Eq. (4.31) and summing with respect to the normalized probabilities p_a, we see that the bound (4.31) is valid for mixed states, too. (Note that the Wigner function $W_n(q,p)$ for Fock states $|n\rangle$ equals $(-1)^n/\pi$ at the origin $q = p = 0$; see Eq. (4.89).) The constraint (4.31) shows that Wigner functions cannot be highly peaked; the quantum "phase space density" cannot be arbitrarily high and so Wigner functions cannot approach delta functions $\delta(q - q_0)\delta(p - p_0)$. Of course, according to Heisenberg's uncertainty principle, position and momentum must fluctuate statistically, and this intrinsic uncertainty is reflected in the uniform bound (4.31). Note that other constraints on Wigner functions were given by Tatarskii (1983) and Lieb (1990). However, no golden rule decides whether a given function is a Wigner function, apart from the Solomonic statement that any density matrix derived from a proper Wigner function should be a density matrix, a Hermitian matrix with non-negative eigenvalues that sum up to unity. Equivalently, all main diagonal elements $\langle a| \hat{\rho} |a\rangle$ derived according to formula (4.30) must be non-negative. Deviations from this law indicate imperfections in experimentally reconstructed Wigner functions, for instance.

In addition to formula (4.26) we can formulate another equivalent way of making quantum-mechanical predictions using the Wigner function, that is, of calculating expectation values via the Wigner function. Consider, for arbitrary constants λ and μ,

$$
\mathrm{tr}\left\{\hat{\rho}\left(\lambda\hat{q}+\mu\hat{p}\right)^k\right\} = \mathrm{i}^k \left. \frac{\partial^k}{\partial\xi^k}\, \widetilde{W}(\xi\lambda,\xi\mu)\right|_{\xi=0}
$$
$$
= \int_{-\infty}^{+\infty}\int_{-\infty}^{+\infty} W(q,p)\,(\lambda q+\mu p)^k\, \mathrm{d}q\,\mathrm{d}p. \qquad (4.34)
$$

In the first line we have used formula (4.12) for the characteristic function $\widetilde{W}(u,v)$, whereas in the second line we have applied the Fourier relationship (4.4) between $\widetilde{W}(u,v)$ and the Wigner function. Comparing the powers of λ and μ, we see that

$$
\mathrm{tr}\left\{\hat{\rho}\, \mathcal{S}\,\hat{q}^m\hat{p}^n\right\} = \int_{-\infty}^{+\infty}\int_{-\infty}^{+\infty} W(q,p)\,q^m p^n\, \mathrm{d}q\,\mathrm{d}p. \qquad (4.35)
$$

The symbol $\mathcal{S}$ means that we should symmetrize all possible products of the m $\hat{q}$ operators and the n $\hat{p}$ operators, that is, we should take the average over all products with the right amount of $\hat{q}$ and $\hat{p}$. For example, $\left(\hat{q}\hat{p}+\hat{p}\hat{q}\right)/2$ corresponds to qp and $\left(\hat{q}^2\hat{p}+\hat{q}\hat{p}\hat{q}+\hat{p}\hat{q}^2\right)/3$ to q^2p. This *Weyl correspondence* (Weyl, 1950) is also a convenient way of making quantum-mechanical predictions in a classical-like fashion. (Note that the Weyl correspondence is equivalent to formula (4.26).) Given a symmetrized operator $\hat{F}$, we can calculate quantum-mechanical averages as if $\hat{F}$ were a classical quantity. However, this pleasing property is Janus-faced. The square of $\hat{F}$, which describes the fluctuations of $\hat{F}$, is not necessarily symmetrized. For example, the square of $\hat{q}\hat{p}+\hat{p}\hat{q}$ does not contain the terms $\hat{q}^2\hat{p}^2$ and $\hat{p}^2\hat{q}^2$. We should use the commutator between $\hat{q}$ and $\hat{p}$ to express $\hat{F}^2$ in terms of symmetrized operators for getting meaningful results. This route is another way in which the mutual exclusion of certain observables sneaks in; here via the commutator relations of position and momentum. So we must not forget that the algebraic structures of quantum mechanics and classical physics are different, despite many similarities. This difference causes a problem in the very concept of a quantum phase space even more serious than negative "probabilities".

4.1.3 Examples

How do typical Wigner functions look? Are they similar to classical phase space densities? Probably the simplest example is the Wigner function for the vacuum state. We insert the quadrature wave function (3.26) in Wigner's formula (4.17), perform the integration and see that the Wigner function for a vacuum is the Gaussian

$$W_0(q,p) = \frac{1}{\pi} \exp\left(-q^2 - p^2\right). \tag{4.36}$$

Classically, this function would correspond to the phase space density of an ensemble of electromagnetic oscillators fluctuating statistically around the origin in phase space with isotropic variances of $1/2$ in our units. Quantum mechanically, these statistical fluctuations occur even if the spatial temporal mode is in a pure vacuum. Figure 4.2 shows the experimentally reconstructed Wigner function for a vacuum, illustrating beautifully the isotropic character of the vacuum fluctuations.

How does a squeezed vacuum look? Let us study the general effect of squeezing in phase space first. We obtain from Wigner's formula (4.17)

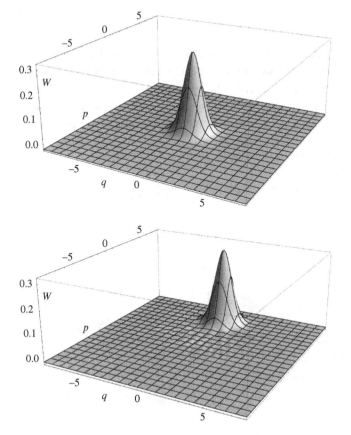

Fig. 4.2. Wigner function of the vacuum (top) and a coherent state (bottom), illustrating that a coherent state is just a "displaced vacuum". Optical homodyne tomography (Section 5.2.2) was used to reconstruct the Wigner functions from the quadratures of Figs. 3.2 and 3.4. (Data: Akira Furusawa and Hidehiro Yonezawa.)

$$W_s(q,p) = \frac{1}{2\pi} \int_{-\infty}^{+\infty} \exp(ipx) \left\langle q - \frac{x}{2} \middle| \hat{S} \hat{\rho} \hat{S}^\dagger \middle| q + \frac{x}{2} \right\rangle dx$$

$$= \frac{1}{2\pi} \int_{-\infty}^{+\infty} \exp(ipx) \, e^{\zeta} \left\langle e^{\zeta} \left(q - \frac{x}{2} \right) \middle| \hat{\rho} \middle| e^{\zeta} \left(q + \frac{x}{2} \right) \right\rangle dx, \quad (4.37)$$

because the squeezing operator $\hat{S}$ defined in Eq. (3.87) rescales the position wave function according to Eq. (3.84). We substitute $e^{\zeta} x$ with x' and get the simple result

$$W_s(q,p) = W(e^{\zeta} q, e^{-\zeta} p). \quad (4.38)$$

The Wigner function for a squeezed state is squeezed in one quadrature direction and stretched accordingly in the orthogonal line, in order to preserve the area in phase space. In this way, the quadrature fluctuations displayed in the Wigner function are redistributed from one quadrature to the canonically conjugate quantity. This redistribution is exactly what we would expect from squeezing in phase space. Using the general result (4.38), we obtain directly from Eq. (4.36) the Wigner function of a squeezed vacuum

$$W_s(q,p) = \frac{1}{\pi} \exp\left(-e^{2\zeta} q^2 - e^{-2\zeta} p^2 \right). \quad (4.39)$$

As for a vacuum, the Wigner function is a Gaussian distribution with, however, unbalanced variances (3.83) indicating the effect of quadrature squeezing. Figure 4.3 shows the experimentally reconstructed Wigner function of a significantly squeezed vacuum generated by parametric amplification. It is an easy exercise to calculate the distribution $\mathrm{pr}(q,\theta)$ for phase-shifted quadratures from the Wigner function (4.39) via the Radon transformation (4.1). We obtain the result

$$\mathrm{pr}(q,\theta) = \left(2\pi \Delta_\theta^2 q \right)^{-1/2} \exp\left(-\frac{q^2}{2\Delta_\theta^2 q} \right) \quad (4.40)$$

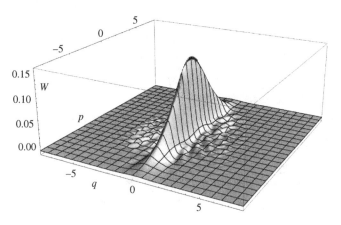

Fig. 4.3. Wigner function of a squeezed vacuum. Optical homodyne tomography (Section 5.2.2) was used to reconstruct the Wigner function from the quadratures of Fig. 3.6. (Data: Akira Furusawa and Hidehiro Yonezawa.)

with the phase-dependent variance

$$\Delta_\theta^2 q = \frac{1}{2}\left(e^{2\varsigma}\sin^2\theta + e^{-2\varsigma}\cos^2\theta\right). \qquad (4.41)$$

The quadrature fluctuations of a squeezed vacuum are Gaussian and, of course, phase dependent. Their variances $\Delta_\theta^2 q$ vary from $e^{-2\varsigma}/2$ to $e^{+2\varsigma}/2$ with a period of π.

What is the Wigner function of a coherent state? Coherent states are displaced vacua, so we would expect that their Wigner functions are displaced vacuum Wigner functions, too, with a displacement given by the complex coherent amplitude $\alpha = (q_0 + ip_0)/\sqrt{2}$. That this expectation is correct is easily seen, considering the general effect of the displacement operator $\hat{D}$ in phase space,

$$\begin{aligned}
W_D(q,p) &= \frac{1}{2\pi}\int_{-\infty}^{+\infty} \exp(ipx)\left\langle q - \frac{x}{2}\Big|\hat{D}\hat{\rho}\hat{D}^\dagger\Big|q + \frac{x}{2}\right\rangle dx \\
&= \frac{1}{2\pi}\int_{-\infty}^{+\infty} \exp\left(i(p-p_0)x\right) \\
&\quad \times \left\langle q - \frac{x}{2} - q_0\Big|\hat{\rho}\Big|q + \frac{x}{2} - q_0\right\rangle dx \qquad (4.42)
\end{aligned}$$

according to the general rule (3.63) and (3.65) for displacing position wave functions. We find that Wigner functions of displaced states are indeed just displaced Wigner functions

$$W_D(q,p) = W(q - q_0, p - p_0) \qquad (4.43)$$

and, consequently, the Wigner function of a coherent state is given by the displaced Gaussian distribution

$$W(q,p) = \frac{1}{\pi}\exp\left(-(q-q_0)^2 - (p-p_0)^2\right). \qquad (4.44)$$

Again, the Wigner function displays the typical features of the quantum state: a coherent state, as produced by a high-quality laser, has a stable coherent amplitude, apart from the absolutely unavoidable quantum fluctuations; coherent light is only contaminated by vacuum noise.

According to the fundamental superposition principle of quantum mechanics, we are entitled to think of quantum superpositions of coherent states. These are states that contain simultaneously two coherent components, one pointing in one direction in phase space and the other pointing in another, for example the even and odd superpositions

$$|\psi_\pm\rangle = \frac{1}{\sqrt{2\eta}}\left(|\alpha_0\rangle \pm |-\alpha_0\rangle\right), \quad \eta = 1 \pm \exp\left(-2|\alpha_0|^2\right), \qquad (4.45)$$

or, represented as density matrices,

$$\rho_\pm = \frac{1}{2\eta}\left(|\alpha_0\rangle\langle\alpha_0| + |-\alpha_0\rangle\langle-\alpha_0| \pm |\alpha_0\rangle\langle-\alpha_0| \pm |-\alpha_0\rangle\langle\alpha_0|\right). \qquad (4.46)$$

Fig. 4.4. Wigner function of a single photon. The picture shows the Wigner function of a Fock state (4.89) with $n = 1$. This Wigner function is partially negative (bottom picture). Optical homodyne tomography (Section 5.2.2) was used to reconstruct the Wigner function. The Fock state was created (Zavatta et al., 2004) by heralded down-conversion discussed in Section 7.1.3. (Data: Marco Bellini.)

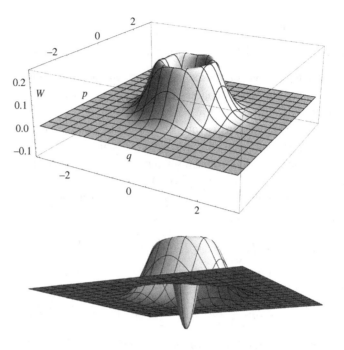

The normalization factor η follows from the scalar product (3.71) of the coherent states. For simplicity we consider a real $\alpha_0 = q_0/\sqrt{2}$. In this case, the quantum state (4.45) corresponds to a superposition of the quadrature wave function (3.66),

$$\psi(q) \propto \exp\left(-\frac{1}{2}(q - q_0)^2\right) \pm \exp\left(-\frac{1}{2}(q + q_0)^2\right) \qquad (4.47)$$

where we omitted any normalization factors. The wave function shows two peaks, one at q_0 and the other at $-q_0$ according to the superimposed coherent amplitudes. Note that this quantum superposition has nothing to do with optical interference. When two fields interfere, their amplitude may be enhanced or cancelled, producing, for example, coherent states of enhanced or zero amplitude (vacuum). The quantum superposition (4.45) still contains both coherent amplitudes $\pm q_0$. It is also very different from an incoherent superposition of $\pm q_0$, where the field has either the amplitude $+q_0$ or the amplitude $-q_0$ with certain probabilities. Such an incoherent statistical mixture corresponds to the density matrix

$$\hat{\rho} = \frac{1}{2}\left(|\alpha_0\rangle\langle\alpha_0| + |-\alpha_0\rangle\langle-\alpha_0|\right), \qquad (4.48)$$

leaving out the quantum superposition terms in the state (4.46). The quadrature amplitude of the superposition state is $+q_0$ *as well as* $-q_0$, with a resolution given by the vacuum fluctuations. This strange behaviour of being at $+q_0$ as well as at $-q_0$ resembles Schrödinger's famous Gedanken experiment about a quantum cat being simultaneously alive and dead (Schrödinger, 1935b). Therefore, states such as (4.45) are named *Schrödinger cat states*. Macroscopic quantum superposition states of light are difficult to make and even more difficult to maintain, because they are easily destroyed by losses or other interactions with the environment, as we discuss in Sections 6.2.2 and 7.1.2. Nevertheless, Schrödinger cat states with small amplitudes α_0 (in the order of 1), Schrödinger kittens, have been made (Neergaard Nielsen et al., 2006; Ourjoumtsev et al., 2006) by a method (Dakna et al., 1997) we briefly explain in Section 5.1.3. The Schrödinger kittens are approximations of the even and odd cat states for small amplitudes. They are truncations of the Fock expansions (3.69) of the coherent states in the superposition (4.45),

$$|\psi_+\rangle = \frac{1}{\sqrt{\cosh |\alpha_0|^2}} \sum_{m=0}^{\infty} \frac{\alpha_0^{2m}}{\sqrt{2m!}} |2m\rangle, \tag{4.49}$$

$$|\psi_-\rangle = \frac{1}{\sqrt{\sinh |\alpha_0|^2}} \sum_{m=0}^{\infty} \frac{\alpha_0^{2m+1}}{\sqrt{(2m+1)!}} |2m+1\rangle. \tag{4.50}$$

Superposition states that are significantly more macroscopic have been made in other areas of physics, for example quantum superpositions of superconducting currents (Friedman et al., 2000).

Which observable phenomena of the Schrödinger cat state (4.45) do we expect? Let us calculate the Wigner function. We see from the density matrix (4.46) that the Wigner function for the Schrödinger cat state is the sum of the Wigner functions for the coherent states $|\alpha_0\rangle$ and $|-\alpha_0\rangle$ plus or minus the Wigner functions for the interference terms $|\alpha_0\rangle\langle-\alpha_0|$ and $|-\alpha_0\rangle\langle\alpha_0|$. Again, we use a real amplitude $\alpha_0 = q_0/\sqrt{2}$. For the interference terms, we obtain from Wigner's formula (4.17) and from our expression (3.66) for the quadrature wave functions of coherent states

$$W_{\alpha_0,-\alpha_0} = \frac{1}{\pi} \exp\left(-q^2 - p^2 - 2iq_0 p\right). \tag{4.51}$$

For the remaining term $W_{-\alpha_0,\alpha_0}$ we replace q_0 by $-q_0$ and so we arrive at the Wigner function for the Schrödinger cat state

$$W_\pm(q,p) \propto \exp\left(-(q-q_0)^2 - p^2\right) + \exp\left(-(q+q_0)^2 - p^2\right)$$
$$\pm\, 2 \exp\left(-q^2 - p^2\right) \cos(2pq_0). \tag{4.52}$$

Like the wave function, the Wigner function exhibits two peaks at the coherent amplitudes $\pm q_0$. However, the interference structure halfway between the peaks displays the quantum superposition of both amplitudes, showing rapid oscillations with a frequency given by the distance $2q_0$ of the superimposed amplitudes, as shown in Figs. 4.5 and 4.6. The Wigner function becomes negative, indicating the non-classical behaviour of the Schrödinger cat state (Schleich et al., 1991; Bužek and Knight, 1995). To predict observable effects of the quantum superposition state (4.45) we calculate the quadrature distributions $\mathrm{pr}(q, \theta)$ via Radon transformation (4.1) of the Wigner function (4.52) and get

$$\mathrm{pr}_\pm(q, \theta) \propto \exp\left(-(q - q_0 \cos\theta)^2\right) + \exp\left(-(q + q_0 \cos\theta)^2\right)$$
$$\pm\, 2 \exp\left(-q^2 - q_0^2 \cos^2\theta\right) \cos(2qq_0 \sin\theta). \qquad (4.53)$$

The position quadrature distribution at $\theta = 0$ shows peaks at $\pm q_0$,

$$\mathrm{pr}_\pm(q) \approx \frac{1}{2} \exp\left(-(q - q_0)^2\right) + \frac{1}{2} \exp\left(-(q + q_0)^2\right). \qquad (4.54)$$

Shifting the reference phase to $\theta = \pi/2$ and replacing q by p turns the quadrature distributions $\mathrm{pr}_\pm(q, \theta)$ into the momentum distributions

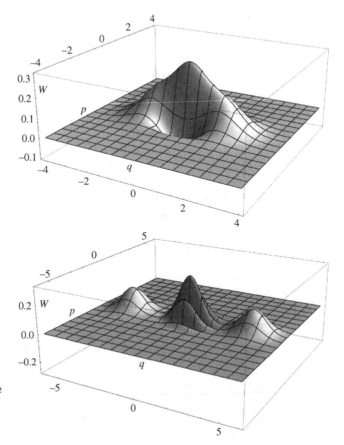

Fig. 4.5. Wigner function of even Schrödinger cat states, the quantum superpositions (4.49) of two coherent states with amplitudes $\pm q_0$. Top: $q_0 = 1.5$, the amplitudes are hardly separated compared with the vacuum noise. Bottom: $q_0 = 4.0$, the amplitudes $\pm q_0$ are clearly distinguishable while a typical interference structure is visible that generates the interference fringes (4.55) in the momentum distribution.

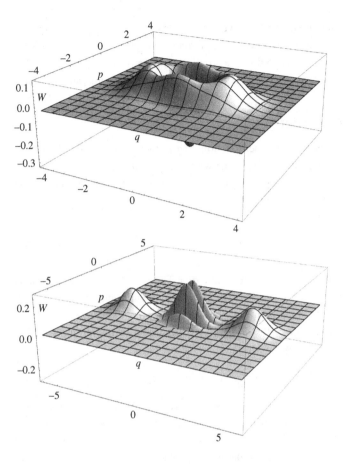

Fig. 4.6. Wigner function of odd Schrödinger cat states, the quantum superpositions (4.50) of two coherent states with amplitudes $\pm q_0$. Same description as in Fig. 4.5.

$$\mathrm{pr}_+(p) \propto \exp\left(-p^2\right)\cos^2(pq_0),$$
$$\mathrm{pr}_-(p) \propto \exp\left(-p^2\right)\sin^2(pq_0),$$

(4.55)

displaying typical interference fringes. The fringes depend on the phase between the superimposed coherent states, the $\pm$ in our case. This interference pattern is contained in the oscillating Wigner function of a Schrödinger cat state.

We have seen that Wigner functions are useful to visualize the phase space properties of quantum states. Wigner functions display quadrature amplitudes, their fluctuations, and possible interferences.

4.2 Other quasiprobability distributions

In many respects the Wigner representation appears as the best compromise between a classical phase space density and the correct quantum-mechanical behaviour. The Wigner function generates the right marginal distributions and it obeys the overlap relation (4.24) for calculating expectation values in a classical-like fashion. Yet the Wigner function may be negative. Is there a way to define a strictly non-negative quasiprobability distribution? Which other useful quasiprobability distributions can we define?

4.2.1 Q function

We may smooth the Wigner function $W(q,p)$ by convolving it with a Gaussian distribution having the same width as vacuum, by taking averages of W around each phase space point (q,p) with a width that corresponds to the vacuum noise, for example as

$$Q(q,p) \equiv \frac{1}{\pi} \int_{-\infty}^{+\infty} \int_{-\infty}^{+\infty} W(q',p') \exp\left(-(q-q')^2 - (p-p')^2\right) dq' \, dp'.$$
(4.56)

This function is known as the Q function. The kernel used for averaging is the Wigner function (4.44) of a coherent state with amplitude $\alpha = (q + ip)/\sqrt{2}$. From the overlap relation (4.24) we immediately see that the Q function simply gives the probability distribution for finding the coherent states $|\alpha\rangle$ in the state $\hat{\rho}$, because

$$Q(q,p) = \frac{1}{2\pi} \text{tr}\{\hat{\rho} |\alpha\rangle\langle\alpha|\} = \frac{1}{2\pi} \langle\alpha| \hat{\rho} |\alpha\rangle.$$
(4.57)

Note that also $\pi^{-1} \langle\alpha| \hat{\rho} |\alpha\rangle$ is frequently called the Q function. Clearly, the Q function is non-negative and normalized to unity, as is easily seen from the completeness relation (3.73) of the coherent states. Consequently, the Q function can be regarded as describing probability densities. We discuss in Section 5.2.4 a scheme to measure the Q function directly as a probability distribution. We also see that the negative regions of the Wigner function cannot be extended over areas significantly wider than $1/2$; otherwise, the Q function could be negative. (This property puts yet another constraint on Wigner functions.) Roughly speaking, the Gaussian smoothing (4.56) means taking the average of $W(q',p')$ values in a circle area around (q,p) with a radius given by the vacuum fluctuations. Because any negativities disappear after this procedure, they must be concentrated in small regions of the Wigner function. The resolution of these negative "probabilities" thus requires an accuracy in the order of the vacuum fluctuations.

The smoothing of the Wigner function is also clearly seen in the Fourier-transformed Q function $\tilde{Q}(u,v)$. In fact, we obtain from the definition (4.56)

$$\tilde{Q}(u,v) \equiv \int_{-\infty}^{+\infty} \int_{-\infty}^{+\infty} Q(q,p) \exp(-iuq - ivp) \, dq \, dp$$

$$= \tilde{W}(u,v) \exp\left(-\frac{u^2 + v^2}{4}\right).$$
(4.58)

Because details of the Wigner function correspond to high-frequency components (u, v), these details are suppressed in the Q representation.

Equation (4.58) also reveals another important property of the Q function. We use the formula (4.12) for the characteristic function $\widetilde{W}(u, v)$ and obtain

$$\widetilde{Q} = \text{tr}\left\{\hat{\rho} \, \exp\left(-\frac{u^2 + v^2}{4}\right) \exp\left(-iu\hat{q} - iv\hat{p}\right)\right\} \tag{4.59}$$

or, introducing the complex notation $\beta = (u + iv)/\sqrt{2}$,

$$\widetilde{Q} = \text{tr}\left\{\hat{\rho} \, \exp\left(-\frac{1}{2}|\beta|^2\right) \exp\left(-i\hat{a}\beta^* - i\hat{a}^\dagger \beta\right)\right\} \tag{4.60}$$

because $\hat{a} = (\hat{q} + i\hat{p})/\sqrt{2}$. We employ the Baker–Hausdorff formula (3.62) and get

$$\widetilde{Q} = \text{tr}\left\{\hat{\rho} \, \exp\left(-i\hat{a}\beta^*\right) \exp\left(-i\hat{a}^\dagger \beta\right)\right\}. \tag{4.61}$$

Consequently,

$$\text{tr}\left\{\hat{\rho} \, \hat{a}^\nu \, \hat{a}^{\dagger\mu}\right\} = i^{\nu+\mu} \frac{\partial^\nu}{\partial\beta^{*\nu}} \frac{\partial^\mu}{\partial\beta^\mu} \widetilde{Q}\bigg|_{\beta=\beta^*=0}$$

$$= \int_{-\infty}^{+\infty} \int_{-\infty}^{+\infty} Q(q, p) \, \alpha^\nu \alpha^{*\mu} \, dq \, dp, \tag{4.62}$$

using the definition (4.58) of the Fourier-transformed Q function in the complex notation $\alpha = (q + ip)/\sqrt{2}$. Expectation values of the form $\text{tr}\{\hat{\rho} \, \hat{a}^\nu \, \hat{a}^{\dagger\mu}\}$ are called *anti-normally ordered*. We have seen that we can express these quantities in terms of the Q function as if $\hat{a}$ and $\hat{a}^\dagger$ were classical amplitudes and not operators. We note, however, that this property relies critically on the ordering $\hat{a}^\nu \, \hat{a}^{\dagger\mu}$, and it is clearly lost when powers $(\hat{a}^\nu \, \hat{a}^{\dagger\mu})^\lambda$ are concerned. (Remember the discussion at the end of Section 4.1.2.)

4.2.2 *P* function

In the theory of photodetection, see for instance Mandel and Wolf (1995, Chap. 12), *normally ordered* expectation values $\text{tr}\{\hat{\rho} \, \hat{a}^{\dagger\mu} \, \hat{a}^\nu\}$ play a distinguished role. How can we find the quasiprobability distribution for normal ordering? We simply reverse the order of the exponentials $\exp(-i\hat{a}\beta^*)$ and $\exp(-i\hat{a}^\dagger \beta)$ in the expression (4.61) to define a new function

$$\widetilde{P}(u, v) \equiv \text{tr}\left\{\hat{\rho} \, \exp\left(-i\hat{a}^\dagger \beta\right) \exp\left(-i\hat{a}\beta^*\right)\right\} \tag{4.63}$$

with $\beta = (u + iv)/\sqrt{2}$. Using the same arguments as in the previous section we see that the *P function*

$$P(q, p) \equiv \frac{1}{(2\pi)^2} \int_{-\infty}^{+\infty} \int_{-\infty}^{+\infty} \widetilde{P}(u, v) \, \exp(iuq + ivp) \, du \, dv \tag{4.64}$$

corresponds to the normal ordering

$$\text{tr}\left\{\hat{\rho} \, \hat{a}^{\dagger\mu} \, \hat{a}^\nu\right\} = \int_{-\infty}^{+\infty} \int_{-\infty}^{+\infty} P(q, p) \, \alpha^{*\mu} \alpha^\nu \, dq \, dp \tag{4.65}$$

with $\alpha = (q + ip)/\sqrt{2}$. Because normally ordered quantities are quite fundamental in quantum optics, the property (4.65) is one reason that the P function, also called the *Glauber–Sudarshan function* (Glauber, 1963; Sudarshan, 1963), is a widely used phase space distribution. Yet another, more important reason is that the P function diagonalizes the density operator in terms of coherent states. To prove this property, we argue along similar lines as in the previous section where we started from (4.58) and arrived at (4.61). Here we do the necessary algebra in reversed order – we start from the definition (4.63) and arrive via the Baker–Hausdorff formula (3.62) at

$$\widetilde{W}(u, v) = \widetilde{P}(u, v) \exp\left(-\frac{u^2 + v^2}{4}\right). \tag{4.66}$$

Consequently,

$$W(q, p) = \int_{-\infty}^{+\infty} \int_{-\infty}^{+\infty} P(q_0, p_0) \frac{1}{\pi} \exp\left(-(q - q_0)^2 - (p - p_0)^2\right) dq_0 \, dp_0. \tag{4.67}$$

Let us read the Gaussian $\pi^{-1} \exp\left(-(q - q_0)^2 - (p - p_0)^2\right)$ in the right-hand side of this equation as the Wigner function (4.44) of a coherent state with density matrix $|\alpha\rangle\langle\alpha|$ and $\alpha = (q_0 + ip_0)/\sqrt{2}$. The left-hand side is the Wigner function of the quantum state $\hat{\rho}$. As Wigner functions are one-to-one linear transformations (4.17) of the density matrix, we obtain the same relationship for the density matrices,

$$\hat{\rho} = \int_{-\infty}^{+\infty} \int_{-\infty}^{+\infty} P(q, p) |\alpha\rangle\langle\alpha| \, dq \, dp, \tag{4.68}$$

having replaced q_0 and p_0 by q and p. This famous result (Sudarshan, 1963) is called the *optical equivalence theorem* (Klauder, 1966). At first glance, formula (4.68) appears as a representation of the quantum state in terms of a distribution of coherent states, that is, as a statistical mixture of classical amplitudes. This is impossible! There is no way to represent a pure state $|\psi\rangle$ as a mixture of coherent states, unless $|\psi\rangle$ itself is a coherent state. Yet the algebra to arrive at the result (4.68) is correct. What is wrong? The answer is that the P function might be very ill-behaved. For instance, we see from Eq. (4.66) that the Wigner function is a smoothed P function. Because the Wigner function can display negative "probabilities" the P function might behave even worse; it might be negative or it might not even exist as a well-tempered distribution. States having such ill-behaved P functions are called *non-classical states*; see the discussion at the beginning of Section 3.2.3.

4.2.3 *s*-parameterized quasiprobability distributions

We may also convolve the Wigner function with Gaussian distributions having a width different from the scale of the vacuum noise. In this way, we obtain a whole family of distributions called the *s-parameterized quasiprobability distributions* $W(q, p; s)$ (Cahill, 1965; Cahill and Glauber, 1969b; Leonhardt and Paul, 1993c). First, we define the characteristic functions

$$\tilde{W}(u, v; s) \equiv \tilde{W}(u, v) \exp\left(\frac{s}{4}\left(u^2 + v^2\right)\right). \tag{4.69}$$

For historical reasons (Cahill and Glauber, 1969a, 1969b) the real parameter *s* happens to be negative when the Wigner function is smoothed. We see from definition (4.69) that in this case the Gaussian distribution is indeed suppressing high frequencies (u, v) describing details in the Wigner function. The s-parameterized quasiprobability distributions are obtained from the characteristic function via inverse Fourier transformation

$$W(q, p; s) \equiv \frac{1}{(2\pi)^2} \int_{-\infty}^{+\infty} \int_{-\infty}^{+\infty} \tilde{W}(u, v; s) \exp(iuq + ivp)\, du\, dv. \tag{4.70}$$

Obviously, all previously studied quasiprobability distributions are included in this family of functions, because they correspond to the parameters

$$s = \begin{cases} +1: & P \text{ function,} \\ 0: & \text{Wigner function,} \\ -1: & Q \text{ function,} \end{cases} \tag{4.71}$$

respectively. The defined distributions interpolate between the P, the Wigner, and the Q function. The range of s, however, is the whole real axis. Note that it is also possible to define quasiprobability distributions corresponding to complex s parameters (Wünsche, 1996).

Let us study some general properties of s-parameterized quasiprobability distributions. Of course, they are normalized to unity, because

$$\int_{-\infty}^{+\infty} \int_{-\infty}^{+\infty} W(q, p; s)\, dq\, dp = \tilde{W}(0, 0; s) = \tilde{W}(0, 0) = 1 \tag{4.72}$$

according to Eq. (4.23). We obtain from the relation

$$\tilde{W}(u, v; s) = \tilde{W}\left(u, v; s'\right) \exp\left(\frac{1}{4}\left(s - s'\right)\left(u^2 + v^2\right)\right) \tag{4.73}$$

the formula

$$W(q, p; s) = \frac{1}{\pi\left(s' - s\right)} \int_{-\infty}^{+\infty} \int_{-\infty}^{+\infty} W(q', p'; s')$$

$$\times \exp\left(-\frac{(q - q')^2 + (p - p')^2}{(s' - s)}\right) dq'\, dp', \tag{4.74}$$

provided that $s < s'$ so that the integral converges. This relation shows that there is a smoothing hierarchy among the s-parameterized quasiprobability distributions. The smaller the parameter s is, the more the distribution is smoothed. Moreover, the marginal distributions $\mathrm{pr}(q, \theta; s)$ of smoothed Wigner functions ($s < 0$) are smoothed accordingly, that is,

$$
\mathrm{pr}(q, \theta; s) \equiv \int_{-\infty}^{+\infty} W(q \cos\theta - p \sin\theta, q \sin\theta + p \cos\theta; s)\, dp
$$

$$
= \frac{1}{\sqrt{\pi |s|}} \int_{-\infty}^{+\infty} \mathrm{pr}(q', \theta) \exp\left(-|s|^{-1}(q - q')^2\right) dq', \qquad (4.75)
$$

because the Wigner function has the quantum-mechanically correct marginals $\mathrm{pr}(q, \theta)$. Additionally, the overlap relation (4.24) must be modified for s-parameterized quasiprobability distributions, because

$$
\mathrm{tr}\{\hat{F}_1 \hat{F}_2\} = 2\pi \int_{-\infty}^{+\infty} \int_{-\infty}^{+\infty} W_1(q,p)\, W_2(q,p)\, dq\, dp
$$

$$
= \frac{1}{2\pi} \int_{-\infty}^{+\infty} \int_{-\infty}^{+\infty} \widetilde{W}_1(u,v)\, \widetilde{W}_2(-u, -v)\, du\, dv
$$

$$
= \frac{1}{2\pi} \int_{-\infty}^{+\infty} \int_{-\infty}^{+\infty} \widetilde{W}_1(u,v; s)\, \widetilde{W}_2(-u, -v; -s)\, du\, dv. \qquad (4.76)
$$

Consequently, we obtain

$$
\mathrm{tr}\{\hat{F}_1 \hat{F}_2\} = 2\pi \int_{-\infty}^{+\infty} \int_{-\infty}^{+\infty} W_1(q, p; s)\, W_2(q, p; -s)\, dq\, dp \qquad (4.77)
$$

and in particular

$$
\mathrm{tr}\{\hat{\rho} \hat{F}\} = 2\pi \int_{-\infty}^{+\infty} \int_{-\infty}^{+\infty} W(q, p; s)\, W_F(q, p; -s)\, dq\, dp. \qquad (4.78)
$$

This relation shows that a smoothing of the quasiprobability distribution must be compensated for by an enhancement in the resolution of the filter function $2\pi W_F(q, p)$ to calculate expectation values, and vice versa. This procedure may cause significant problems, because it requires extremely high accuracy for $W(q, p; s)$ and it may involve singular filter functions $2\pi W_F(q, p; s)$. We see that the price to be paid for having a non-negative quasiprobability distribution is the introduction of additional noise in practical applications. This noise appears in the smoothing of the marginal distributions, and it must be compensated for by enhancing filter functions to correctly predict observable quantities. Finally, we note that also a certain s *ordering* of operators (Cahill and Glauber, 1969b) can be defined to calculate expectation values. However, apart from the normal ordering (4.65) for the P function, the symmetric ordering (4.35) for the Wigner representation, and the anti-normal ordering (4.62) corresponding to the Q function, these ordering procedures are relatively complicated and rarely used. The reader is referred to the comprehensive articles by Cahill and Glauber (1969a; 1969b) for the details.

4.3 Examples

How do Q functions look? How smoothed are they compared with Wigner functions? How singular can a P function be? Let us study some examples. The simplest candidate to consider is a Fock state $|n\rangle$. We see immediately from formula (4.57) and the Poissonian photon statistics (3.70) of a coherent state that the Q function of a Fock state is given by

$$
\begin{aligned}
Q(q,p) &= \frac{1}{2\pi} |\langle \alpha | n \rangle|^2 \\
&= \frac{1}{2\pi n!} \exp\left(-|\alpha|^2\right) |\alpha|^{2n} \\
&= \frac{1}{2\pi n!} \exp\left(-\frac{1}{2}(q^2 + p^2)\right) \left(\frac{q^2 + p^2}{2}\right)^n .
\end{aligned}
\tag{4.79}
$$

According to the Gaussian asymptotics for the Poissonian distribution (Schleich et al., 1988; Vogel and Schleich, 1992) we can approximate formula (4.79) for large quantum numbers and get

$$
Q_n \sim \frac{1}{2\pi^{3/2} r} \exp\left(-(r - r_n)^2\right)
\tag{4.80}
$$

with

$$
r = \sqrt{q^2 + p^2}
\tag{4.81}
$$

and the Bohr–Sommerfeld radius (Dowling et al., 1991)

$$
r_n = \sqrt{2n + 1}.
\tag{4.82}
$$

We see that the Q function of a Fock state describes a ring with the Bohr–Sommerfeld radius r_n in phase space. This illustrates that Fock states are typical particle-like states containing exactly n energy quanta and showing no wave-like phase dependence. The P function for a Fock state is obtained from Eq. (4.79) by Fourier transformation according to the general relations (4.69) and (4.71),

$$
P = \frac{1}{n!} \exp\left(\frac{q^2 + p^2}{2}\right) \left(\frac{1}{2}\left(\frac{\partial^2}{\partial q^2} + \frac{\partial^2}{\partial p^2}\right)\right)^n \delta(q)\, \delta(p).
\tag{4.83}
$$

This result indicates that Fock states are indeed non-classical, because their P functions contain derivatives of the two-dimensional delta function $\delta(q)\, \delta(p)$. The only exception is the vacuum state with $n = 0$. This example illustrates the mathematical subtleties involved in the P representation.

Let us consider another example, thermal light. The thermal state of a light mode is described by the statistical mixture (3.49) of Fock states with the Boltzmann distribution (3.41) in terms of the β parameter (3.47). We use expression (4.79) to calculate the Q function of the thermal state

$$Q(q,p) = \frac{1}{2\pi Z} \exp\left(-|\alpha|^2\right) \sum_{n=0}^{\infty} \frac{1}{n!}\left(|\alpha|^2 e^{-\beta}\right)^n$$

$$= \frac{1}{2\pi Z} \exp\left(-|\alpha|^2(1 - e^{-\beta})\right)$$

$$= \frac{1}{2\pi(\bar{n}+1)} \exp\left(-\frac{q^2+p^2}{2\bar{n}+2}\right) \tag{4.84}$$

where we applied the expression (3.50) for the Planck spectrum of the average photon number $\bar{n}$. The Q function is a Gaussian distribution centred at the origin in phase space. In the limiting case of vanishing temperature, we obtain the Q function $Q_0(q,p)$ of a vacuum, whereas for finite temperature the Gaussian distribution (4.84) is accordingly broader. This difference illustrates the additional fluctuations involved in a thermal state. From the Fourier-transformed Q function

$$\tilde{Q}(q,p) = \exp\left(-\frac{\bar{n}+1}{2}(q^2+p^2)\right) \tag{4.85}$$

we obtain via the relationship (4.69) with the definitions (4.71) by inverse Fourier transformation both the Wigner and the P function of the thermal state,

$$W(q,p) = \frac{1}{\pi(2\bar{n}+1)} \exp\left(-\frac{q^2+p^2}{2\bar{n}+1}\right), \tag{4.86}$$

$$P(q,p) = \frac{1}{2\pi\bar{n}} \exp\left(-\frac{q^2+p^2}{2\bar{n}}\right). \tag{4.87}$$

Like the Q function, the Wigner function displays the additional thermal fluctuations. The P function of a thermal state is a well-behaved positive function that can be rightfully regarded as a probability distribution. In this sense thermal states are classical states. According to the optical equivalence theorem (4.68) the P function diagonalizes the density operator in terms of coherent states. So, instead of seeing the thermal state as a statistical mixture of non-classical Fock states, we may unravel the thermal density operator into a Gaussian distribution of coherent states; thermal light appears as a Boltzmann distribution of photons or as a Gaussian distribution of waves. In this way we find a satisfying physical interpretation of thermal states and, simultaneously, a good example to demonstrate the general ambiguity in unravelling a mixed-state density operator mentioned in Section 1.2.2.

Formula (4.86) turns out to reveal the Wigner function $W_n(q,p)$ of Fock states as well. We expand $W(q,p)$ in terms of $e^{-\beta}$

$$W(q,p) = \left(1 - e^{-\beta}\right) \sum_{n=0}^{\infty} W_n(q,p)\, e^{-n\beta} \qquad (4.88)$$

with

$$W_n(q,p) = \frac{(-1)^n}{\pi} \exp\left(-q^2 - p^2\right) L_n\left(2q^2 + 2p^2\right). \qquad (4.89)$$

Here the $L_n(q)$ denote the Laguerre polynomials, and we have utilized their relation

$$\sum_{n=0}^{\infty} L_n(q)\, z^n = (1-z)^{-1} \exp\left(qz(z-1)^{-1}\right), \qquad (4.90)$$

see Erdélyi et al. (1953, Eq. 10.12(17)). Because the thermal density operator (3.49) is expanded in the same way as the expression (4.88), the $W_n(q,p)$ must be indeed the Wigner functions for the Fock states $|n\rangle$. Figure 4.4 illustrates the Wigner function of a single photon. Figure 4.7 shows that in contrast to the Q function, the Wigner function for a Fock state displays a "wavy sea" of rings in the area enclosed by the Bohr–Sommerfeld band (Dowling et al., 1991). This feature illustrates again that Fock states are clearly non-classical. Note that the "wavy sea" is necessary to guarantee the orthogonality of the Fock states, because the overlap of two Wigner functions $W_n(q,p)$ and $W_{n'}(q,p)$ must vanish, according to formula (4.27). The transition to the Q function, however, smoothes out the waves, and only the Bohr–Sommerfeld ring at $(2n+1)^{1/2}$ remains. This result shows strikingly that the Q function displays far less signature of a quantum state in phase space than does the Wigner function.

We can understand this another way. Significantly different quantum states may create similar Q functions. Given a picture of a Q function, it may be difficult to infer the state behind the picture. Probably the best example to demonstrate this is the Schrödinger cat state (4.45). Using the scalar product (3.71) of coherent states, we immediately obtain from formula (4.57) the Q function, omitting the normalization factor,

$$Q(\alpha) \propto \exp\left(-|\alpha - \alpha_0|^2\right) + \exp\left(-|\alpha + \alpha_0|^2\right)$$
$$\pm\ 2 \exp\left(-|\alpha_0|^2 - |\alpha|^2\right) \cos\left(2\,\mathrm{Im}(\alpha_0^*\alpha)\right). \qquad (4.91)$$

All that is left from the beautiful quantum interference structure clearly displayed in the Wigner function (4.52) is an exponentially small feature suppressed by $\exp(-|\alpha_0|^2)$. The more macroscopic the quantum superposition (4.45) is, the smaller is this term. If we neglect this little interference feature we obtain the Q function of the incoherent mixture (4.48). The Q function cannot clearly discriminate between macroscopic superpositions

Fig. 4.7. Wigner function of a Fock state (top) compared with the corresponding Q function (bottom). We used Eqs. (4.79) and (4.89) with $n = 4$. The Wigner function shows significantly more detail than the Q function.

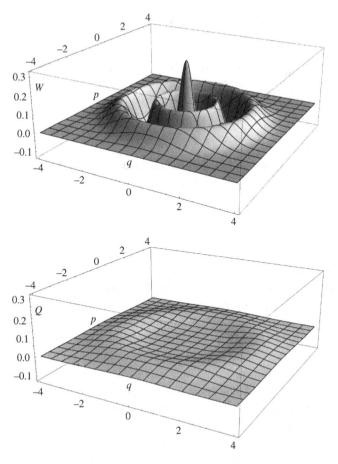

and statistical mixtures, that is, between the classical *either* α_0 *or* $-\alpha_0$ and the quantum mechanical α_0 *as well as* $-\alpha_0$.

Nevertheless, both the Q function and the P function are mathematically equivalent to any other representation for quantum states, and we are entitled to use this equivalence in theoretical calculations. On the other hand, when the Q function is numerically or experimentally given, the retrieval of details hidden in the Q function (but clearly displayed in the Wigner function) takes significant effort in precision. In any case, we must perform a deconvolution procedure that is typically delicate. This fact motivates the conclusion that experimental efforts should be aimed at measuring the Wigner function rather than the Q function to determine the quantum state. The measurement or even the reconstruction of the P function is clearly beyond feasibility in general, because this mathematical object might be ill-behaved, as we have seen in the case of a Fock state. On the other hand, a well-behaved P function can be experimentally determined (Kiesel et al., 2008).

4.4 Questions

4.1 Why are probability distributions of position and momentum plagued by fundamental problems in quantum mechanics?

4.2 How is the Wigner function defined?

4.3 What are marginal distributions? Show how you obtain the position and momentum distributions from the Wigner function.

4.4 How is the characteristic function connected to the Fourier-transformed quadrature distributions?

4.5 Prove Weyl's formula

$$\tilde{W}(u, v) = \mathrm{tr}\{\hat{\rho} \, \exp(-iu\hat{q} - iv\hat{p})\}.$$

4.6 Derive Wigner's formula

$$W(q, p) = \frac{1}{2\pi} \int_{-\infty}^{+\infty} \exp(ipx) \left\langle q - \frac{x}{2} \right| \hat{\rho} \left| q + \frac{x}{2} \right\rangle dx.$$

4.7 Show that, in momentum representation,

$$W(q, p) = \frac{1}{2\pi} \int_{-\infty}^{+\infty} \exp(iqy) \left\langle p + \frac{y}{2} \right| \hat{\rho} \left| p - \frac{y}{2} \right\rangle dy.$$

4.8 Why is the Wigner function of a quantum state real and normalized?

4.9 Prove the overlap formula

$$\mathrm{tr}\{\hat{F}_1 \hat{F}_2\} = 2\pi \int_{-\infty}^{+\infty} \int_{-\infty}^{+\infty} W_1(q, p) \, W_2(q, p) \, dq \, dp.$$

4.10 Which conclusions can you draw from the overlap formula?

4.11 Show that $|W(q, p)| \leq \pi^{-1}$.

4.12 What is symmetric ordering? Why can you express the expectation values of symmetrically ordered operators in terms of the Wigner function as if they were classical quantities? Why is this analogy not entirely perfect?

4.13 Draw the Wigner functions of squeezed coherent states.

4.14 Show that the quadrature distribution of the squeezed vacuum is

$$\mathrm{pr}(q, \theta) = \left(2\pi \Delta_\theta^2 q \right)^{-1/2} \exp \left(-\frac{q^2}{2\Delta_\theta^2 q} \right)$$

with

$$\Delta_\theta^2 q = \frac{1}{2} \left(e^{2\varsigma} \sin^2 \theta + e^{-2\varsigma} \cos^2 \theta \right).$$

4.15 What are Schrödinger cat states? How do they differ from statistical mixtures of coherent states?

4.16 Deduce their Fock expansions

$$|\psi_+\rangle = \frac{1}{\sqrt{\cosh |\alpha_0|^2}} \sum_{m=0}^{\infty} \frac{\alpha_0^{2m}}{\sqrt{2m!}} \, |2m\rangle,$$

$$|\psi_-\rangle = \frac{1}{\sqrt{\sinh |\alpha_0|^2}} \sum_{m=0}^{\infty} \frac{\alpha_0^{2m+1}}{\sqrt{(2m+1)!}} \, |2m+1\rangle.$$

4.17 Calculate the Wigner functions of Schrödinger cat states.

4.18 What are the characteristic features of their quadrature distributions?

4.19 How is the Q function defined? Explain the concept of the smoothing of probability distributions.

4.20 Why is $Q(q,p) = (2\pi)^{-1} \langle \alpha | \hat{\rho} | \alpha \rangle$? Why is this expression non-negative?

4.21 Explain why you can calculate anti-normally ordered operators with the help of the Q function as if they were classical quantities.

4.22 How is the P function defined?

4.23 Why can you calculate normally ordered expectation values with the P function as if they were classical quantities?

4.24 What is the optical equivalence theorem? Prove the theorem.

4.25 What is non-classical light?

4.26 What are s-parameterized quasiprobability distributions? How do the Wigner, Q and P belong to this family of quasiprobability distributions?

4.27 Why are all s-parameterized quasiprobability distributions of quantum states real and normalized?

4.28 How are quasiprobability distributions with different s parameters connected to each other? How are the marginal distributions connected?

4.29 Derive the overlap formula for s-parameterized quasiprobability distributions.

4.30 Draw the Q and the Wigner function of a Fock state.

4.31 Deduce the Q, Wigner and P function of a thermal state.

4.32 Give an interpretation for the P function of the thermal state.

4.5 Homework problem

In addition to the traditional Wigner function $W(q,p)$, the quasiprobability distribution for position and momentum, we can also introduce Wigner functions for other canonically conjugate variables. One example is frequency ω and time t, because, according to Fourier analysis, $\omega = i\partial/\partial t$ and hence

$$\left[\omega, t\right] = i.$$

For a complex signal $\psi(t)$ with spectrum $\tilde{\psi}(\omega)$ we define the Wigner function $W(\omega, t)$ for frequency and time as

$$W(\omega, t) = \frac{1}{2\pi} \int_{-\infty}^{+\infty} \exp(i t \Omega)\, \tilde{\psi}^*(\omega + \frac{\Omega}{2})\, \tilde{\psi}(\omega - \frac{\Omega}{2})\, d\Omega$$

$$= \frac{1}{2\pi} \int_{-\infty}^{+\infty} \exp(i \omega \tau)\, \psi^*(t - \frac{\tau}{2})\, \psi(t + \frac{\tau}{2})\, d\tau.$$

(a) Suppose

$$\psi(t) = \psi_0(t) e^{-i\omega_0 t}.$$

Show that

$$W(\omega, t) = W(\omega - \omega_0, t).$$

(b) Suppose the signal $\psi(t)$ starts at some specific time, say $t = 0$, such that $\psi = 0$ for $t < 0$. What are the integration limits in the time representation of Wigner's formula? Why is $W(\omega, t) = 0$ for $t < 0$?

(c) Consider an oscillating causal signal that exponentially fades away,

$$\psi(t) = \Theta(t)e^{-i\omega_0 t - \gamma t}.$$

Calculate the frequency–time Wigner function $W(\omega, t)$ for this ψ.

(d) Why is the $W(\omega, t)$ of part (c) a quasidistribution?

(e) Calculate the spectral density

$$w(\omega) = \int_{-\infty}^{+\infty} W(\omega, t)\, dt.$$

(f) Plot $w(\omega)$ and discuss it. How is the width of $w(\omega)$ related to the characteristic time scale γ^{-1} of the exponential decay?

4.6 Further reading

Apart from the Wigner function defined in the phase space of position and momentum, other possible Wigner functions exist for different systems. For instance, spin systems may be described by continuous Wigner functions (Agarwal, 1981; Dowling, Agarwal and Schleich, 1994) or Q functions (Luis, 2002). The discrete Wigner functions are another intriguing class of quasiprobability distributions for finite-dimensional systems. See the interesting paper by Wootters (1987) for prime-dimensional state spaces and the extension to odd-dimensional systems by Cohendet et al. (1988, 1990). Wigner functions for even dimensions are a bit odd, and they, together with the odd-dimensional ones, have been considered by Leonhardt (1995, 1996). See also Vourdas's review on discrete quantum mechanics with its connections to number theory (Vourdas, 2004). Wigner functions have been applied in classical physics as well, as quasi-distributions in frequency and time (Paye, 1992; Almeida, 1994) and in classical optics (Dragoman, 1997, 2002).

Chapter 5
Simple optical instruments

5.1 Beam splitter

Compared with other fundamental experiments in physics, optical tests of quantum mechanics are often distinguished by their simplicity. Most quantum optical experiments do not require a whole industry – an optical table of equipment and a few people are often sufficient. Good ideas, good research problems are more important. "Research is to see what everybody has seen and to think what nobody has thought" (Jammer, 1989). A simple optical beam splitter, for instance, is already a nice device to demonstrate the quantum nature of light. Quite a number of puzzling quantum effects have been seen by splitting or recombining photons at a small cube of glass. Additionally, the beam splitter serves as a theoretical paradigm for other linear optical devices. Interferometers, semitransparent mirrors, dielectric interfaces, wave guide couplers, and polarizers are all described sufficiently well by a simple beam splitter model. The quantum effects of almost all passive optical devices can be understood assuming appropriate beam splitter models. (It's all done with mirrors.)

5.1.1 Heisenberg picture

An ideal beam splitter is a reversible, lossless device in which two incident beams may interfere to produce two outgoing beams. For instance, a dielectric interface inside a cube or plate of glass splits a light beam into two, see Fig. 5.1. We may reverse this situation by sending the two beams back to the cube, where they interfere constructively to restore the original beam. However, if we change the phases of the two

Fig. 5.1. Beam splitter.
(Photo: Oliver Glöckl and
Natalia Korolkova.)

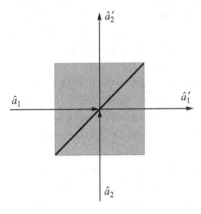

Fig. 5.2. Diagram of a lossless
beam splitter.

beams, their mutual interference generates two outgoing beams in general. So four beams might be involved, two incident and two outgoing light modes, and the splitting of just one beam is a special case. Most polarizing beam splitters use anisotropic media (for instance, calcite) to split the two polarizations of the incident field into two spatially separate beams. Wave guide couplers apply the optical tunnelling effect for mixing the light fields travelling in two fibres. All these simple optical instruments act like beam splitters. To make our theoretical beam splitter model as general as possible, let us describe the device as a *four-port*, that is, as a black box with two input and two output ports having certain properties, see Fig. 5.2.

What happens if two coherent light beams with complex amplitudes α_1 and α_2 interfere? In classical optics the amplitudes are simply superimposed according to the linear transformation

$$\begin{pmatrix} \alpha_1' \\ \alpha_2' \end{pmatrix} = B \begin{pmatrix} \alpha_1 \\ \alpha_2 \end{pmatrix} \tag{5.1}$$

described by the matrix

$$B = \begin{pmatrix} B_{11} & B_{12} \\ B_{21} & B_{22} \end{pmatrix}. \tag{5.2}$$

In quantum optics the complex amplitudes α_1 and α_2 correspond to the annihilation operators $\hat{a}_1$ and $\hat{a}_2$ of the incident fields, and the outgoing beams are characterized by the operators $\hat{a}'_1$ and $\hat{a}'_2$. We assume that the linear interference (5.1) is also valid for quantum fields, that is,

$$\begin{pmatrix} \hat{a}'_1 \\ \hat{a}'_2 \end{pmatrix} = B \begin{pmatrix} \hat{a}_1 \\ \hat{a}_2 \end{pmatrix}. \tag{5.3}$$

This simple model describes very well most passive and lossless devices in which two input beams interfere to produce two outgoing fields.[1] Note that our model (5.3) is a special case of the multi-mode transformation (2.30) that describes scattering in a material or the controlled distribution of light modes in an optical *multiport* (Mattle et al., 1995; Törmä et al., 1995; Törmä, 1998; Törmä and Jex, 1999; Törmä et al., 2002) also called a *passive quantum optical network* (Leonhardt, 2003) or a *Gaussian quantum channel* (Eisert and Wolf, 2007). Here we assume that essentially only two modes interact with each other. On the other hand, any multiport, however complicated, can be constructed from elementary beam splitters (Reck et al., 1994).

As the model (5.3) describes the change of the operators $\hat{a}_1$ and $\hat{a}_2$ after beam splitting, it corresponds to the Heisenberg picture of the process. We easily determine the general properties of the beam splitter matrix B using this picture. Assuming that the incoming and the outgoing beams are both independent bosonic modes, their annihilation operators must satisfy

$$[\hat{a}'_\nu, \hat{a}'^\dagger_\mu] = [\hat{a}_\nu, \hat{a}^\dagger_\mu] = \delta_{\nu\mu} ,$$
$$[\hat{a}'_\nu, \hat{a}'_\mu] = [\hat{a}_\nu, \hat{a}_\mu] = 0. \tag{5.4}$$

Consequently, the beam splitter matrix B must obey

$$|B_{11}|^2 + |B_{12}|^2 = |B_{21}|^2 + |B_{22}|^2 = 1 , \quad B_{11}B^*_{21} + B_{12}B^*_{22} = 0 \tag{5.5}$$

or, in other words, B is unitary

$$B^{-1} = B^\dagger. \tag{5.6}$$

---

[1] It is quite typical that classical laws rule interference phenomena in the classical as well as in the quantum domain. Usually quantum effects change only the visibility of the interference but not the phenomenon itself.

From this condition it follows that the total intensity $\hat{a}_1^\dagger \hat{a}_1 + \hat{a}_2^\dagger \hat{a}_2$ is invariant; the lossless beam splitter conserves energy. Apart from the unitarity (5.6) the beam splitter coefficients are free parameters that depend on the particular experimental situation. They comprise the material properties of an interferometer, a linear coupler, or a polarizer. As in our general simplified model for light beams, the electromagnetic oscillator (see Section 3.1), we have reduced the wealth of the classical optics involved in the device to a few material parameters. Addition-ally, we have assumed that we are dealing with only four well-defined optical modes. Our simplified model allows us to focus on the essential quantum properties of four ports and to be general.

To proceed further we recall the well-known mathematical structure of two-dimensional unitary matrices. Any unitary B can be represented as the matrix product

$$B = e^{i\Lambda/2} \begin{pmatrix} e^{i\Phi/2} & 0 \\ 0 & e^{-i\Phi/2} \end{pmatrix} \begin{pmatrix} \cos(\Theta/2) & \sin(\Theta/2) \\ -\sin(\Theta/2) & \cos(\Theta/2) \end{pmatrix} \begin{pmatrix} e^{i\Psi/2} & 0 \\ 0 & e^{-i\Psi/2} \end{pmatrix}$$

(5.7)

with the real numbers Λ, Θ, Ψ and Φ or, expressed explicitly,

$$B = e^{i\Lambda/2} \begin{pmatrix} \cos(\Theta/2)\, e^{i(\Phi+\Psi)/2} & \sin(\Theta/2)\, e^{i(\Phi-\Psi)/2} \\ -\sin(\Theta/2)\, e^{i(-\Phi+\Psi)/2} & \cos(\Theta/2)\, e^{i(-\Phi-\Psi)/2} \end{pmatrix}.$$

(5.8)

Equation (5.7) shows that any four-port can be considered as acting in three steps. First the phases of the incident modes are changed, then the amplitudes are mixed (rotated), and finally the phases are changed again. In many cases we can incorporate the phase shifts in the definition of the reference phases of the incoming or outgoing beams. The rotation of the mode operators, however, remains the key feature of four-ports. In most later calculations concerning beam splitters, we will consider only real rotation matrices B. In this case we may represent B in terms of the *transmissivity* τ and the *reflectivity* ϱ as

$$B = \begin{pmatrix} \tau & -\varrho \\ \varrho & \tau \end{pmatrix}.$$

(5.9)

In this notation τ equals $\cos(\Theta/2)$, ϱ means $-\sin(\Theta/2)$, and the relation

$$\tau^2 + \varrho^2 = 1$$

(5.10)

accounts for the energy conservation of the lossless four-port.

Note that we have silently smuggled in one essential quantum fea-ture of beam splitters. A beam splitter is a four-port not only in the case of two incoming fields interfering to produce two outgoing beams; a beam splitter is always a four-port. Even if only one beam is split into two, if literally nothing behind the semitransparent mirror is interfering with the incident field, quantum mechanically this nothing means a vacuum state. The sheer possibility that the second mode behind the mirror might be excited makes a difference. The vacuum fluctuations carried by the empty mode (and entering the apparatus via the so-called *unused port* of the beam splitter) do cause physical effects. In particular, we will see in Section 5.2.4 that this picture of the vacuum fluctuations behind the mirror is useful for understanding a fundamental quantum optical experiment. From a formal point of view the explanation of this intriguing quantum feature is elementary in the Heisenberg picture. Suppose that only one beam described by the annihilation operator $\hat{a}_1$ splits into two modes corresponding to the operators $\hat{a}'_1$ and $\hat{a}'_2$ according to the linear transformations $\hat{a}'_1 = B_1\hat{a}_1$ and $\hat{a}'_2 = B_2\hat{a}_1$. Because the commutator $[\hat{a}'_1, \hat{a}'^{\dagger}_2]$ gives $B_1 B_2^*$ and not zero, the outgoing modes cannot be independent quantum systems. The introduction of the second beam, $\hat{a}_2$, however, and the unitarity of the beam splitter matrix guarantee that the outgoing fields can be considered independent bosonic modes.

5.1.2 Schrödinger picture

In the Heisenberg picture the state of the incident beams is invariant, whereas the mode operators $\hat{a}_1$ and $\hat{a}_2$ are changed according to the linear transformation (5.3). In the Schrödinger picture we encounter the opposite situation – the operators are invariant while the states are changed. The standard way of deducing the behaviour in the Schrödinger representation from the Heisenberg picture (and vice versa) is to find an evolution operator (1.33). This operator, here denoted by $\hat{U}_B$, performs the transformation (5.3) of the observables in the Heisenberg representation

$$\begin{pmatrix} \hat{a}'_1 \\ \hat{a}'_2 \end{pmatrix} = \hat{U}^{\dagger}_B \begin{pmatrix} \hat{a}_1 \\ \hat{a}_2 \end{pmatrix} \hat{U}_B. \tag{5.11}$$

In the Schrödinger picture the density operator $\hat{\rho}$ is changed accordingly

$$\hat{\rho}' = \hat{U}_B \hat{\rho} \, \hat{U}^{\dagger}_B. \tag{5.12}$$

Both formulae (5.11) and (5.12) are designed to produce identical expectation values $\mathrm{tr}\{\hat{\rho}F(\hat{a}_1, \hat{a}_2)\}$, and hence both pictures are considered physically equivalent. In the case of a pure state, the state vector $|\psi\rangle$ is transformed as

$$|\psi\rangle' = \hat{U}_B|\psi\rangle. \tag{5.13}$$

A convenient trick for finding the desired operator $\hat{U}^{\dagger}_B$ of the beam splitting is to employ the Jordan–Schwinger representation (Jordan,

1935; Schwinger, 1952) of an angular momentum operator in terms of two bosonic operators, one for each mode. The Jordan–Schwinger operators are defined as

$$\hat{L}_t = \frac{1}{2} \left(\hat{a}_1^\dagger, \hat{a}_2^\dagger \right) \mathbb{1}_2 \begin{pmatrix} \hat{a}_1 \\ \hat{a}_2 \end{pmatrix} = \frac{1}{2} \left(\hat{a}_1^\dagger \hat{a}_1 + \hat{a}_2^\dagger \hat{a}_2 \right),$$

$$\hat{L}_x = \frac{1}{2} \left(\hat{a}_1^\dagger, \hat{a}_2^\dagger \right) \sigma_x \begin{pmatrix} \hat{a}_1 \\ \hat{a}_2 \end{pmatrix} = \frac{1}{2} \left(\hat{a}_1^\dagger \hat{a}_2 + \hat{a}_2^\dagger \hat{a}_1 \right),$$

$$\hat{L}_y = \frac{1}{2} \left(\hat{a}_1^\dagger, \hat{a}_2^\dagger \right) \sigma_y \begin{pmatrix} \hat{a}_1 \\ \hat{a}_2 \end{pmatrix} = \frac{i}{2} \left(\hat{a}_2^\dagger \hat{a}_1 - \hat{a}_1^\dagger \hat{a}_2 \right),$$

$$\hat{L}_z = \frac{1}{2} \left(\hat{a}_1^\dagger, \hat{a}_2^\dagger \right) \sigma_z \begin{pmatrix} \hat{a}_1 \\ \hat{a}_2 \end{pmatrix} = \frac{1}{2} \left(\hat{a}_1^\dagger \hat{a}_1 - \hat{a}_2^\dagger \hat{a}_2 \right)$$

(5.14)

in terms of the unity matrix in two dimensions $\mathbb{1}_2$ and the *Pauli matrices*

$$\sigma_x = \begin{pmatrix} 0 & 1 \\ 1 & 0 \end{pmatrix}, \quad \sigma_y = \begin{pmatrix} 0 & -i \\ i & 0 \end{pmatrix}, \quad \sigma_z = \begin{pmatrix} 1 & 0 \\ 0 & -1 \end{pmatrix}. \tag{5.15}$$

An easy exercise verifies that $\hat{L}_x$, $\hat{L}_y$, and $\hat{L}_z$ obey the commutation relations of angular-momentum components,

$$[\hat{L}_x, \hat{L}_y] = i\hat{L}_z, \quad [\hat{L}_y, \hat{L}_z] = i\hat{L}_x, \quad [\hat{L}_z, \hat{L}_x] = i\hat{L}_y. \tag{5.16}$$

The operator $\hat{L}_t$ commutes with all other $\hat{L}_k$. We also see that

$$\hat{L}_t(\hat{L}_t + 1) = \hat{L}_x^2 + \hat{L}_y^2 + \hat{L}_z^2 \tag{5.17}$$

represents the total angular momentum $\hat{L}^2$. As $\hat{L}_t$ describes half the total photon number, the total angular momentum depends on the total number of photons in the two modes. These properties show that the Jordan Schwinger operators (5.14) behave indeed like angular momentum components.

We can give a simple physical interpretation to the Jordan–Schwinger operators in the case when the two modes are the two polarizations of a light beam. Suppose $\hat{a}_1$ describes the quantum amplitude of one linearly polarized component, say the vertical polarization, while $\hat{a}_2$ refers to the horizontal polarization. In this case, the beam splitter is a device that changes the polarization or splits a light beam into its polarization components (a polarizing beam splitter). The Jordan–Schwinger operators play the role of *quantum Stokes parameters* that characterize the degree and type of polarization. In particular, we identify the classical *Stokes parameters* (Born and Wolf, 1999) with the expectation values

$$s_0 = \langle 2\hat{L}_t \rangle, \quad s_1 = \langle 2\hat{L}_z \rangle, \quad s_2 = \langle 2\hat{L}_x \rangle, \quad s_3 = -\langle 2\hat{L}_y \rangle. \tag{5.18}$$

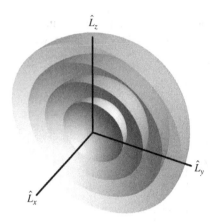

Fig. 5.3. Quantum Poincaré spheres. The quantum Stokes parameters $\hat{L}_x$, $\hat{L}_y$, $\hat{L}_z$ lie on spheres with radius (5.17) where $\hat{L}_t$ assumes half-integer values The figure illustrates the first four of such spheres (shown as half spheres).

The classical Stokes parameters lie on a sphere, the *Poincaré sphere*. As we have seen in Eq. (5.17), the quantum Stokes parameters lie on a sphere as well, see Fig. 5.3. The radius of this *quantum Poincaré sphere* is determined by the total photon number $\hat{a}_1^\dagger \hat{a}_1 + \hat{a}_2^\dagger \hat{a}_2 = 2\hat{L}_t$. Suppose the total photon number is a fixed number n, for example 1 for a single photon distributed over two polarization modes. In this case, the subspace over which the Jordan–Schwinger operators act is $(n+1)$-dimensional. So the Jordan–Schwinger representation describes spin systems of total spin (5.17) $n/2$, unifying the formalism of integer and half-integer spins. If the total photon number is not fixed, the total intensity fluctuates from observation to observation and the quantum Poincaré sphere is fuzzy. But also here the Jordan–Schwinger operators serve as the starting point for the quantum theory of polarized light (Luis, 2002; Korolkova and Loudon, 2005; Luis and Korolkova, 2006; Korolkova, 2007) and unpolarized light (Lehner et al., 1996; Klimov et al., 2005).

Let us return to the beam splitter in the Schrödinger picture. How do the Jordan–Schwinger components transform the mode operators? We note that, for a constant Λ,

$$\exp(-i\Lambda\hat{L}_t)\begin{pmatrix}\hat{a}_1\\\hat{a}_2\end{pmatrix}\exp(i\Lambda\hat{L}_t) = \exp(i\Lambda/2)\begin{pmatrix}\hat{a}_1\\\hat{a}_2\end{pmatrix} \tag{5.19}$$

and, for a constant Φ,

$$\exp(-i\Phi\hat{L}_z)\begin{pmatrix}\hat{a}_1\\\hat{a}_2\end{pmatrix}\exp(i\Phi\hat{L}_z) = \begin{pmatrix}\exp(i\Phi/2)\hat{a}_1\\\exp(-i\Phi/2)\hat{a}_2\end{pmatrix}, \tag{5.20}$$

as is easily seen from the basic property (3.4) of the phase shifting operator $\hat{U}$ defined in Eq. (3.3). We also note that, for another constant Θ,

$$\exp(-i\Theta\hat{L}_y) \begin{pmatrix} \hat{a}_1 \\ \hat{a}_2 \end{pmatrix} \exp(i\Theta\hat{L}_y) = \begin{pmatrix} \cos(\Theta/2) & \sin(\Theta/2) \\ -\sin(\Theta/2) & \cos(\Theta/2) \end{pmatrix} \begin{pmatrix} \hat{a}_1 \\ \hat{a}_2 \end{pmatrix}. \quad (5.21)$$

The easiest way of verifying this formula is to show that both sides obey the same differential equation with respect to Θ, as we see by differentiating both sides and using the commutation relations of $\hat{L}_y$ and the annihilation operators $\hat{a}_1$ and $\hat{a}_2$ for the left-hand side. Because both sides of Eq. (5.21) agree for $\Theta = 0$, they must be identical.

Now we have everything on hand to obtain from the representation (5.7) of the B matrix the evolution operator $\hat{U}_B$ as

$$\hat{U}_B = \exp(i\Lambda\hat{L}_t) \exp(i\Phi\hat{L}_z) \exp(i\Theta\hat{L}_y) \exp(i\Psi\hat{L}_z) \quad (5.22)$$

for the constants Φ, Θ, Ψ, and Λ. It is instructive to visualize the effect of a real-valued beam splitter transformation (5.9) on the Schrödinger wave function for the quadratures q_1 and q_2. As pointed out in the previous section, the case of a real B matrix contains the physical essence of any beam splitter transformation. According to rule (5.13) and representation (5.22), the two-mode wave function

$$\psi(q_1, q_2) = \langle q_1 | \langle q_2 | | \psi \rangle \quad (5.23)$$

of the incident beams is changed to

$$\psi'(q_1, q_2) = \langle q_1 | \langle q_2 | \exp(i\Theta\hat{L}_y) | \psi \rangle \quad (5.24)$$

to produce the wave function $\psi'(q_1, q_2)$ of the outgoing states. Here and throughout the rest of this chapter we use primes to identify the quantum states of the outgoing beams. We express $\hat{L}_y$ of the Jordan–Schwinger representation (5.14) in terms of the quadratures (3.9),

$$\hat{L}_y = \frac{1}{2}\left(\hat{q}_1\hat{p}_2 - \hat{q}_2\hat{p}_1\right), \quad (5.25)$$

and obtain in the Schrödinger representation

$$\psi'(q_1, q_2) = \exp\left(\frac{\Theta}{2}\left(q_1\frac{\partial}{\partial q_2} - q_2\frac{\partial}{\partial q_1}\right)\right)\psi(q_1, q_2)$$
$$= \psi(q_1', q_2') \quad (5.26)$$

with

$$\begin{pmatrix} q_1' \\ q_2' \end{pmatrix} = \begin{pmatrix} \cos(\Theta/2) & -\sin(\Theta/2) \\ \sin(\Theta/2) & \cos(\Theta/2) \end{pmatrix} \begin{pmatrix} q_1 \\ q_2 \end{pmatrix}. \quad (5.27)$$

In the last step we have used the fact that formula (5.26) obeys the same differential equation with respect to Θ as the expression (5.26) does, because

$$\frac{\partial \psi(q_1', q_2')}{\partial \Theta} = \left(\frac{\partial \psi}{\partial q_1} \frac{\partial q_1}{\partial q_1'} + \frac{\partial \psi}{\partial q_2} \frac{\partial q_2}{\partial q_1'} \right) \frac{\partial q_1'}{\partial \Theta} + \left(\frac{\partial \psi}{\partial q_1} \frac{\partial q_1}{\partial q_2'} + \frac{\partial \psi}{\partial q_2} \frac{\partial q_2}{\partial q_2'} \right) \frac{\partial q_2'}{\partial \Theta}$$
$$= \frac{1}{2} \left(q_1 \frac{\partial}{\partial q_2} - q_2 \frac{\partial}{\partial q_1} \right) \psi(q_1', q_2'). \tag{5.28}$$

As the formulae (5.26) and (5.26) agree for $\Theta = 0$, they must be identical for all Θ values. The wave function is simply rotated. In other words, a beam splitter transforms the quadrature wave function in a classical way, according to the transmittivity $\tau = \cos(\Theta/2)$ and reflectivity $\varrho = -\sin(\Theta/2)$ of light at the beam splitter:

$$\psi'(q_1, q_2) = \psi(\tau q_1 + \varrho q_2, -\varrho q_1 + \tau q_2). \tag{5.29}$$

In the momentum representation, we obtain the same result

$$\widetilde{\psi}'(p_1, p_2) = \widetilde{\psi}(\tau p_1 + \varrho p_2, -\varrho p_1 + \tau p_2). \tag{5.30}$$

In many cases, the rotation of the wave function gives a simple intuitive picture (Leonhardt, 1993) for the quantum effects of the beam splitter and related passive optical devices, see, e.g., Fig. 5.4. For example, if two beams in equally squeezed vacuum states interfere according

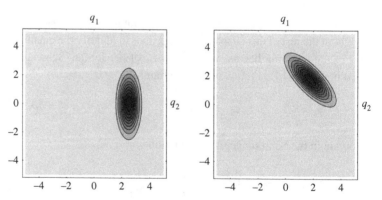

Fig. 5.4. Optical interference of two squeezed states with opposite squeezing parameters. Left: total position quadrature probability distribution $|\psi(q_1, q_2)|^2$ of the incident light. Right: quadrature distribution $|\psi(q_1, q_2)'|^2$ of the outgoing light. The first incident beam has a squeezing parameter ζ of 0.5 and a coherent amplitude q_0 of 2.5; the second beam is in a squeezed vacuum with $\zeta = -0.5$. The rotated quadrature distribution after the optical interference at the beam splitter visualizes the quantum entanglement of the outgoing light.

to the real transformation (5.9), they leave the apparatus unchanged, simply because their total wave function is isotropic,

$$\psi(q_1, q_2) = \frac{e^\zeta}{\sqrt{\pi}} \exp\left(-\frac{1}{2}.e^{2\zeta}\left(q_1^2 + q_2^2\right)\right). \tag{5.31}$$

(We have used Eqs. (3.26) and (3.84) for the wave functions of the squeezed vacuum states.) On the other hand, if two oppositely squeezed vacuum states

$$\psi_1(q_1) = e^{\zeta/2}\pi^{-1/4} \exp\left(-\frac{1}{2}e^{2\zeta}q_1^2\right),$$
$$\psi_2(q_2) = e^{-\zeta/2}\pi^{-1/4} \exp\left(-\frac{1}{2}e^{-2\zeta}q_2^2\right), \tag{5.32}$$

interfere at a real 50:50 beam splitter (where $\tau = \varrho = 1/\sqrt{2}$), they produce a *two-mode squeezed vacuum*

$$\psi'(q_1, q_2) = \frac{1}{\sqrt{\pi}} \exp\left(-\frac{1}{4}e^{2\zeta}(q_1 + q_2)^2 - \frac{1}{4}e^{-2\zeta}(q_1 - q_2)^2\right). \tag{5.33}$$

This wave function describes an entangled state, because the quadratures are correlated, see Fig. 5.5, in particular for strong squeezing $\zeta \to \infty$ when the wave function nearly vanishes unless $q_1 = q_2$.

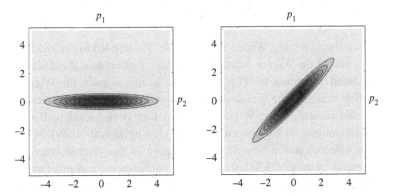

Fig. 5.5. Creation of an Einstein–Podolski–Rosen state by optical interference of two oppositely squeezed vacuum states. Left: total momentum quadrature probability distribution $|\widetilde{\psi}(p_1, p_2)|^2$ of the incident light. Right: quadrature distribution $|\widetilde{\psi}(p_1, p_2)'|^2$ of the outgoing light. The first incident beam is in a squeezing vacuum with ζ of -1.0 while the second beam is in a squeezed vacuum with $\zeta = 1.0$. The outgoing light is in a two-mode squeezed vacuum (5.33), also known as the Einstein–Podolski–Rosen state (7.20). The rotated quadrature distribution after the optical interference at the beam splitter shows the significant correlations of the Einstein–Podolski–Rosen state.

We argue in Section 7.1.3 that the two-mode squeezed vacuum is the strongest entangled state of two light modes with given average total energy (Barnett and Phoenix, 1989, 1991). So the squeezing directions determine whether the two interfering modes produce a perfectly disentangled or a maximally correlated pair of beams (Böhmer and Leonhardt, 1995). This and many other quantum properties of polarizers and beam splitters can be easily explained with the simple picture of rotating wave functions (Leonhardt, 1993).

How is the Wigner function transformed under the action of a beam splitter? Clearly, two modes are involved, and so we describe the quantum state of the incident beams by a two-mode Wigner function extending Wigner's formula (5.34) to two degrees of freedom. For understanding the action of a beam splitter in the Wigner representation, let us focus on the mode rotation (5.9) only. (The effect of the phase shifts has been considered already in the definition (4.1) of the Wigner function.) Because quadrature wave functions are simply rotated, the Wigner function is rotated as well,

$$W'(q_1, p_1, q_2, p_2) = W(q'_1, p'_1, q'_2, p'_2) \tag{5.34}$$

in terms of the transformed quadratures

$$\begin{pmatrix} q'_1 \\ q'_2 \end{pmatrix} = \begin{pmatrix} \tau & \varrho \\ -\varrho & \tau \end{pmatrix} \begin{pmatrix} q_1 \\ q_2 \end{pmatrix}, \quad \begin{pmatrix} p'_1 \\ p'_2 \end{pmatrix} = \begin{pmatrix} \tau & \varrho \\ -\varrho & \tau \end{pmatrix} \begin{pmatrix} p_1 \\ p_2 \end{pmatrix}, \tag{5.35}$$

as is easily seen using Wigner's formula (4.17) extended to two modes. The points of the Wigner function move exactly like classical quantities distributed according to $W(q_1, p_1, q_2, p_2)$. In this respect, the Wigner function behaves like a classical phase space density. Note, however, that this feature of the Wigner representation is restricted to linear transformations, as has been noticed early on by Moyal (1949). Nonlinear effects cause a complicated reshaping of the Wigner function in general. Formula (5.34), though a simple mathematical relation, is the key for understanding the behaviour of beam splitters whenever phase space quantities are concerned. We also notice immediately from the expression (4.44) of the Wigner function for coherent states that the optical interference of coherent light produces just coherent light. To be more precise, if the incident beams are in the coherent states $|\alpha_1\rangle$ and $|\alpha_2\rangle$, the outgoing fields are in the coherent states $|\alpha'_1\rangle$ and $|\alpha'_2\rangle$, with α'_1 and α'_2 given by the classical interference law (5.1). This is consistent with our motivation for the quantum theory of a beam splitter.

5.1.3 Fock representation and wave–particle dualism

We have considered beam splitting in the Heisenberg and Schrödinger picture and using Wigner functions. What happens in the photon number representation? How are photons distributed from an incident light beam to the outgoing field modes? We express the density operator in the Fock basis

$$\hat{\rho} = \sum_{n_1,n_2=0}^{\infty} \sum_{m_1,m_2=0}^{\infty} \rho(n_1,n_2;m_1,m_2) |n_1,n_2\rangle\langle m_1,m_2| \tag{5.36}$$

with the density matrix

$$\rho(n_1,n_2;m_1,m_2) = \langle n_1,n_2| \hat{\rho} |m_1,m_2\rangle. \tag{5.37}$$

Here $|n_1,n_2\rangle$ denotes a two-mode Fock state $|n_1\rangle \otimes |n_2\rangle$. According to relation (3.28) we may mathematically generate this state from the vacuum $|0,0\rangle$ by the procedure

$$|n_1,n_2\rangle = \frac{1}{\sqrt{n_1!\,n_2!}}\,\hat{a}_1^{\dagger n_1}\hat{a}_2^{\dagger n_2} |0,0\rangle. \tag{5.38}$$

In the Schrödinger picture the observables are invariant, whereas the state is changed by the transformation operator $\hat{U}_B$ according to Eq. (5.13). To find the transformed density operator

$$\hat{\rho}' = \sum_{n_1,n_2=0}^{\infty} \sum_{m_1,m_2=0}^{\infty} \rho(n_1,n_2;m_1,m_2)\, \hat{U}_B |n_1,n_2\rangle\langle m_1,m_2| \hat{U}_B^{\dagger}, \tag{5.39}$$

describing the quantum state of the outgoing beams, we need to consider the beam splitting of the Fock states

$$|n_1,n_2\rangle' \equiv \hat{U}_B |n_1,n_2\rangle = \frac{1}{\sqrt{n_1!\,n_2!}}\,\hat{U}_B\,\hat{a}_1^{\dagger n_1}\hat{a}_2^{\dagger n_2}\,\hat{U}_B^{\dagger} |0,0\rangle. \tag{5.40}$$

Note that this equation contains a subtlety: we have replaced $|0,0\rangle$ by $\hat{U}_B^{\dagger}|0,0\rangle$, using the fact that an incident two-mode vacuum produces just vacuum, because the beam splitter conserves energy. To proceed, we express in Eq. (5.40) the "old" mode operators $\hat{a}_1$ and $\hat{a}_2$ in terms of the "new" operators $\hat{a}_1'$ and $\hat{a}_2'$. We simply invert the basic transformation (5.3), which is easily done

$$\begin{pmatrix} \hat{a}_1 \\ \hat{a}_2 \end{pmatrix} = B^{\dagger} \begin{pmatrix} \hat{a}_1' \\ \hat{a}_2' \end{pmatrix} = \begin{pmatrix} B_{11}^* & B_{21}^* \\ B_{12}^* & B_{22}^* \end{pmatrix} \begin{pmatrix} \hat{a}_1' \\ \hat{a}_2' \end{pmatrix}, \tag{5.41}$$

because the beam splitter matrix B is unitary. Consequently, we obtain via the transformation (5.11) in the Heisenberg picture

$$|n_1,n_2\rangle' = \frac{1}{\sqrt{n_1!\,n_2!}} \left(B_{11}\hat{a}_1^{\dagger} + B_{21}\hat{a}_2^{\dagger}\right)^{n_1} \left(B_{12}\hat{a}_1^{\dagger} + B_{22}\hat{a}_2^{\dagger}\right)^{n_2} |0,0\rangle. \tag{5.42}$$

So, in essence, we have simply replaced the "old" mode operators by the "new" mode operators in definition (5.38) to find an expression for the optically mixed Fock states. We expand the binomials in Eq. (5.42), use definition (5.38) again, and arrive at the lengthy formula

$$|n_1, n_2\rangle' = \frac{1}{\sqrt{n_1! n_2!}} \sum_{k_1, k_2 = 0}^{n_1, n_2} \binom{n_1}{k_1} \binom{n_2}{k_2} (B_{11})^{k_1} (B_{21})^{n_1 - k_1}$$
$$\times (B_{12})^{k_2} (B_{22})^{n_2 - k_2} \sqrt{(k_1 + k_2)! \, (n_1 + n_2 - k_1 - k_2)!}$$
$$\times |k_1 + k_2, n_1 + n_2 - k_1 - k_2\rangle \tag{5.43}$$

in terms of the binomial coefficients (3.99). Quite in contrast to the quadrature representation, the beam splitter transformation in the Fock basis is rather involved. A few instructive special cases of the general formula (5.43) are, however, worth considering in detail.

What happens if precisely n photons are "split"? In this case the first beam is in the Fock state $|n\rangle$, whereas the second mode is a vacuum. We obtain from formula (5.43), assuming a real beam splitter matrix,

$$|n, 0\rangle' = \sum_{k=0}^{n} \sqrt{\binom{n}{k}} \, \tau^k \varrho^{n-k} |k, n - k\rangle. \tag{5.44}$$

Of course, photons are not split to fractions. Rather, they are distributed by the beam splitter to the two outgoing fields. The probability p_k of finding k photons in beam one (and consequently $n - k$ photons in beam two) is given by

$$p_k = \binom{n}{k} (\tau^2)^k (\varrho^2)^{n-k} \tag{5.45}$$

with the binomial coefficient (3.99). This rule turns out to be the statistical law for distributing n distinguishable classical particles to two channels (Tijms, 2007), the law of a photon lottery. Commonly, photons are said to be indistinguishable particles. So why do they appear as being distinguishable here? The distinction between distinguishable and indistinguishable particles is a subtle issue in statistics. In statistical physics, we do not discriminate between the individual particles or microstates that constitute a macrostate, but it matters a great deal whether the microstates are distinguishable or indistinguishable in principle. Suppose n distinguishable particles are distributed to two channels, the case related to the splitting of n photons. Let τ^2 be the probability for each particle to end up in channel one and ϱ^2 the probability of leaving in channel two. The sum $\tau^2 + \varrho^2$ gives unity, of course. As all particles are independent, the probability of finding k selected particles in channel one and $n - k$ individual particles in channel two is

given by the product $(\tau^2)^k(\varrho^2)^{n-k}$. The particles are distinguishable –
we could label them in principle as individuals – but we cannot or
choose not to discriminate between them. In the macrostate we consider, any k particles are in channel one and any $n-k$ particles in channel
two. The probability for the macrostate is thus enhanced by the number
of ways of choosing k particles from n. In combinatorics, this number
was found to be the binomial coefficient (3.99), because we can label n
objects in $n!$ different ways, but do not discriminate between the $k!$ possible labellings in channel one and the $(n - k)!$ possibilities in channel
two. Hence we must divide $n!$ by both $k!$ and $(n - k)!$ to get the number
of choices (3.99). Equation (5.45) thus describes the probability for distributing distinguishable particles into two channels. Now imagine the
particles were fundamentally indistinguishable. In this case, there is no
difference between, say, particle 1 in channel one, particle 2 in channel
two and particle 2 in channel one, particle 1 in channel two. Statistically, both situations correspond to the same microstate, whereas for
distinguishable particles they would count as two microstates. In general, the binomial coefficient (3.99) in the probability (5.45) accounts
for the difference between the statistics of distinguishable and indistinguishable particles in sorting them into two channels. So, contrary
to common wisdom, photons count as distinguishable particles in the
statistics of beam splitting, a fact confirmed by experiment (Brendel
et al., 1988).

Formula (5.44) also describes *photon subtraction* (Dakna et al.,
1997), a method that has been applied to generate Schrödinger cat states
(4.45) of small amplitudes, Schrödinger kittens (Neergaard Nielsen
et al., 2006; Ourjoumtsev et al., 2006). What is photon subtraction?
Suppose we wish to extract exactly one photon from a light mode.
We could impinge the mode onto a highly transmissive beam splitter where only a small fraction of the incident light is reflected. If, on
the reflected light, a photodetector registers one click it will most certainly have registered only one photon. This photon is subtracted from
the incident light. Alternatively, one could use a highly reflective beam
splitter and monitor the transmitted photons. In this case the reflected
light is the photon subtracted light. In order to determine the result of
photon subtraction, we represent the incident light as a superposition of
Fock states $|n\rangle$. We use our formula (5.44) for the splitting of photons
and project it onto a one-photon Fock state in the second mode. The
resulting state is $\sqrt{n}\,|n - 1\rangle$: photon subtraction acts like the annihi-
lation operator (3.20). If we counted two photons in the second beam
and subtracted two photons from the first beam, the device would act
as the square of the annihilation operator. The action of the annihilation
operator on the squeezed vacuum turns out to generate a state that is

remarkably similar to the even Schödinger cat state (4.49) for small amplitude α_0 (resulting from a squeezed vacuum with a relatively small squeezing parameter). If we subtracted two photons we would obtain a state that resembles the odd Schödinger cat state (4.50). Returning to photon subtraction in general, the prefactor $\sqrt{n}$ in the probability amplitude indicates that there are n different ways of singling out one photon from the n photons in the Fock state $|n\rangle$. Again, photons behave as distinguishable particles.

Splitting precisely n photons reveals the particle nature of light. Photons are indivisible energy quanta, and so the beam splitter faces no other choice than distributing them by chance. On the other hand, what happens if two photons interfere optically? To be precise, suppose that both incident beams carry exactly one photon each. What would we expect? Individual photons cannot have definite optical phases. However, if they are forced to interfere either constructively or destructively, they might consider both options as equally likely and emerge in pairs in beam one or in beam two. That this is indeed true is easily seen from formula (5.43) assuming a real 50:50 beam splitter with $\tau = \varrho = 1/\sqrt{2}$. In this case we obtain

$$|1, 1\rangle' = \frac{1}{\sqrt{2}}\left(|0, 2\rangle - |2, 0\rangle\right). \tag{5.46}$$

Photons appear in pairs. To be more precise, the outgoing photons are in the quantum superposition state of being both together in one beam or in the other. Consequently, if one photon is detected in one beam, then no photon should be in the other. The count correlation of the two outgoing beams should exhibit a pronounced minimum, an effect that was experimentally demonstrated by Hong et al. (1987) for the first time and that has served as a benchmark for the quality of photon pairs ever since. We have thus seen that the splitting or the interference of single photons is a wonderful way of demonstrating the fundamental wave–particle dualism of light.

5.2 Detection

Making sense of the quantum measurement process is the root of our troubled relationship with quantum mechanics. How do physical quantities make the transition from being certain possibilities to being uncertain facts? But quantum measurements not only pose intellectual puzzles; understanding the restrictions of quantum mechanics has led to ingenious experimental techniques to get around them. For instance, we cannot see quantum objects as they really are, because as soon as we observe them we disturb them. But, as we explain in this section, one can reconstruct faithful images of quantum states from the results of many observations, probing their different quantum aspects separately. We cannot measure position and momentum at the same time and with arbitrary precision, but, as we also show, one can simultaneously measure such conjugate variables with some degree of imprecision set by quantum mechanics. Of course, such quantum measurements not only lead to new techniques, they also rely on the latest technology. Photodetection is the very basis of quantum optical measurements. Experiments with non-classical light require highly efficient detectors at the limits of present-day technology. These detectors must have low electronic noise, and they should have single-quantum resolution. So any progress in detector design widens the scope of measuring interesting quantum effects. How do detectors work, and which types are commonly used?

5.2.1 Photodetector

Most photodetectors (Silberhorn, 2007) apply a version of the *photoelectric effect* to operate (that was discovered by Heinrich Hertz (Hertz, 1887) in 1887 and has been important for quantum mechanics since Einstein's Nobel prize-winning 1905 paper (Einstein, 1905a)). Radiation ionizes a piece of photosensitive material and produces freely moving electrons, that is, an electric current that can be amplified and handled by electronic means. Most modern photodetectors are photodiodes. A diode is an electronic device that conducts electric currents in only one direction. Figure 5.6 shows the principal scheme of a semiconductor photodiode. The diode is made from (positively) p-doped and (negatively) n-doped material. At the junction between these materials, electrons tend to diffuse into the p-doped material, while holes drift into the n-doped layer until the created voltage counterbalances the diffusion. The junction is starved of charge carriers and therefore not conducting. An applied forward-bias voltage (positive in the p-type region and negative in the n-type layer) may overcome the blockade of the diode by pushing the carriers back into the junction such that a

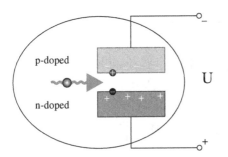

Fig. 5.6. Photodiode. A photon creates an electron–hole pair in the carrier-starved region between the p-doped and n-doped layers such that a current can flow. This photocurrent is proportional to the photon flux. Reproduced with permission from (Silberhorn, 2007).

current flows. On the other hand, a reverse-bias voltage has the opposite effect: the diode indeed conducts only in one direction. Photodiodes are often, but not always, operated in reverse-bias mode. Due to the photoelectric effect electrons and holes are created in the carrier starved region between the p-doped and n-doped layers. This photocurrent is proportional to the number of created carriers; they, in turn, are proportional to the number of photons: the photodiode measured the photon flux. In practice, not all photons create charge carriers; so photodetectors are not perfectly efficient, but state-of-the-art photodiodes may reach efficiencies over 95%. However, the electronic noise of the photodiode is typically much larger than the contribution of a single photon to the photocurrent. Therefore, linear response photodiodes do not resolve individual photons.

Avalanche photodiodes, on the other hand, do respond to single photons. They act like the Geiger–Müller counters used to detect nuclear radiation. A high voltage is applied across the so-called multiplication region of the diode where the charge carriers knock out further carriers and therefore exponentially multiply. A single photon may trigger an avalanche of carriers that then appears as a distinct electronic detector click. Avalanche photodiodes are very sensitive and can be made highly efficient, but they are over-sensitive when more than one photon is initially detected, because the size of the avalanche does not much depend on the number of carriers initially set free by the photoelectric effect. One or more photons may trigger the same detector click. A way around this problem is *multiplexing* (Paul et al., 1996). Figure 5.7 shows

Fig. 5.7. Multiplexed detection schemes. Avalanche photodiodes respond to single photons, but cannot discriminate between one or more detected photons. Multiplexing is a way around this problem. The incident light is distributed to an array of avalanche photodiodes by an arrangement of beam splitters. Each individual detector rarely encounters more than one photon; so the whole device accurately measures the number of photons. Reproduced with permission from Silberhorn (2007).

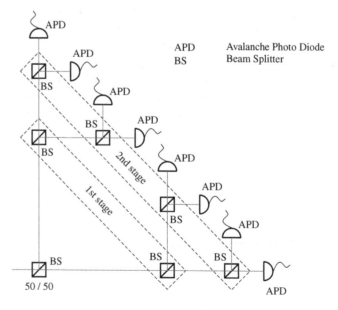

APD   Avalanche Photo Diode
BS   Beam Splitter

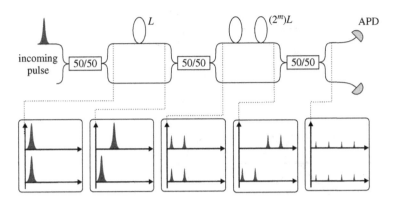

Fig. 5.8. Time-multiplexed detector. The fibre couplers and loops multiplex a pulse of light in time such that two avalanche photodiodes are sufficient for measuring the photon number. This simple scheme replaces the experimentally more complicated one of Fig. 5.7. Reproduced with permission from (Silberhorn, 2007).

an arrangement of beam splitters that distribute the incident light onto an array of avalanche photodiodes. If the light is sufficiently diluted such that each photodiode will hardly encounter more than one photon, the multiplexed detector indeed measures the accurate photon number and can be used to retrieve the photon statistics (Paul et al., 1996). An elegant way of implementing this scheme is *time multiplexing* (Achilles et al., 2004; Silberhorn, 2007), as shown in Fig. 5.8. A sequence of fibre couplers and fibre loops splits and delays light pulses in a controlled way such that rarely more than one photon arrives on the other end where two avalanche photodiodes sit. The couplers and loops act like the spatial multiplexing shown in Fig. 5.7, but only two photodetectors are needed and the loopy fibres are flexible but stable, which considerably simplifies the scheme. Also the effect of detection losses may be taken into account in the obtained photon statistics (Achilles et al., 2004).

In practice, we must often face inefficiencies and noise in realistic photodetection. A convenient model to understand the effect of these experimental problems is provided by imagining a fictitious beam splitter placed in front of an ideal detector, see Section 6.2.2 and, in particular, Fig. 6.1. Only the transmitted photons are counted, so that the transmissivity of the beam splitter corresponds to the detection efficiency. Dissipation is always accompanied by fluctuations. These degrade the quantum noise properties of the detected light. The fluctuations are modelled by a vacuum entering the unused port of the fictitious beam splitter. In Section 6.2.2 we show that this beam splitter model is typical for all loss processes, not only for detection losses. In Section 6.2.1 we discuss how the Wigner function of the signal is smoothed during absorption. Nonclassical features of light are easily lost in inefficient detection.

Fig. 5.9. Balanced homodyne detector. The signal is mixed with coherent light, called the local oscillator, at a balanced beam splitter and the outgoing beams are detected. The difference of the photocurrents is proportional to the quadrature $\hat{q}_\theta$ with a reference phase θ set by the phase of the local oscillator.

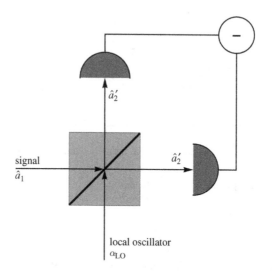

5.2.2 Balanced homodyne detection

Under idealized circumstances the photon number is measured in direct photodetection. However, another method of detection has been invented, in which the field amplitudes (the quadrature components) are measured instead of the quantized intensity. As we have seen in Chapter 3, intensity (photon number) and field amplitude (quadrature) are distinct quantities. There is no simple relationship between the photon statistics and the quadrature distributions in the quantum regime. (We show in Appendix B how the two are related to each other.) Additionally, the field amplitudes contain phase information, and so they are dependent on phase. Quadrature components $\hat{q}_\theta$ are defined (3.12) with respect to a certain reference phase θ that can be varied experimentally. How do we measure the quadratures?

In 1983 Yuen and Chan proposed (Yuen and Chan, 1983) and subsequently Abbas, Chan, and Yee (Abbas et al., 1983) first demonstrated balanced homodyne detection, a method designed for measuring the degree of quadrature squeezing.[2] The principal scheme of the balanced homodyne detector is depicted in Fig. 5.9. The signal interferes with a coherent laser beam at a well-balanced 50:50 beam splitter. The laser

---

[2] Note that there are also other ways of quadrature measurement based on earlier versions of homodyne detection (Yuen and Shapiro, 1978a, 1978b; Shapiro et al., 1979; Yuen and Shapiro, 1980). The balanced version of homodyne detection has the great practical advantage of cancelling technical noise and the classical instabilities of the reference field. Note also that the idea of homodyne detection was born in the microwave technology developed during World War II. The precursor of the optical balanced homodyne detector was the microwave balanced mixer radiometer (Dicke, 1946).

The local oscillator not only provides a phase reference, it selects from the quantum field of light the mode that is actually measured in homodyne detection. Only the mode that matches the spatial temporal beam shape of the local oscillator is multiplied by the large amplitude α_{LO} in Eq. (5.49) and so brought above the noise floor of the photodiodes. In this way the observer separates the quantum object, here a single optical mode, from the "rest of the world". If the interesting bit of the quantum state of light happens to be in modes that do not completely match the local oscillator mode the interesting features are partially lost by mode mismatch. Designing the mode of the local oscillator such that it matches the detected light as perfectly as possible is crucial in applications of homodyne detection on non-classical light, in particular for optical homodyne tomography that we describe in Section 5.2.3.

Although our analysis is rather crude, the final result (5.50) is indeed correct and has been verified by more sophisticated theories of homodyne detection (Carmichael, 1987; Walker, 1987; Braunstein, 1990; Vogel and Grabow, 1993; Ou and Kimble, 1995; Raymer et al., 1995; Banaszek and Wódkiewicz, 1997; Tyc and Sanders, 2004; Vogel and Welsch, 2006). The reason is that in the regime of a strong local oscillator, classical optics is well-justified for understanding the behaviour of the local oscillator light (in the sense of Bohr's correspondence principle). Our simple model is typical for many other theories of quantum measurement. A quantum object, here the signal mode, encounters a classical apparatus, here the beam splitter, the local oscillator, and the detectors. The theoretical model of the measurement is a hybrid of quantum mechanics and classical physics. We may certainly push the limits between the quantum world and the classical apparatus a bit further. We may quantize the local oscillator, for instance, but there is always a border between measuring device and quantum system. It is wise to make the cut between the quantum world and the classical apparatus as early as possible, yet without losing essential features.

The balanced homodyne detector shares another characteristic aspect of many other quantum measurement devices – the detector is an amplifier. The local oscillator amplifies the signal by the mutual optical mixing of the two. Or, seen from a different point of view, the homodyne detector is an interferometer that can be measurably imbalanced by a single photon in the signal mode, because the reference field is very intense. The coherent amplification by the local oscillator has an important technical advantage. The so-amplified signal is well above the electronic noise floor of the photodiodes. The signal amplitude is enhanced so that even the noisy linear response photodiodes can detect the quantum features of the signal. On the other hand, these photodetectors may reach nearly 100% efficiency (Polzik et al., 1992). In

SOBALD DIE ATOMIS BEOBACHTET WERDEN,
BENEHMEN SIE SICH AUF EINMAL GANZ ANDERS

Fig. 5.10. "As soon as we watch the atoms they behave differently." Quantum objects are significantly changed by measurements, in general. They are showing us only their particular aspects, the wave or the particle aspect, for instance, but not what they really are. (Reproduced with friendly permission of P. Evers.)

this way the balanced homodyne detector takes advantage of the high efficiency of photodiodes and at the same time can determine signals with single-photon resolution – a nearly perfect technical solution! Moreover, apart from technical advantages, balanced homodyne detection has been given a profound physical application in measuring the quantum state of light (Leonhardt, 1997a).

5.2.3 Quantum tomography

But how can we observe quantum states? As a fundamental feature of quantum mechanics, we cannot see physical objects *as they are*. According to Heisenberg's uncertainty principle, for instance, we cannot measure position and momentum simultaneously *and* precisely. While observing the position of a mechanical system we are losing the momentum information. We cannot see the things as they are, because as soon as we watch them they behave differently, see Fig. 5.10. Instead, we see only the various aspects of the physical objects, such as either position or momentum or either the wave or particle aspect, that depend on the particular kind of observation. Moreover, these features are complementary; they exclude each other, yet they are only different sides of the same coin. As we cannot see the true nature of "quantum things", they might rather resemble abstract ideas than things we call visible and real. What we do see are only the different aspects of a quantum object, the "quantum shadows" in the sense of Plato's famous cave parable (Plato, 1935); Plato compared people to prisoners who were chained in a cave and forced to see only the shadows of the things outside projected onto the wall of their cave.

However, although as macroscopic observers we may play the role of Plato's prisoners, we still have some advantages compared to them. Although we are chained to the macroscopic world and see only mere shadows of the real, quantum world, we may get insights into quantum objects by changing the directions under which the shadows are projected. For example, we could first observe the position of a particle and then the momentum of a second particle prepared in the same state as the first one. If we repeat this procedure on many identically prepared quantum states we may reconstruct the true picture. Remember that in some cases in medical imaging we cannot see an "object" directly without harming it or we are simply restricted for practical reasons, and yet we can tomographically reconstruct the shape of the hidden object. In transmission computerized tomography, for instance, a cross-section of the human body is scanned by a thin X-ray beam whose attenuated intensity is recorded by a detector. The object (or the apparatus) is rotated to yield many intensity distributions at different angles. Finally, a computer processes the data to build a picture of the object in the form of a spatial distribution of the absorption coefficient.

All we need to do for understanding the key idea of quantum state tomography is to translate this procedure to the language of quantum mechanics, see Fig. 5.11. We picture the "shape" of a quantum object in

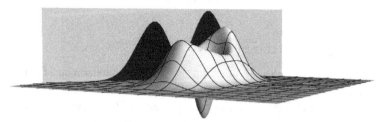

Fig. 5.11. Tomography, from the Greek word $\tau o\mu o\varsigma = slice$, is a method to infer the shape of a hidden object from its shadows (projections) under various angles. Quantum tomography is the application of this idea to quantum mechanics. For instance, in optical homodyne tomography the Wigner function plays the role of the hidden object. The observable "quantum shadows" are the quadrature distributions and are measured using homodyne detection. From these distributions the Wigner function or, more generally, the quantum state is reconstructed. The figure shows the Wigner function of a Schrödinger kitten state (4.50) reconstructed from experiment. (Data: Akira Furusawa and Hidehiro Yonezawa.)

phase space using the Wigner representation. The transmission profiles correspond to the marginal distributions

$$\text{pr}(q,\theta) = \int_{-\infty}^{+\infty} W(q\cos\theta - p\sin\theta, q\sin\theta + p\cos\theta)\,dp \qquad (5.51)$$

of the Wigner function $W(q,p)$, that is, to shadows projected onto a line in quantum phase space. Certainly, because we cannot measure simultaneously and precisely the position q and the momentum p, we cannot observe the Wigner function directly as a probability distribution. Yet we can measure the quadrature histograms

$$\text{pr}(q,\theta) = \langle q|\,\hat{U}(\theta)\,\hat{\rho}\,\hat{U}^{\dagger}(\theta)\,|q\rangle, \qquad (5.52)$$

and by varying the phase θ we see the quantum object under different angles. (As usual $\hat{U}(\theta)$ denotes the phase-shifting operator defined in Eq. (3.3).) Given the distributions $\text{pr}(q,\theta)$, we apply the mathematics of computerized tomography to infer the Wigner function.

In quantum optics, the quadratures q_θ of a spatial temporal mode can be precisely measured using balanced homodyne detection. Here the angle θ is defined by the phase of the local oscillator with respect to the signal. The phase θ can be easily varied using a piezo-electric translator. To measure the quadrature distributions, we may fix the phase angle θ and perform a series of homodyne measurements at this particular phase to build up a quadrature histogram $\text{pr}(q,\theta)$. Then the local oscillator phase should be changed in order to repeat the procedure at a new phase, and so on (Smithey et al., 1993). Another possibility is to monitor the phase while it drifts or to sweep it in a known way (Breitenbach et al., 1995), see Fig. 5.12. In any case,

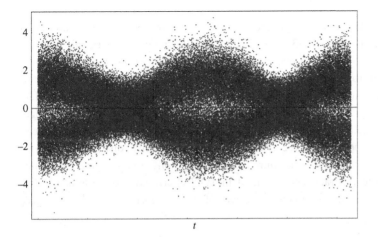

Fig. 5.12. Quadrature measurements. The quadrature q_θ of light in a Schrödinger kitten state (4.50) was measured using homodyne detection while the phase θ was varied. The data were used to reconstruct the Wigner function shown in Fig. 5.11. (Data: Akira Furusawa and Hidehiro Yonezawa.)

Fig. 5.13. The first "quantum pictures". Reconstructed Wigner functions of a squeezed vacuum (a, b) and a vacuum (c, d) viewed in 3D and as contour plots, with equal numbers of constant height contours. The figures are reproduced with permission from the pioneering paper of Smithey et al. (1993). Copyright of the American Physical Society. These are the first pictures of the vacuum fluctuations and of the anisotropic quantum noise of a squeezed vacuum.

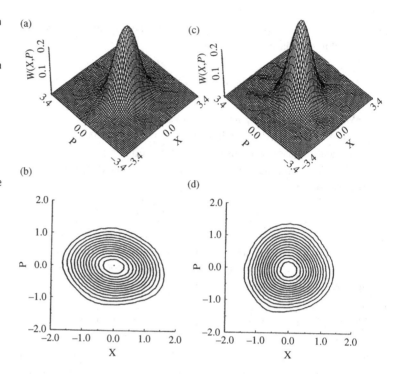

the homodyne measurement must be repeated many times on identically prepared light modes (or on a continuous flux of light in a steady state) to gain sufficient statistical information about the quadrature values at a sufficient number of reference phases (Leonhardt et al., 1996; Leonhardt, 1997a). Finally, the Wigner function is tomographically reconstructed from the experimental data. Alternatively, one can also sample the density matrix directly, as we discuss in Appendix B, or use maximum likelihood reconstruction (Hradil et al., 2004; Lvovsky, 2004).

The pioneering experimental demonstration of this method was published in 1993 by Smithey, Beck, Raymer, and Faridani (Smithey et al., 1993). They coined the name *optical homodyne tomography* for this fundamental measurement technique. Figure 5.13 shows the first Wigner functions reconstructed from experimental data using this scheme. The work was (indirectly) inspired by Vogel and Risken (1989), who first noticed that homodyne detection can be applied for reconstructing quasiprobability distributions. Probably the first who introduced tomography to quantum mechanics were Bertrand and Bertrand in 1987 (Bertrand and Bertrand, 1987). However, the main objective of their paper was rather to give a more convincing approach to Wigner's formula (4.17) than to propose an experimental technique. (Our definition (4.1) of the Wigner function was inspired by their idea.) The mathematics of tomography dates back to 1917, when Johann Radon published in the article (Radon, 1917) "Über die Bestimmung von Funktionen durch ihre Integralwerte längs gewisser Mannigfaltigkeiten" an inversion formula for the later so-called *Radon transformation* (5.51). This theory was rediscovered in the 1960s and early 1970s for medical imaging, a technique Cormack and Hounsfield shared the Nobel Prize for. Like any other idea of importance, quantum tomography has roots in the development of science.

How does tomography work? The ground has been already prepared in our approach to Wigner's formula (4.17) in Section 4.1. According to relation (4.9) the Fourier transformed quadrature distribution gives the characteristic function in polar coordinates, that is, the Fourier-transformed Wigner function. We simply need to perform the Fourier inversion in polar coordinates and obtain

$$
\begin{aligned}
W&(q,p) \\
&= \frac{1}{(2\pi)^2} \int_{-\infty}^{+\infty} \int_0^\pi \widetilde{W}(\xi\cos\theta, \xi\sin\theta) |\xi| \exp\left(i\xi(q\cos\theta + p\sin\theta)\right) \\
&\quad \times d\theta \, d\xi \\
&= \frac{1}{(2\pi)^2} \int_{-\infty}^{+\infty} \int_0^\pi \int_{-\infty}^{+\infty} \mathrm{pr}(x,\theta) |\xi| \exp\left(i\xi(q\cos\theta + p\sin\theta - x)\right) \\
&\quad \times dx \, d\theta \, d\xi
\end{aligned}
\tag{5.53}
$$

using Eq. (4.9) and definition (4.5). In the last line the ξ integration does not concern the quadrature distributions $\mathrm{pr}(x,\theta)$. We simplify formula (5.53), introducing the kernel

$$
K(x) = \frac{1}{2} \int_{-\infty}^{+\infty} |\xi| \exp(i\xi x) \, d\xi,
\tag{5.54}
$$

and obtain

$$
W(q,p) = \frac{1}{2\pi^2} \int_0^\pi \int_{-\infty}^{+\infty} \mathrm{pr}(x,\theta) K(q\cos\theta + p\sin\theta - x) \, dx \, d\theta.
\tag{5.55}
$$

The kernel $K(x)$ exists in the sense of a generalized function (Gel'fand and Shilov, 1965) like Dirac's delta function, which took decades to become accepted in mathematics. Physicists are excused, and so we can regularize the generalized function $K(x)$ in a simple way. First we express the kernel (5.54) as

$$K(x) = \frac{1}{2} \left(\int_0^{+\infty} \exp(i\xi x)\, \xi\, d\xi - \int_{-\infty}^0 \exp(i\xi x)\, \xi\, d\xi \right)$$

$$= \frac{1}{2i} \frac{\partial}{\partial x} \left(\int_0^{+\infty} \exp(i\xi x)\, d\xi - \int_{-\infty}^0 \exp(i\xi x)\, d\xi \right)$$

$$= \frac{1}{2i} \frac{\partial}{\partial x} \left(\int_0^{\infty} \exp(i\xi x)\, d\xi - \int_0^{\infty} \exp(-i\xi x)\, d\xi \right)$$

$$= \frac{\partial}{\partial x} \operatorname{Im} \int_0^{\infty} \exp(i\xi x)\, d\xi . \tag{5.56}$$

To regularize the integral (5.56), we add an infinitely small yet positive imaginary part $+i\varepsilon$ to x so that

$$K(x) = \frac{\partial}{\partial x} \operatorname{Im} \int_0^{\infty} \exp\left(i\xi(x+i\varepsilon)\right) d\xi \tag{5.57}$$

converges and obtain the result

$$K(x) = \frac{\partial}{\partial x} \operatorname{Re} \frac{1}{x+i\varepsilon}. \tag{5.58}$$

The infinitesimal imaginary part $+i\varepsilon$ added to x means that we should slightly move the integration contour of (5.55) below the real axis. This regularization is equivalent (Gel'fand and Shilov, 1965) to *Cauchy's principal value* of the integral (5.55). We abbreviate this value by the symbol $\mathcal{P}$ and write

$$K(x) = \frac{\partial}{\partial x} \frac{\mathcal{P}}{x} \equiv -\frac{\mathcal{P}}{x^2}. \tag{5.59}$$

As a typical generalized function, $K(x)$ only makes sense as a kernel in integrations with respect to well-behaved functions. (Strictly speaking, $-\mathcal{P}x^{-2}$ serves only as a convenient abbreviation for the derivative of a principal value integral involving $\mathcal{P}x^{-1}$.) Finally, we obtain the compact formula

$$W(q,p) = -\frac{\mathcal{P}}{2\pi^2} \int_0^{\pi} \int_{-\infty}^{+\infty} \frac{\operatorname{pr}(x,\theta)\, dx\, d\theta}{(q\cos\theta + p\sin\theta - x)^2} \tag{5.60}$$

for the *inverse Radon transformation*. This expression shows how the Wigner function $W(q,p)$ can be calculated from a mathematically given set of quadrature distributions $\operatorname{pr}(x,\theta)$. A real-world numerical application of this formula requires, however, a certain filtering of the kernel (5.54). The inverse Radon transformation (5.60) shows how quantum states can be reconstructed from "quantum shadows" in principle. The sampling of the density matrix described in Appendix B or maximum likelihood fits (Badurek et al., 2004; Hradil et al., 2004, 2006; Řeháček et al., 2008), in particular the elegant method (Lvovsky, 2004), provide alternatives with advantages in practice.

5.2.4 Simultaneous measurement of conjugate variables

Quantum systems manage to combine classically contradicting features – the wave *and* the particle aspects, for instance. Other features they may comprise are mutually exclusive because of quantum mechanics only. For instance, as a consequence of Heisenberg's uncertainty principle the position q and the momentum p cannot be observed both simultaneously *and* precisely. On the other hand, position and momentum define the very state of a mechanical system in classical physics (or, more generally, the statistical distribution of q and p describes an ensemble of mechanical objects). And, in classical mechanics, one can, in principle, simultaneously measure position and momentum to an arbitrary degree of precision. In quantum mechanics, these *canonically conjugate* quantities obey the commutation relation (3.10) with $\hbar$ set to unity.[3] Commutators are connected to uncertainty relations, because for any two observables $\hat{F}$ and $\hat{G}$ Robertson's *general uncertainty relation* predicts (Davydov, 1991)

$$\Delta^2 \hat{F} \, \Delta^2 \hat{G} \geq -\frac{1}{4} \langle [\hat{F}, \hat{G}] \rangle^2 \qquad (5.61)$$

for the standard deviations of F and G from their mean values ($\Delta^2 F \equiv \langle F^2 \rangle - \langle F \rangle^2$ and $\Delta^2 G \equiv \langle G^2 \rangle - \langle G \rangle^2$). Note that the commutator of two Hermitian operators is anti-Hermitian with purely imaginary expectation values. So the right-hand side of relation (5.61) is non-negative. The uncertainty relation (5.61) quantifies the uncertainty principle by giving an absolute bound on the uncertainty product $\Delta F \Delta G$.

What happens when we *attempt* to measure simultaneously the position *and* the momentum of a mechanical system? In 1965 Arthurs and Kelly published a remarkable brief report in the *Bell System Technical Journal* (Arthurs and Kelly, 1965) entitled "On the Simultaneous

---

[3] The relation (3.10) was discovered by Max Born after Heisenberg's first flash of insight into the magnificent structure of quantum mechanics in 1925 (Heisenberg, 1969). As Born declared (Born, 1956), "I shall never forget the thrill I experienced when I succeeded in condensing Heisenberg's ideas on quantum conditions in the mysterious equation $\hat{p}\hat{q} - \hat{q}\hat{p} = h/(2\pi i)$". Wolfgang Pauli (Pauli, 1933) proved Heisenberg's uncertainty relation (3.80); Howard P. Robertson (Robertson, 1929) showed that the uncertainty relation follows from the commutation rule (3.10).

Measurement of a Pair of Conjugate Observables". They described such an attempt – a Gedanken experiment on a one-dimensional wave packet. Let us understand what is going on there. However, we will not consider Arthurs and Kelly's particular example here. (The reader is referred to Stenholm's excellent article (Stenholm, 1992) for the details.) We will sketch only the general idea behind this and other schemes for measuring jointly q and p.

Certainly, we cannot measure the position and the momentum simultaneously *and* precisely. Yet literally taken, this fact does not exclude the possibility to observe q and p at the same time with, however, limited accuracy. We could allow for some extra quantum noise to be involved in the measurement process. How can we put this idea into precise terms? Let us assume that we have two meters attached to our system – one for observing q and the other for p, described by the relations

$$\hat{Q}_1 = \hat{q}_s + \hat{F}, \quad \hat{P}_2 = \hat{p}_s + \hat{G}. \tag{5.62}$$

The subscript s refers to the signal, whereas $\hat{Q}_1$ denotes the position read by one meter and $\hat{P}_2$ the momentum read by the other meter. The operators $\hat{F}$ and $\hat{G}$ describe the extra quantum fluctuations necessary for measuring simultaneously the position and the momentum of the signal. We do not need to specify the state of the fluctuations nor the particulars of the operators. First, we require only that they carry no pre-existing amplitude, that is,

$$\langle \hat{F} \rangle = \langle \hat{G} \rangle = 0. \tag{5.63}$$

Although we cannot measure the intrinsic $\hat{q}_s$ and $\hat{p}_s$ simultaneously, we can read the meters $\hat{Q}_1$ and $\hat{P}_2$ at the same time. For this, we require that

$$[\hat{Q}_1, \hat{P}_2] = 0. \tag{5.64}$$

In addition, we assume that the signal state and the fluctuations $\hat{F}$ and $\hat{G}$ are separate physical systems. Consequently, the commutators $[\hat{q}_s, \hat{G}]$ and $[\hat{F}, \hat{p}_s]$ vanish and we obtain from the canonical commutator relation (3.80)

$$-[\hat{F}, \hat{G}] = i. \tag{5.65}$$

This commutation relation is the only other constraint on the extra fluctuation and shows on a formal level that the introduction of two non-commuting operators $\hat{F}$ and $\hat{G}$ is indeed necessary to satisfy the condition (5.64) for the simultaneous measurement, that is, to compensate the mutual exclusion of $\hat{q}_s$ and $\hat{p}_s$. In classical physics (or in the "classical regime" of a large position and momentum scale) we could neglect the extra fluctuations, and we would obtain the simplest possible description of a joint measurement of q and p. So we can agree that the operators $\hat{Q}_1$ and $\hat{P}_2$ defined in Eq. (5.62) describe the outcome of a simultaneous yet imprecise measurement of position and momentum in quantum mechanics.

How precisely do $\hat{Q}_1$ and $\hat{P}_2$ correspond to the actual position $\hat{q}_s$ and momentum $\hat{p}_s$ of the signal? Which limit does Heisenberg's uncertainty relation set? Arthurs and Kelly (Arthurs and Kelly, 1965) answered this question in a general and elegant way: they considered the effect of the extra fluctuations on the uncertainty product $\Delta Q_1 \Delta P_2$ of the measured position and momentum values compared with the intrinsic product $\Delta q_s \Delta p_s$. We obtain from definition (5.62) and from Eq. (5.63) that the product $\Delta^2 Q_1 \Delta^2 P_2$ satisfies

$$\Delta^2 Q_1 \Delta^2 P_2 = \Delta^2 q_s \Delta^2 p_s + \langle \hat{F}^2 \rangle \langle \hat{G}^2 \rangle + \Delta^2 q_s \langle \hat{G}^2 \rangle + \Delta^2 p_s \langle \hat{F}^2 \rangle$$

$$\geq \left(\Delta q_s \Delta p_s + \sqrt{\langle \hat{F}^2 \rangle \langle \hat{G}^2 \rangle} \right)^2 . \tag{5.66}$$

In the last line we estimated the arithmetic mean $(\Delta^2 q_s \langle \hat{G}^2 \rangle + \Delta^2 p_s \langle \hat{F}^2 \rangle)/2$ by the geometric mean $\Delta q_s \Delta p_s (\langle \hat{F}^2 \rangle \langle \hat{G}^2 \rangle)^{1/2}$. (We have used the fact that $(a + b)/2 \geq (ab)^{1/2}$ for all real a and b.) We apply Heisenberg's uncertainty relation for the intrinsic position q_s and momentum p_s to arrive at

$$\Delta Q_1 \Delta P_2 \geq \frac{1}{2} + \sqrt{\langle \hat{F}^2 \rangle \langle \hat{G}^2 \rangle}. \tag{5.67}$$

Let us estimate the product of the fluctuations $\langle \hat{F}^2 \rangle$ and $\langle \hat{G}^2 \rangle$. As their average amplitudes vanish, Eq. (5.63), the expectation values $\langle \hat{F}^2 \rangle$ and $\langle \hat{G}^2 \rangle$ describe the variances. We use Robertson's general uncertainty relation (5.61) to express the fluctuations in terms of the commutator (5.65) and obtain the Heisenberg bound

$$\langle \hat{F}^2 \rangle \langle \hat{G}^2 \rangle \geq \frac{1}{4}. \tag{5.68}$$

We use this estimation in relation (5.67) and arrive at the famous result (Arthurs and Kelly, 1965)

$$\Delta Q_1 \Delta P_2 \geq 1. \tag{5.69}$$

This simple relation quantifies the effect of the extra noise involved in a simultaneous yet imprecise measurement of position and momentum. The uncertainty product of the measured Q_1 and P_2 values exceeds the Heisenberg limit (3.80) by a factor of two. As we have seen, this result is rather general and requires few, quite natural assumptions.

Given fixed variances $\langle \hat{F}^2 \rangle$ and $\langle \hat{G}^2 \rangle$ of the extra fluctuations in a simultaneous measurement of position and momentum, what are the minimum uncertainty states (with respect to the observed quantities) (Leonhardt et al., 1995a)? Remember that in our analysis we have estimated the arithmetic mean $(\Delta^2 q_S \langle \hat{G}^2 \rangle + \Delta^2 p_S \langle \hat{F}^2 \rangle)/2$ by the corresponding geometric mean, and then we have solely used estimations for the fluctuations $\langle \hat{F}^2 \rangle$ and $\langle \hat{G}^2 \rangle$. Given the latter quantities, the uncertainty product is minimized if the geometric mean equals the arithmetic one, that is, if $\Delta^2 q_S \langle \hat{G}^2 \rangle$ equals $\Delta^2 p_S \langle \hat{F}^2 \rangle$. Also, we must minimize the intrinsic uncertainty product $\Delta q_S \Delta p_S$. According to Pauli's proof from Section 3.3, only the squeezed states (3.89) have minimal uncertainty in their intrinsic position and momentum fluctuations. Consequently, the minimum uncertainty states for the observed joined position and momentum values are the squeezed states with (Leonhardt et al., 1995a)

$$\frac{\Delta^2 q_S}{\Delta^2 p_S} = \frac{\langle \hat{F}^2 \rangle}{\langle \hat{G}^2 \rangle} \tag{5.70}$$

or, in terms of the squeezing parameter ζ of Eq. (3.83),

$$\zeta = -\frac{1}{4} \ln \left(\langle \hat{F}^2 \rangle / \langle \hat{G}^2 \rangle \right). \tag{5.71}$$

The ratio of the extra fluctuations $\langle \hat{F}^2 \rangle$ and $\langle \hat{G}^2 \rangle$ determines the squeezing for the best adapted state. The less the extra fluctuation of one of the observables $\hat{F}$ or $\hat{G}$ is, the higher is the influence of the intrinsic uncertainty and the higher must be the squeezing of the position or the momentum variance, respectively, for minimizing the uncertainty product.

Quantum optics is the field in which most modern experimental tests of the fundamentals of quantum physics have been performed. How do we implement the idea of Arthurs and Kelly? How do we measure simultaneously position and momentum in quantum optics? Let us recall what we mean by measurements of position and momentum. The in-phase and out-of-phase quadrature components $\hat{q}$ and $\hat{p}$ obey the canonical commutation relation (3.10), and, consequently, they share all algebraic properties of mechanical position and momentum operators. We have seen in Section 5.2.2 how the quadratures can be measured via balanced homodyne detection. Yet in addition, we would like to have a device for making two "copies" of a light beam so that we can measure separately the position quadrature of the first beam and the momentum quadrature of the second "copy". What about using a simple beam splitter? It could split the incident spatial temporal mode into two parts. We could guide each outgoing beam to a homodyne detector, one for measuring $\hat{q}$ on the first beam and the other for measuring $\hat{p}$ on the second one, see Fig. 5.14. We must only ensure that the local oscillators of the two homodyne detectors have a phase difference of $\pi/2$. This is readily

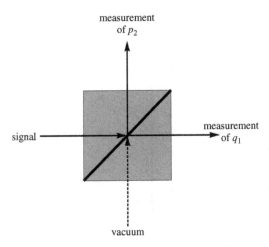

measurement
of p_2

signal

measurement
of q_1

vacuum

Fig. 5.14. Simultaneous measurement of position and momentum quadratures. The incident signal is split into two emerging beams. Each beam represents an independent system. In this case quantum mechanics does not raise any objections to measurements of the position quadrature on one beam and of the momentum on the other. However, the vacuum field entering the apparatus via the "unused" second port of the beam splitter introduces extra noise. The uncertainty principle is not violated but taken literally, that is, we can measure the position and the momentum simultaneously but not precisely.

achieved using a common local oscillator that is split into two parts at a second beam splitter. One partial beam is directed to the first homodyne detector, and the other is phase-shifted via a $\lambda/4$ plate and directed to the second homodyne detector. The scheme appears like two entangled homodyne apparata; see Fig. 5.15. Strictly speaking, four input fields are involved – the signal and a vacuum at the first beam splitter and the local oscillator and a vacuum at the second. Additionally, four output beams are travelling toward the four photodetectors. In view of this, the apparatus is called an *eight-port homodyne detector*.

This device was used by Walker and Carroll (Walker and Carroll, 1984; Walker, 1987) to perform the first genuine simultaneous measurement of position and momentum. Although the scheme had precursors in microwave technology (Engen and Hoer, 1972), these devices had never operated on the quantum level until the pioneering work of Walker and Carroll. Why is quantum mechanics not violated in the joint measurement of canonically conjugate quadratures? How is this scheme related to the general idea of Arthurs and Kelly (Arthurs and Kelly, 1965)? What does the eight-port homodyne detector actually measure? To answer all these questions we need to understand only the action of the first beam splitter, where the signal is divided into two parts. The rest of the device serves to perform only the homodyne measurements

Fig. 5.15. Eight-port homodyne detector.

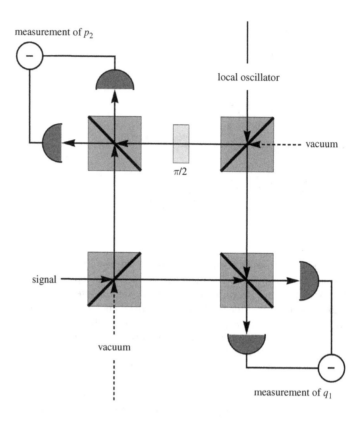

on the two outgoing beams (provided, of course, that the local oscillator is strong).

Roughly speaking, beam splitting is always "noisy". The incident photons are distributed as whole energy quanta to the two outgoing beams. Yet the average intensity ratio of the transmitted and the reflected beam is a constant given by the ratio of the transmittance τ and the reflectance ϱ of the beam splitter. To resolve this conflict between the wave-like distribution of the average intensities and the appearance of discrete particles, the signal photons are distributed as single units, but at random. Each photon is transmitted with the probability τ^2 and reflected with the probability ϱ^2. This randomness causes some additional noise in the simultaneously measured quadratures. Alternatively, we may understand the extra detection noise as being caused by the vacuum entering the second, "unused" port of the beam splitter. These vacuum fluctuations contaminate the signal field, so that the jointly measured quadratures are only fuzzy pictures of the intrinsic position and momentum quantities. In this way the violation of Heisenberg's uncertainty principle is avoided.

To put these words into precise terms, let us apply the simple quantum theory of beam splitting developed in this chapter. Beam splitting is based on optical interference, that is, on the superposition of the incident fields. In the Heisenberg picture the annihilation operators $\hat{a}_1'$ and $\hat{a}_2'$ of the emerging beams are linear transformations of the operators $\hat{a}_1$ and $\hat{a}_2$ for the incident modes

$$\begin{pmatrix} \hat{a}_1' \\ \hat{a}_2' \end{pmatrix} = \begin{pmatrix} \tau & -\varrho \\ \varrho & \tau \end{pmatrix} \begin{pmatrix} \hat{a}_1 \\ \hat{a}_2 \end{pmatrix}. \tag{5.72}$$

For simplicity we have assumed that the beam-splitting matrix is real, a situation that can always be achieved by redefining the reference phases of the mode operators $\hat{a}_1$, $\hat{a}_2$ and $\hat{a}_1'$, $\hat{a}_2'$ if required. The relation (5.10) between the transmittance τ and the reflectance ϱ accounts for the fact that photons are either transmitted (with probability τ^2) or reflected (with probability ϱ^2). The eight-port homodyne detector measures the q quadrature of the first outgoing beam and the p quadrature of the second. As $\hat{a}$ equals $(\hat{q} + i\hat{p})/\sqrt{2}$, these quantities are given by the simple expressions[4]

$$\hat{q}_1' = \tau\hat{q}_1 - \varrho\hat{q}_2, \quad \hat{p}_2' = \varrho\hat{p}_1 + \tau\hat{p}_2. \tag{5.73}$$

As a consequence of the canonical commutation relation (3.10) for the operator pairs $\hat{q}_1$, $\hat{p}_1$ and $\hat{q}_2$, $\hat{p}_2$, the measured quadratures $\hat{q}_1'$ and $\hat{p}_2'$ do commute. This fact is not surprising, for otherwise we simply could not measure the two quantities simultaneously. Formula (5.73) shows how the eight-port homodyne detector brings into being the central idea (5.62) of Arthurs and Kelly. The measured q quadrature is proportional to the intrinsic position $\hat{q}_1$, apart from the noise term $-\varrho\hat{q}_2$. As only a fraction τ^2 of the incident intensity reaches the q detector the measured position quadrature is reduced by the factor of τ. The extra quantum noise originates from the field entering the second port of the beam splitter. This noise is enhanced for a low transmittance τ and reduced for a highly transmitting beam splitter. Similarly, the p detector measures $\varrho\hat{p}_1$, apart from the noise contribution $\tau\hat{p}_2$.

So the quantum optical version of Arthurs and Kelly's Gedanken experiment is as simple as this: split a beam into two parts and measure simultaneously the position quadrature on one beam and the momentum quadrature on the other. The measured quantities are proportional to the operators

$$\hat{Q}_1 = \hat{q}_1 - \frac{\varrho}{\tau}\hat{q}_2, \quad \hat{P}_2 = \hat{p}_1 + \frac{\tau}{\varrho}\hat{p}_2, \tag{5.74}$$

---

[4] Note (Freyberger et al., 1993) that $\hat{q}_1'$ and $\hat{p}_2'$ are the observables in the Einstein–Podolsky–Rosen Gedanken experiment (Bohr, 1935; Einstein et al., 1935), that we briefly discuss in Section 7.1.1.

that is, to the Arthurs and Kelly variables (5.62). The mode entering the second, "unused" port of the beam splitter introduces just the right quantity of extra quantum fluctuations required for not violating Heisenberg's uncertainty principle. We have seen that one effect of this extra noise is the doubling (5.69) of the uncertainty product.

What is the probability distribution $\mathrm{pr}(Q_1, P_2)$ for the simultaneously measured position and momentum values? From this distribution we could gain much more detailed information about the effect of the extra quantum noise involved than quantified in the uncertainty product only. Again, we use our simple quantum theory of beam splitting to calculate $\mathrm{pr}(Q_1, P_2)$.

Let us describe the state of the incident signal by the density operator $\hat{\rho}$ and the associated Wigner function $W(q_1, p_1)$. Additionally, we must take into account the light beam entering the second input port of the beam splitter. In most experiments this beam would just "not exist" classically, meaning in quantum optics that the second incident mode is a vacuum with the Wigner function (4.36)

$$W_0(q_2, p_2) = \frac{1}{\pi} \exp\left(-q_2^2 - p_2^2\right). \tag{5.75}$$

According to Eq. (5.34) the beam splitter transforms the total Wigner function $W(q_1, p_1, q_2, p_2)$ of the two incident beams as if $W(q_1, p_1, q_2, p_2)$ were a classical probability distribution,

$$W'(q_1, p_1, q_2, p_2) = W(q_1', p_1') \, W_0(q_2', p_2') \tag{5.76}$$

with the changed variables (5.35). As a fundamental property (4.1) of the Wigner function, the probability distribution $\mathrm{pr}(q_1, p_2)$ is given by integrating $W'(q_1, p_1, q_2, p_2)$ with respect to the unobserved quantities p_1 and q_2, that is, by

$$\mathrm{pr}(q_1, p_2) = \int_{-\infty}^{+\infty} \int_{-\infty}^{+\infty} W'(q_1, p_1, q_2, p_2) \, dp_1 \, dq_2. \tag{5.77}$$

To find an instructive expression for $\mathrm{pr}(q_1, p_2)$ we simply change the variables in the integration (5.77). We use

$$q = q_1' = \tau q_1 + \varrho q_2, \quad p = p_1' = \tau p_1 + \varrho p_2 \tag{5.78}$$

instead of p_1 and q_2. Obviously,

$$q_2' = -\varrho q_1 + \tau q_2 = \frac{1}{\varrho}\left(-q_1 + \tau q\right), \tag{5.79}$$

where we have applied the photon probability conservation (5.10) at the beam splitter. In a similar way we get

$$p_2' = \frac{1}{\tau}\left(p_2 - \varrho p\right). \tag{5.80}$$

Splitting a beam into two parts means distributing the incident intensity to the two outgoing beams (with a ratio of τ/ϱ). To compensate for this intensity loss we rescale the position and momentum variables

$$Q_1 = \frac{q_1}{\tau}, \quad P_2 = \frac{p_2}{\varrho} \tag{5.81}$$

and their probability distribution

$$\mathrm{pr}(Q_1, P_2) = (\varrho\tau)^{-1}\,\mathrm{pr}(q_1, p_2). \tag{5.82}$$

The prefactor $(\varrho\tau)^{-1}$ appears because $\mathrm{pr}(Q_1, P_2)$ is a probability density. The distribution $\mathrm{pr}(Q_1, P_2)$ of the scaled variables Q_1 and P_2 describes the information we infer about the intrinsic phase space density.

Taking all these calculations and definitions (5.75)–(5.82) into account, we obtain the simple result

$$\mathrm{pr}(Q_1, P_2) = \int_{-\infty}^{+\infty}\int_{-\infty}^{+\infty} W(q,p)$$
$$\times \frac{1}{\pi}\exp\left(-\frac{\tau^2}{\varrho^2}(q - Q_1)^2 - \frac{\varrho^2}{\tau^2}(p - P_2)^2\right)dq\,dp. \tag{5.83}$$

The probability distribution $\mathrm{pr}(Q_1, P_2)$ for the simultaneously measured conjugate variables is a filtered Wigner function. It resolves $W(q,p)$ within an ellipse in phase space. Let us consider a balanced beam splitter with $\tau = \varrho$. From the definition (4.56) of the Q function we notice immediately that the measured phase space density is exactly this quasiprobability distribution. *The eight-port homodyne detector measures the Q function!* Note that this feature is shared by a remarkable number of other schemes for measuring simultaneously position and momentum (Shapiro and Wagner, 1984; Schleich et al., 1992; Leonhardt and Paul, 1993a), and in particular by the original example of Arthurs and Kelly (Arthurs and Kelly, 1965; Stenholm, 1992) (for balanced meters).

In the case of the unbalanced beam splitter, the Q function is squeezed. This squeezed Q function is called a *Husimi function* (Husimi, 1940). To quantify the measurement induced squeezing, we take advantage of the overlap formula (4.24) and the Wigner representation (4.38) and (4.43) of the squeezing (3.87) and the displacement operator (3.55), respectively. We see that the measured phase space density is the quantum overlap (Walker, 1987; Leonhardt, 1993)

$$\mathrm{pr}(Q_1, P_2) = \frac{1}{2\pi}\,\mathrm{tr}\{\hat{\rho}\,|\alpha,\zeta\rangle\langle\alpha,\zeta|\} \tag{5.84}$$

with the squeezed state

$$|\alpha,\zeta\rangle = \hat{D}(\alpha)\,\hat{S}(\zeta)\,|0\rangle \tag{5.85}$$

characterized by the displacement

$$\alpha = \frac{1}{\sqrt{2}}\left(Q_1 + iP_2\right) \tag{5.86}$$

and the squeezing parameter

$$\zeta = \ln(\tau/\varrho). \tag{5.87}$$

This squeezing effect is easily explained: for a large transmittance τ the precision of the Q_1 quadrature is enhanced, whereas the precision of P_2 is reduced. The reason is simply that in this case the intrinsic position $\hat{q}_1$ in $\hat{Q}_1$ is less contaminated by the vacuum fluctuations entering the "unused" port of the beam splitter, whereas p_1 suffers accordingly. And of course, for a small transmittance τ the situation is reversed. To be more quantitative, we calculate the measured position distribution $\mathrm{pr}(Q_1)$ by integrating $\mathrm{pr}(Q_1, P_2)$ with respect to P_2. We obtain immediately from Eq. (5.83)

$$\mathrm{pr}(Q_1) = \frac{|\tau/\varrho|}{\sqrt{\pi}} \int_{-\infty}^{+\infty} \mathrm{pr}(q)\,\exp\left(-(\tau/\varrho)^2(q - Q_1)^2\right) dq, \tag{5.88}$$

because the marginals of the Wigner function are the quadrature distributions (4.1). The squeezed position fluctuations of the incident vacuum filter the distribution $\mathrm{pr}(q)$ and limit the resolution of the intrinsic position quadrature. Evidently, for the momentum distribution we obtain the similar result

$$\mathrm{pr}(P_2) = \frac{|\varrho/\tau|}{\sqrt{\pi}} \int_{-\infty}^{+\infty} \mathrm{pr}(p)\,\exp\left(-(\varrho/\tau)^2(p - P_2)^2\right) dp. \tag{5.89}$$

If the precision of the measured Q_1 quadrature is high, the precision of P_2 is accordingly poor. In an extreme case the beam splitter is completely transmitting, so that the Q_1 homodyne detector receives the full signal. The Q_1 resolution is perfect, but, on the other hand, the P_2 detector is detached from the signal and, consequently, measures no signal information any more. By changing the transmittance we could squeeze the resolution of the signal Wigner function, but we can never make the total effect of the extra quantum fluctuations arbitrarily small.

Yes, we can see experimentally an overall phase space picture of a quantum system, but the picture is fuzzy. This is the price to be paid if one dares to measure simultaneously the position and the momentum, that is, all phase space properties, in a single experiment. However, if we prefer to observe the phase space aspects one at a time, in form of phase shifted quadratures, we circumvent this problem. In this case we can indeed tomographically infer the Wigner function, as we have seen in Section 5.2.3.

So far we have assumed that we split the incident signal to measure the position quadrature on one of the outgoing beams and the momentum quadrature on the other. Although "unused", the second input port of the beam splitter allows the vacuum fluctuations to sneak in. This picture has helped us to understand why the apparatus does not violate Heisenberg's uncertainty principle. What happens if we let the signal interfere with a second incident beam instead of just splitting it? This second field might be a squeezed vacuum or a coherent beam of laser light, for instance. To describe the effect of the interference with "something" instead of "nothing" we can, fortunately, perform the same calculations as in the previous case. The only difference is that we do not specify the state $\hat{\rho}_R$ of the second incident field and the associated Wigner function $W_R(q_2, p_2)$. To calculate the measured phase space density we simply repeat our previous procedure and obtain the result

$$\mathrm{pr}(Q_1, P_2) = \int_{-\infty}^{+\infty} \int_{-\infty}^{+\infty} W(q, p) \, W_{SQR}(q - Q_1, p - P_2) \, dq \, dp \qquad (5.90)$$

with

$$W_{SQR}(q, p) = W_{QR}\left(\frac{\tau}{\varrho} q, \frac{\varrho}{\tau} p\right) \qquad (5.91)$$

and

$$W_{QR}(q, p) = W_R(q, -p). \qquad (5.92)$$

So quite generally, the probability distribution $\mathrm{pr}(Q_1, P_2)$ for the simultaneously measured position and momentum values is a filtered Wigner function. Following Popper (1982), Wódkiewicz (1984a; 1984b) called the expression (5.90) a *propensity*. Two operations relate the filter function $W_{SQR}(q, p)$ to the Wigner function $W_R(q, p)$ of the second incident beam. The first one is the already familiar squeezing (5.91) brought about by using an unbalanced beam splitter. Additionally, the momenta p are inverted in Eq. (5.92). We have not noticed this feature in the case of beam splitting, because the vacuum Wigner function is inversion invariant. In classical optics the inversion of the p quadrature components means a *phase conjugation* because p is proportional to the imaginary part of the complex wave amplitude α. The signal interferes optically with the second beam, and so the phase conjugated wave amplitude of the latter enters the interference pattern. We would expect that $W_{QR}(q, p)$ is the Wigner function for the complex conjugate of the density matrix

$$\hat{\rho}_{QR} \equiv \hat{\rho}_R^* \qquad (5.93)$$

in position representation. (We mean by this definition that the density matrix $\langle q | \hat{\rho}_{QR} | q \rangle$ should be the complex conjugate of $\langle q | \hat{\rho}_R | q \rangle$.)

That Eq. (5.93) is indeed correct is easily seen using Wigner's formula (4.17)

$$
\begin{aligned}
W_{QR}(q,p) &= \frac{1}{2\pi}\int_{-\infty}^{+\infty} \exp(ipx)\langle q - \frac{x}{2}|\hat{\rho}_R|q + \frac{x}{2}\rangle^* \, dx \\
&= \frac{1}{2\pi}\int_{-\infty}^{+\infty} \exp(ipx)\langle q + \frac{x}{2}|\hat{\rho}_R|q - \frac{x}{2}\rangle \, dx \\
&= W_R(q,-p).
\end{aligned}
\tag{5.94}
$$

In the last step we have replaced x by $-x$ in the integration.

Aharonov, Albert, and Au (Aharonov et al., 1981) called the complex conjugate $\hat{\rho}_{QR}$ of $\hat{\rho}_R$ the density matrix of the *quantum ruler*. The ruler, essentially the second incident field, probes the Wigner function of the signal. The filter function in $\mathrm{pr}(Q_1, P_2)$ is the squeezed and displaced Wigner function $W_{QR}(q_2, p_2)$ of the quantum ruler. When the second input port of the beam splitter is "unused", the ruler is in the vacuum state and we obtain Eq. (5.83). As in this special case, we could squeeze the resolution of the signal Wigner function by changing the transmittance of the beam splitter. However, as we have seen in the discussion of Eq. (4.31), any physically meaningful Wigner function cannot be highly peaked and in particular cannot approach a two-dimensional delta function. Of course, this applies also to the Wigner function of the quantum ruler. So the resolution of the filtering (5.90) is always limited, and we can never measure the true Wigner function directly as a probability distribution (besides the fact that $W(q,p)$ might be negative).

Beam splitters are noisy "copy machines" for quantum light fields, and therefore they cause extra quantum fluctuations in a simultaneous measurement of position and momentum quadratures. What would happen, however, if we had already two perfect copies of a light beam and let them interfere (Leonhardt and Paul, 1994)? Imagine that the first incident beam is in a pure state described by Schrödinger's position wave function $\psi(q_1)$, whereas the second beam is just in the complex conjugate state $\psi^*(q_2)$. Both light fields should interfere at a balanced beam splitter with $\tau = -\varrho = 1/\sqrt{2}$, and we measure q on the first and p on the second beam. According to the quantum theory of beam splitting in the Schrödinger picture (Leonhardt, 1993) described in Section 5.1.2, the total wave function $\psi'(q_1, q_2)$ of the outgoing beams is the rotated wave function of the incident beams, that is,

$$
\psi'(q_1, q_2) = \psi\big((q_1 - q_2)/\sqrt{2}\big)\psi^*\big((q_1 + q_2)/\sqrt{2}\big).
\tag{5.95}
$$

Because the momentum is measured on the second beam, it is advantageous to express ψ' in the momentum representation with respect to p_2

$$\tilde{\psi}'(q_1, p_2) = \frac{1}{\sqrt{2\pi}} \int_{-\infty}^{+\infty} \psi'(q_1, q_2) \exp(-i p_2 q_2) \, dq_2 \,. \qquad (5.96)$$

According to Born's interpretation of the wave function, the modulus squared of $\tilde{\psi}'(q_1, p_2)$ gives the probability distribution $\mathrm{pr}(q_1, p_2)$ of the measured q_1 and p_2 values. We use the integration variable $x = q_2\sqrt{2}$ and the scaled quadratures $Q_1 = q_1/\sqrt{2}$ and $P_2 = p_2/\sqrt{2}$ with the corresponding probability distribution (5.82), and obtain

$$\mathrm{pr}(Q_1, P_2) = \frac{1}{2\pi} \left| \int_{-\infty}^{+\infty} \psi\left(Q_1 - \frac{x}{2}\right) \psi^*\left(Q_1 + \frac{x}{2}\right) \exp(-i P_2 x) \, dx \right|^2. \qquad (5.97)$$

We glance at Wigner's formula (4.17) and realize immediately that $\mathrm{pr}(Q_1, P_2)$ is essentially the modulus squared of the Wigner function of the first incident beam. Because the Wigner function is real, the modulus squared is just the square, and we get the result (Leonhardt and Paul, 1994)

$$\mathrm{pr}(Q_1, P_2) = 2\pi \, W^2(Q_1, -P_2). \qquad (5.98)$$

If we have two copies of the light beam, one in the state $\psi(q_1)$ and the other in the conjugate state $\psi^*(q_2)$ and let them interfere, the eight-port homodyne detector measures the square of the Wigner function for the state $\psi(q_1)$. Although we cannot measure the Wigner function directly as a probability distribution, we can, in principle, measure its square!

5.3 Questions

5.1 Describe the beam splitter in the Heisenberg picture.

5.2 Use the Bose commutation relations between the incident and the emerging modes in order to show that the beam splitter matrix B is unitary.

5.3 Why is $\hat{a}_1^\dagger \hat{a}_1 + \hat{a}_2^\dagger \hat{a}_2$ a conserved quantity? What is the physical meaning of this conservation law?

5.4 Explain the structure of the beam splitter matrix. Why is the mode mixing by a rotation the most important feature of the beam splitter transformation?

5.5 What is the meaning of the Schrödinger picture for the beam splitter?

5.6 Explain why $\hat{\rho}' = \hat{U}_B \hat{\rho} \, \hat{U}_B^\dagger$ in the Schrödinger picture if you can write the mode transformation of the beam splitter in the Heisenberg picture as

$$\begin{pmatrix} \hat{a}_1' \\ \hat{a}_2' \end{pmatrix} = \hat{U}_B^\dagger \begin{pmatrix} \hat{a}_1 \\ \hat{a}_2 \end{pmatrix} \hat{U}_B.$$

5.7 Show that the Jordan–Schwinger operators obey the commutation rules of angular momentum components.

5.8 What is the physical meaning of the Jordan–Schwinger operators?

5.9 Why do you get

$$\hat{U}_B = \exp(i\Lambda\hat{L}_t)\,\exp(i\Phi\hat{L}_z)\,\exp(i\Theta\hat{L}_y)\,\exp(i\Psi\hat{L}_z)\,?$$

5.10 Why does the beam splitter rotate the quadrature wave function?

5.11 Use the rotation of the quadrature wave function to show that the beam splitter transforms the Wigner function like a probability distribution of classical amplitudes.

5.12 Discuss the interference of various squeezed vacua at the beam splitter using the wave function representation. When are the emerging modes uncorrelated and when do you get entangled states?

5.13 Derive the result

$$\hat{U}_B \,|n, 0\rangle = \sum_{k=0}^{n} \sqrt{\binom{n}{k}}\, \tau^k \varrho^{n-k} |k, n - k\rangle.$$

5.14 Why do photons behave like distinguishable particles in beam splitting?

5.15 What happens when two photons interfere at a 50:50 beam splitter?

5.16 Compare the interference of two photons with the interference of coherent states.

5.17 How does a photodetector work? What is a photodiode? What is an avalanche photodiode?

5.18 What is a homodyne detector?

5.19 Show that the homodyne detector measures the quadrature with respect to the phase of the local oscillator.

5.20 How does the homodyne detecter select the mode it measures?

5.21 What is optical homodyne tomography? Why is it significant?

5.22 How is the Wigner function reconstructed from quadrature histograms?

5.23 Why are simultaneous measurements of position and momentum not in conflict with quantum mechanics?

5.24 Prove that Heisenberg's uncertainty relation is doubled in such measurements.

5.25 What are the corresponding minimum uncertainty states?

5.26 Sketch the quantum optical implementation of simultaneous measurements of position and momentum.

5.27 Where is the inevitable noise source in such measurements?

5.28 Show that the quadratures $\hat{q}_1' = \tau\hat{q}_1 - \varrho\hat{q}_2$ and $\hat{p}_2' = \varrho\hat{p}_1 + \tau\hat{p}_2$ commute.

5.29 How is the measured phase space probability distribution related to the Wigner function?

5.30 In which case is the Q function measured?

5.31 How is the square of the Wigner function measured?

5.4 Homework problem

The beam splitter transmits light with transmissivity τ and reflects with reflectivity ϱ. Imagine the reflected light is lost. In this case, the beam splitter may serve as an elementary model for a scatterer, describing loss of light. Here you will study this model of loss. As we derived in Section 5.1.3, the beam splitter transforms a Fock state $|n\rangle$ into

$$|n,0\rangle' = \sum_{k=0}^{n} \sqrt{\frac{n!}{k!\,(n-k)!}}\, \tau^k \varrho^{n-k}\, |k, n+k\rangle.$$

(a) Suppose the incident light is in the coherent state

$$|\alpha\rangle = \exp\left(-\frac{|\alpha|^2}{2}\right) \sum_{n=0}^{\infty} \frac{\alpha^n}{\sqrt{n!}}\, |n\rangle.$$

The outgoing light is transformed into the two-mode state $|\alpha, 0\rangle'$ with $|n\rangle$ replaced by $|n, 0\rangle'$. Show that

$$|\alpha,0\rangle' = |\tau\alpha\rangle_1\, |\varrho\alpha\rangle_2.$$

(b) Give an interpretation for the result of part (a).

(c) Consider the reduced state

$$\hat\rho = \text{tr}_2\{|\alpha,0\rangle'\,{}'\langle\alpha,0|\},$$

dropping the index 1. Show that

$$\hat\rho = |\tau\alpha\rangle\langle\tau\alpha|.$$

(d) Consider multiple scattering when the transmissivity τ is exponentially reduced as

$$\tau = e^{-\gamma t}$$

with t being a parameter, for example the propagation time of light through the scattering medium. The goal is to derive a compact equation of motion for the density matrix $\hat\rho$. As a first preparatory step, show that

$$\hat a\,|\alpha\rangle = \left(\frac{\alpha}{2} - \frac{\partial}{\partial\alpha^*}\right)|\alpha\rangle, \quad \hat a^\dagger\,|\alpha\rangle = \left(\frac{\alpha^*}{2} + \frac{\partial}{\partial\alpha}\right)|\alpha\rangle,$$

regarding α and α^* as independent variables.

(e) Use the result of part (d) and the definition of the coherent state, $\hat a\,|\alpha\rangle = \alpha|\alpha\rangle$, to show that

$$\frac{d\hat\rho}{dt} = -\gamma\left(\hat a^\dagger \hat a \hat\rho - 2\hat a\hat\rho\hat a^\dagger + \hat\rho\hat a^\dagger\hat a\right).$$

(f) In part (e) you derived the equation of motion for coherent states with density matrix $|\alpha\rangle\langle\alpha|$. Use your knowledge of the P function to give a brief argument why the same equation describes the evolution of an arbitrary state in scattering.

(g) Show that the photon number distribution obeys the master equation

$$\frac{dp_n}{dt} = -2\gamma n p_n + 2\gamma (n+1) p_{n+1}.$$

(h) Give an interpretation for the master equation deduced in part (g).

5.5 Further reading

The first quantum theory describing the action of a lossless beam splitter was developed by Brunner et al. (1965) starting from a microscopic model; for a review see (Paul, 1982). From the large literature on beam splitters we mention (Campos et al., 1989; Leonhardt, 1993) and the references reviewed by Leonhardt (1993, 1997a, 2003). For multi-ports and quantum optical networks see Reck *et al.* (1994); Mattle *et al.* (1995); Törmä, Jex, and Stenholm (1995); Törmä (1998); Törmä and Jex (1999); Törmä, Jex, and Schleich (2002); Leonhardt (2003). In the context of quantum information and quantum communication, linear transformations of mode operators applied to states with Gaussian Wigner functions are called Gaussian quantum channels (Eisert and Wolf, 2007).

The quantum theory of homodyne detection is developed and explained in Carmichael (1987); Walker (1987); Braunstein (1990); Vogel and Grabow (1993); Ou and Kimble (1995); Raymer et al. (1995); Banaszek and Wódkiewicz (1997); Tyc and Sanders (2004); Vogel and Welsch (2006).

For a review on optical homodyne tomography, including recent applications, see Lvovsky and Raymer (2009). The sampling method of state reconstruction is developed in its most user friendly form in Leonhardt et al. (1996) and Leonhardt, (1997a). For maximum likelihood state reconstruction, see Hradil et al. (2004) Lvovsky, 2004).

For the theory of multiport homodyning, see the early yet excellent paper by Walker (1987). The eight-port homodyne detector became widely known in quantum optics when it was used to define the phase of quantum light in an operational way (Noh et al., 1991; 1992a; 1992b; 1993a; 1993b), see Leonhardt and Paul (1993b; 1993c); Leonhardt et al. (1995); Leonhardt (1997a). For an illuminating discussion on the notorious problem of defining a quantum optical phase, see Lynch's critical review (1995).

For the theory of simultaneous measurements of position and momentum, see Stenholm's classic article (1992).

Chapter 6
Irreversible processes

6.1 Lindblad's theorem

Although the fundamental laws of physics explain a great deal of the material world, they often seem to contradict our immediate experience. For instance, as physicists we believe that all elementary interactions are reversible, but as people we are immersed in irreversibility. The collision of two atoms may well be reversible and predictable, but for us time flows in a definite direction and life is sometimes unpredictable. "What then is time? If no one asks me, I know what it is. If I wish to explain it to him who asks, I do not know." (Attributed to Augustine.) Nevertheless, as we try to explain in this section, one can include irreversibility, the true arrow of time, into quantum mechanics in a concise and elegant form put on solid mathematical ground in Lindblad's theorem (Lindblad, 1976). Having then understood the basics of irreversible quantum dynamics, life may not be quite the same any more.

6.1.1 Irreversibility

A beam splitter is a reversible optical instrument: one could collect the outgoing light beams and bring them together again in a second beam splitter where they constructively interfere to re-establish the light in its original state. But now imagine a light beam is scattered many times. For example, light entering biological tissue is interacting with each of the cells along its path. Some of the light is scattered and only a part of the light continues in the original direction. After some distance in the material the light is almost completely scattered; it is lost. We could model each cell as a miniature beam splitter. Each individual act of scattering is completely reversible, but we lost control over all the outgoing light beams. In principle, we could bring all the light back together again, but in practice it is gone. Many processes in our everyday experience are irreversible; the one we experience most directly, life, surely is. However, it is believed that at a fundamental level all elementary interactions are as reversible as the beam-splitting of light. If we could control all elementary processes we could run them backwards.

Yet there is one type of process that is fundamentally irreversible, and this is the quantum measurement process. Whenever Nature makes the decision that a quantum state $|\psi\rangle$ realizes its observable property a and so turns into an eigenstate $|a\rangle$, that decision is irrevocable. Which one of the eigenstates is selected is truly random, although the probability p_a of materializing the physical quantity a is completely determined by the quantum state $|\psi\rangle$ as $p_a = |\langle a|\psi\rangle|^2$. If we prepare the quantum system in the state $|\psi\rangle$ many times and repeat the observation, the statistical frequency of getting the result a approaches p_a, but each individual act of materialization from possibility to fact is random and irreversible.

Both types of irreversible processes, loss of control in a complex system and the quantum measurement, turn out to obey the same type of dynamics, the same principal equation of motion for the quantum state, known as the *quantum master equation*. This is not a coincidence, for the following reason. Suppose that the irreversibility of a process is caused by our lack of control over all the sub-processes involved, like in the scattering of light. The quantum states of the lost light modes will materialize as physical quantities somewhere, the scattered light will eventually appear as photons, but we simply do not monitor them. If we wish to describe the quantum system we can monitor, the remaining light in our case, we must average over all possible observational results of the scattered light. Now, suppose we observe a direct quantum measurement process. For instance, imagine we measure the quadratures of two light modes. The individual results are random, but after many runs of the experiment a pattern emerges in the quadrature histograms. In order to say something reproducible about a quantum system we must average. For example, we could average over the measurement results of one mode and monitor the quantum state of the other. In this case, we deliberately disregard information we have, the individual measurement results for one mode. In short, whether we choose to ignore data or are ignorant about the fundamental processes involved, the result is the same. As we will see, we can elegantly describe all irreversible quantum processes in one equation.

6.1.2 Reversible dynamics

Before we turn to the quantitative description of irreversible dynamics, it is wise to recapitulate the essence of reversible processes. Consider a quantum system with density matrix

$$\hat{\rho} = \sum_n \rho_n |\psi_n\rangle\langle\psi_n|. \tag{6.1}$$

The ρ_n describe the probability of having the quantum state $|\psi_n\rangle$ in a statistical ensemble. If the dynamics is reversible the ρ_n cannot change; they express our initial degree of ignorance about the particular quantum state. The state $|\psi_n\rangle$ evolves according to the Schrödinger equation

$$i\frac{d|\psi\rangle}{dt} = \hat{H}|\psi\rangle. \tag{6.2}$$

Here t denotes a parameter of the evolution, t may be the propagation time or distance, for example. The operator $\hat{H}$ is the Hamiltonian given here in units of $\hbar$, for simplicity, as if

$$\hbar = 1. \tag{6.3}$$

The Schrödinger dynamics conserves the normalization of the state $|\psi\rangle$, because

$$\frac{d\langle\psi|\psi\rangle}{dt} = \frac{d\langle\psi|}{dt}|\psi\rangle + \langle\psi|\frac{d|\psi\rangle}{dt} = \left(\langle\psi|i\hat{H}\right)|\psi\rangle - \langle\psi|\left(i\hat{H}|\psi\rangle\right) = 0. \tag{6.4}$$

Since the probabilities ρ_n do not evolve, we obtain for the density matrix (6.1) by differentiation

$$\frac{d\hat{\rho}}{dt} = i\left(\hat{\rho}\hat{H} - \hat{H}\hat{\rho}\right). \tag{6.5}$$

This equation, called the *von Neumann equation*, describes reversible quantum processes in general.

6.1.3 Irreversible dynamics

Göran Lindblad (Lindblad, 1976) derived the most general equation of motion for irreversible processes, assuming very little about the evolution of the density matrix. Essentially, he only postulated that the density matrix remains a density matrix, a Hermitian operator $\hat{\rho}$ with eigenvalues between 0 and 1 that one can interpret as probabilities. Here we will not reproduce Lindblad's mathematical proof (Lindblad, 1976), but rather motivate Lindblad's result on physical grounds. Suppose, for a moment, we can describe an irreversible process by the effective Hamiltonian $\hat{H}_{eff}$. This Hamiltonian must not be Hermitian, because otherwise we would obtain the ordinary reversible dynamics (6.5). So $\hat{H}_{\textit{eff}}$ should consist of a Hermitian and an anti-Hermitian part similar to the real and imaginary part of a complex number,

$$\hat{H}_{\textit{eff}} = \hat{H} - i\hat{V} \tag{6.6}$$

where $\hat{V}$ is a Hermitian operator. If $\hat{H}_{\textit{eff}}$ were to act on the quantum state $|\psi\rangle$ as in the Schrödinger equation (6.2), the Hamiltonian would alter the normalization of $|\psi\rangle$, because

$$\frac{d\langle\psi|\psi\rangle}{dt} = -2\,\langle\psi|\hat{V}|\psi\rangle. \tag{6.7}$$

We can turn this feature of non-Hermitian Hamiltonians to our advantage in the description of irreversible processes. Suppose we interpret $\langle\psi|\psi\rangle$ as the probability of remaining in the quantum state $|\psi\rangle$. By state we mean the normalized Hilbert space vector $|\psi\rangle$ divided by $\sqrt{\langle\psi|\psi\rangle}$. In the density matrix (6.1) we can include the probabilities ρ_n in the modulus square of the vectors $|\psi\rangle$ such that $\hat{\rho} = \sum_n |\psi_n\rangle\langle\psi_n|$. Equation (6.7) describes how the probability of a quantum state evolves. But a probability cannot grow beyond the maximal value of 1, the probability of statistical certainty. Consequently, the right-hand side of Eq. (6.7) must be negative or zero for $|\psi\rangle$ with $\langle\psi|\psi\rangle = 1$. A natural choice for $\hat{V}$ with this property is

$$\hat{V} = \sum_a \gamma_a \hat{T}_a^\dagger \hat{T}_a \tag{6.8}$$

with some operators $\hat{T}_a$ and positive numbers γ_a, because

$$\langle\psi|\hat{T}_a^\dagger \hat{T}_a|\psi\rangle = \left(\hat{T}_a|\psi\rangle\right)^\dagger \left(\hat{T}_a|\psi\rangle\right) \geq 0. \tag{6.9}$$

Note that the structure (6.8) is not only the natural, but the only choice, for the following reason: if $\langle\psi|\hat{V}|\psi\rangle$ is non-negative for all normalized $|\psi\rangle$ then the eigenvalues V_a of $\hat{V}$ are non-negative. The eigen representation $\hat{V} = \sum_a V_a |a\rangle\langle a|$ agrees with Eq. (6.8) if we set $\gamma_a = V_a$ and $\hat{T}_a = |a\rangle\langle a|$. In this way we have determined the general structure of all physically meaningful irreversible Hamiltonians. The $\hat{T}_a$ describe the irreversible sub-processes that occur at the rates γ_a.

Yet non-Hermitian Hamiltonian dynamics is not everything that happens in irreversible processes. The Hamiltonian dynamics may establish the probability for remaining in the quantum state $|\psi\rangle$ during the evolution of the system, but it does not account for stochastic changes of $|\psi\rangle$. Such stochastic transitions must occur with probability $1 - \langle\psi|\psi\rangle$ if the state is to remain in $|\psi\rangle$ with probability $\langle\psi|\psi\rangle$. We may imagine them as sudden changes triggered by the randomness inherent to irreversible processes, as *quantum jumps*. The quantum jumps cause fluctuations that are necessarily connected to the dissipative process; fluctuation and dissipation always go hand in hand, a fact described in statistical physics as the *fluctuation–dissipation theorem* (Callen and Welton, 1951; Landau and Lifshitz, Vol. V, 1996). A well-known classical example is Brownian motion: a small object immersed in a viscous fluid experiences both the viscosity and the relentless random kicks by the molecules of the fluid.

Let us model the quantum jumps by transition operators $\hat{T}_a$ that are supposed to suddenly transform $|\psi\rangle$ into $\hat{T}_a|\psi\rangle$. In this way the density matrix (6.1) would be transformed into $\hat{T}_a \hat{\rho} \hat{T}_a^\dagger$. Although the quantum jumps are random, on average they should occur at certain rates Γ_a. To describe the effect of both fluctuations and dissipations on the density matrix, averaged over many runs, we add to the non-Hermitian Hamiltonian dynamics the quantum jumps into different states,

$$\frac{d\hat{\rho}}{dt} = i\left(\hat{\rho}\hat{H}_{\text{eff}}^\dagger - \hat{H}_{\text{eff}}\hat{\rho}\right) + \sum_a \Gamma_a \hat{T}_a \hat{\rho} \hat{T}_a^\dagger. \tag{6.10}$$

It remains to identify the transition operators and their rates. The fluctuations must balance the dissipative processes, because the total probability for the quantum state, $\text{tr}\{\hat{\rho}\}$, remains constant (and unity). We differentiate $\text{tr}\{\hat{\rho}\}$ with respect to t and obtain from the equation of motion (6.10) and the structure (6.8) of the effective Hamiltonian

$$\frac{d\,\text{tr}\{\hat{\rho}\}}{dt} = \sum_a \text{tr}\left\{-2\gamma_a \hat{T}_a \hat{\rho} \hat{T}_a^\dagger + \Gamma_a \hat{T}_a \hat{\rho} \hat{T}_a^\dagger\right\}. \tag{6.11}$$

It is reasonable to postulate that the balance between fluctuations and dissipations holds *in detail*, for each sub-process,

$$\hat{T}_a = \hat{T}_a, \quad \Gamma_a = 2\gamma_a, \tag{6.12}$$

because arbitrary density operators $\hat{\rho}$ may experience any one of the individual transitions $\hat{T}_a$. Taking all these considerations into account, we finally arrive at a consistent equation of motion for irreversible dynamics, the *quantum master equation*

$$\frac{d\hat{\rho}}{dt} = i[\hat{\rho}, \hat{H}] - \sum_a \gamma_a \left(\hat{T}_a^\dagger \hat{T}_a \hat{\rho} - 2\hat{T}_a \hat{\rho} \hat{T}_a^\dagger + \hat{\rho} \hat{T}_a^\dagger \hat{T}_a\right). \tag{6.13}$$

Now we can relax our tentative assumptions. For example, we can once again adopt the traditional interpretation of the density matrix (6.1) as an ensemble of normalized pure states. What might have appeared as guesswork in our argument has a solid basis in mathematics: Lindblad (1976) has proved that the quantum master equation (6.13) describes all physically allowed irreversible processes. One can also turn our argument around and represent processes of Lindblad form (6.13) as sequences of coherent evolution stages randomly interrupted by quantum jumps. This stochastic representation of the master equation (6.13) is useful in computer simulations. Here one can investigate irreversible processes by running several such *quantum trajectories* (Plenio and Knight, 1998) in sequence. Our argument also illustrates the general ambiguity between irreversible quantum processes and quantum measurements. We can interpret quantum jumps as the collapses of the wave function in observations, but we can also understand them as random events triggered by the environment our system is interacting with.

Apart from being food for thought, the Lindblad form (6.13) of the master equation is of considerable practical advantage in modelling irreversible processes. It allows us to shortcut tedious calculations of microscopic details by the application of physical insight and intuition. All we need to do is to establish the relevant transition operators $\hat{T}_a$ and obtain their rates γ_a from experimental data. To give a concrete example, for an atom with ground state $|g\rangle$ and excited state $|e\rangle$, the transition $|g\rangle\langle e|$ describes the spontaneous emission of light and $|e\rangle\langle g|$ the incoherent excitation of the atom; the transition rates depend on the environment the atom is interacting with and can be calculated or inferred from experiments. For modelling irreversible processes on light itself, for example loss and gain in materials, we may employ annihilation and creation operators as transition operators.

6.2 Loss and gain

In our starting point of quantum optics, the quantum field theory of light explained in Chapter 2, we considered light in loss- and gain-less transparent materials. There we developed the theoretical model of the light mode. Now imagine a light beam is partially absorbed or amplified. When the beam leaves the dissipative material, we can still regard light as a superposition of modes, but their quantum states have experienced the irreversible dynamics described in Lindblad's theorem. Let us pay attention to the fate of only a single mode. In case some of the incident light is scattered in other modes, the scattered light will simply be written off as part of the losses.

6.2.1 Absorption and amplification

A material may absorb light, but it could also amplify or emit light: the material may contain atoms in population-inverted excited states or simply be sufficiently hot. In the first case, incident light is amplified by stimulated emission, in the second case the material spontaneously emits light. Let us describe the absorption process by the annihilation operator $\hat{a}$ and the amplification or emission of light by the creation operator $\hat{a}^\dagger$. We entirely focus on the dissipative dynamics, ignoring anything else,

$$\hat{H} = 0, \tag{6.14}$$

and thus consider two dissipative processes with the transition operators

$$\hat{T}_- = \hat{a}, \quad \hat{T}_+ = \hat{a}^\dagger \tag{6.15}$$

that happen at the rates γ_- and γ_+. The balance between γ_- and γ_+ will determine the net loss or gain along the effective interaction distance or time described by the parameter t.

It is instructive to visualize the dissipation dynamics of the master equation (6.13) by translating it into an equation for the Wigner function. The Wigner function is a quasiprobability distribution of the quadratures; so we express the annihilation and creation operators (6.15) in the master equation (6.13) in terms of our standard quadrature representation (3.8). We obtain

$$
\begin{aligned}
\frac{d\hat{\rho}}{dt} &= -\frac{\gamma_- + \gamma_+}{2} \left(\left(\hat{q}^2 + \hat{p}^2 \right) \hat{\rho} - 2 \left(\hat{q}\hat{\rho}\hat{q} + \hat{p}\hat{\rho}\hat{p} \right) + \hat{\rho} \left(\hat{q}^2 + \hat{p}^2 \right) \right) \\
&\quad + (\gamma_+ - \gamma_-) \left(i\hat{q}\hat{\rho}\hat{p} - i\hat{p}\hat{\rho}\hat{q} - \hat{\rho} \right) \\
&= (\gamma_- - \gamma_+) \left(\hat{\rho} + i\hat{p}\hat{\rho}\hat{q} - i\hat{q}\hat{\rho}\hat{p} \right) \\
&\quad - \frac{\gamma_- + \gamma_+}{2} \left([\hat{q}, [\hat{q}, \hat{\rho}]] + [\hat{p}, [\hat{p}, \hat{\rho}]] \right).
\end{aligned}
\tag{6.16}
$$

Then we apply some correspondence rules between the Wigner function W_F of the operator $\hat{F}$ and products of $\hat{F}$ and the quadratures $\hat{q}$ and $\hat{p}$. In particular, we use the rules

$$
\hat{q}\hat{F} \leftrightarrow \left(q + \frac{i}{2}\frac{\partial}{\partial p} \right) W_F, \quad \hat{F}\hat{q} \leftrightarrow \left(q - \frac{i}{2}\frac{\partial}{\partial p} \right) W_F,
$$
$$
\hat{p}\hat{F} \leftrightarrow \left(p - \frac{i}{2}\frac{\partial}{\partial q} \right) W_F, \quad \hat{F}\hat{p} \leftrightarrow \left(p + \frac{i}{2}\frac{\partial}{\partial q} \right) W_F.
\tag{6.17}
$$

The correspondence rules (6.17) follow from Wigner's formula (4.17) or its momentum representation (4.19) with $\hat{\rho}$ replaced by $\hat{F}$. For example,

$$W_{qF}(q,p) = \frac{1}{2\pi} \int_{-\infty}^{+\infty} \exp(\mathrm{i}px) \left\langle q - \frac{x}{2} | \hat{q}\hat{F} | q + \frac{x}{2} \right\rangle dx$$

$$= \frac{1}{2\pi} \int_{-\infty}^{+\infty} \exp(\mathrm{i}px) \left(q - \frac{x}{2}\right) \left\langle q - \frac{x}{2} | \hat{F} | q + \frac{x}{2} \right\rangle dx$$

$$= \left(q + \frac{\mathrm{i}}{2} \frac{\partial}{\partial p}\right) \frac{1}{2\pi} \int_{-\infty}^{+\infty} \exp(\mathrm{i}px) \left\langle q - \frac{x}{2} | \hat{F} | q + \frac{x}{2} \right\rangle dx,$$

$$(6.18)$$

which gives the first rule, and the others follow suit. In order to translate the master equation (6.16) into the language of the Wigner function, we deduce from the rules (6.17) the correspondence rules of the commutators

$$\left[\hat{q}, \hat{F}\right] \leftrightarrow \mathrm{i}\frac{\partial W_F}{\partial p}, \quad \left[\hat{p}, \hat{F}\right] \leftrightarrow -\mathrm{i}\frac{\partial W_F}{\partial q}, \qquad (6.19)$$

the equivalent of Poisson brackets (Landau and Lifshitz, Vol. I, 1982), and replace the $(\gamma_+ - \gamma_-)$ term in the master equation (6.16) by $\partial(qW)/\partial q + \partial(pW)/\partial p$ according to the rules (6.17). In this way we arrive at

$$\frac{\partial W}{\partial t} = (\gamma_- - \gamma_+)\left(\frac{\partial qW}{\partial q} + \frac{\partial pW}{\partial p}\right) + \frac{\gamma_- + \gamma_+}{2}\left(\frac{\partial^2 W}{\partial q^2} + \frac{\partial^2 W}{\partial p^2}\right), \quad (6.20)$$

the *Fokker–Planck equation* (Risken, 1996) of the Wigner function. As we will see, the first term in Eq. (6.20), called the drift term, describes the growth or loss of the amplitude while the second term, the diffusion term, accounts for amplification and absorption noise.

We solve the Fokker–Planck equation by Fourier transformation. We obtain for the Fourier-transformed Wigner function, the characteristic function $\widetilde{W}(u, v)$, the simpler equation

$$\frac{\partial \widetilde{W}}{\partial t} = (\gamma_+ - \gamma_-)\left(u\frac{\partial \widetilde{W}}{\partial u} + v\frac{\partial \widetilde{W}}{\partial v}\right) - \frac{\gamma_- + \gamma_+}{2}\left(u^2 + v^2\right)\widetilde{W}. \quad (6.21)$$

One verifies that the solution of this equation is

$$\widetilde{W}(u, v, t) = \widetilde{W}(u\sqrt{\eta}, v\sqrt{\eta}, t_0)\exp\left(\frac{\gamma_- + \gamma_+}{\gamma_- - \gamma_+}(\eta - 1)\frac{u^2 + v^2}{4}\right) \qquad (6.22)$$

in terms of the factor

$$\eta = \exp\left(2(\gamma_+ - \gamma_-)(t - t_0)\right) \qquad (6.23)$$

that, as we will show, describes the loss or gain, depending on the rates γ_- and γ_+. We see that the characteristic function is re-scaled by η and that it may have acquired additional noise, which is elegantly described in terms of s-parameterized quasiprobability distributions as follows. We obtain for the Fourier transforms (4.69) the expression

$$\tilde{W}(u,v,t;s) = \tilde{W}(u,v,t)\exp\left(\frac{s}{4}\left(u^2+v^2\right)\right) = \tilde{W}\left(u\sqrt{\eta},v\sqrt{\eta},t_0;s'\right) \quad (6.24)$$

with the transformed s parameter

$$s' = \frac{s}{\eta} + \frac{\gamma_- + \gamma_+}{\gamma_- - \gamma_+}\left(1-\frac{1}{\eta}\right). \quad (6.25)$$

From the characteristic function (6.24) we arrive by Fourier transformation (4.70) at the result

$$W(q,p,t;s) = \eta^{-1}W\left(q/\sqrt{\eta},p/\sqrt{\eta},t_0;s'\right). \quad (6.26)$$

We can draw several conclusions from formula (6.26). First, we see that amplitudes at their original values q_0 and p_0 appear at $q = q_0\sqrt{\eta}$ and $p = p_0\sqrt{\eta}$, they shrink or grow by the factor (6.23) depending on whether loss or gain dominates. Final and original amplitudes are proportional, which means that with our choice (6.15) of the transition operators we describe *linear* absorption or amplification. In non-linear optics (Shen, 1984; Boyd, 1992) loss or gain may depend on the intensity, for example absorbers may become saturated at high intensities. Such cases are not described by the theory explained here. We also see that in our case the loss or gain (6.23) depends exponentially on the interaction time t and the difference between the amplification and absorption rates $\gamma_+ - \gamma_-$. The drift term in the Fokker–Planck equation (6.20) is the one proportional to $\gamma_+ - \gamma_-$. So we conclude that this term indeed describes the amplitude change, whereas the transformed s parameter (6.25) depends on $\gamma_- + \gamma_+$, the prefactor of the diffusion term in Eq. (6.20). According to our understanding of the s parameterized quasiprobability distributions, the transformed s parameter describes additional noise, for example the amplification noise due to spontaneous emission from the excited atoms that otherwise amplify light by stimulated emission.

Let us discuss some specific cases. Consider a pure absorber with zero gain, $\gamma_+ = 0$, that for example, describes losses in homodyne detection due to detector inefficiencies or mode mismatch. Optical homodyne tomography reconstructs the Wigner function that corresponds to $s = 0$. We obtain from Eq. (6.25)

$$s' = 1 - \frac{1}{\eta}. \quad (6.27)$$

So, in effect, an s parameterized quasiprobability distribution (6.27) appears instead of the Wigner function (Leonhardt and Paul, 1993c).

Equation (6.27) shows that 50% losses, $\eta = 1/2$, are sufficient to reduce the Wigner function to the Q function that is always positive. So 50% overall efficiency sets the benchmark for the observation of negative quasiprobability distributions. Equation (6.25) also shows that in the case of a pure absorber the P function, corresponding to $s = 1$, always remains a P function. According to the optical equivalence theorem (4.68) the P function represents the quantum state as a quasi-ensemble of coherent states. In particular, the P function of a coherent state is a delta function. Consequently, coherent states remain coherent states after pure absorption.

Let us turn to the case of pure amplification, $\gamma_- = 0$, describing the ideal laser amplifier. In this case, a gain of 2 is sufficient to turn the P function into a Wigner function. Squeezed states would have lost all their squeezing, because after amplification their P function is no longer singular. Therefore, amplifying squeezed light is not a practical way to achieve higher squeezing, unless one uses a parametric amplifier, an additional squeezer, that selectively amplifies one of the quadratures. Equation (6.25) also shows that in the limit of large gain, $\eta \to \infty$, all quasiprobabilities approach the Q function that is directly observable by simultaneous measurement of position and momentum. In the regime of large gain, we are reaching the macroscopic, classical world.

6.2.2 Absorber

Equation (6.26) contains the complete theory of linear absorption and amplification; we could be satisfied with it, but it is wise to analyse it further and to cast the theory in another form. Suppose that we are dealing with an absorber where loss outweighs gain,

$$\gamma_- > \gamma_+, \tag{6.28}$$

without requiring that γ_+ vanishes. In this case the light mode approaches a stationary state for $t \to \infty$, the state where the light is completely absorbed, or, to be more precise, where the light reaches an absorption and emission equilibrium with the absorber. We obtain the stationary state from the characteristic function (6.22) in the limit $t \to \infty$ or, according to Eq. (6.23), for $\eta \to 0$. We get the Gaussian

$$\widetilde{W} \to \exp\left(-\frac{\gamma_- + \gamma_+}{\gamma_- - \gamma_+} \frac{u^2 + v^2}{4}\right) = \widetilde{W}_{th}. \tag{6.29}$$

We see that the stationary state for $t \to \infty$ does not depend on the initial state, all traces of the initial light are absorbed; $\widetilde{W}_{th}$ depends only on the absorption and emission rates. What is this state? We calculate the

Wigner function, the Fourier transform of the characteristic function. The Fourier transform of a Gaussian is another Gaussian, the Wigner function (4.86) of the thermal state (3.49). Consequently, the equilibrium state of absorption is a state in thermal equilibrium. Comparing our Fourier-transformed Gaussian (6.29) with the Wigner function (4.86) we obtain the average photon number $\bar{n}$ of the thermal state,

$$2\bar{n} + 1 = \frac{\gamma_- + \gamma_+}{\gamma_- - \gamma_+}. \tag{6.30}$$

From this relationship and the Planck distribution (3.50) follows

$$\frac{\gamma_+}{\gamma_-} = \frac{\bar{n}}{\bar{n}+1} = e^{-\beta}, \quad \beta = \frac{\hbar\omega}{k_B T}. \tag{6.31}$$

The temperature T of the light in steady state is set by the ratio of the emission and absorption rates. Normally, T is the temperature of the absorber. Note that in the optical range of the spectrum $\hbar\omega$ exceeds $k_B T$ at room temperature by about a factor of 100. So typical laboratory temperatures are irrelevant here and the thermal equilibrium state of light is simply the vacuum state. In other parts of the spectrum, the temperature of laboratory equipment is crucial. For example, probing the quantum optics of microwaves requires state-of-the-art cryogenics at liquid helium temperatures.

Equation (6.31) shows that the emission and absorption rates for an arbitrary material are not independent; their ratio is set by the Boltzmann factor $\exp(-\beta)$. We may understand this feature as follows: the absorber contains atoms that absorb or emit light of frequency ω, depending on whether they are in their ground or excited states with the fitting energy difference, $\hbar\omega$. The absorption rate γ_- is proportional to the number N_- of ground state atoms, while the emission rate γ_+ is proportional to N_+, the number of atoms in excited states. The proportionality factors correspond to the absorption and emission rates of a single photon by a single atom. These elementary rates are exactly the same, because the elementary act of emitting a photon is the exact reverse of absorbing it – at a fundamental level, the absorption and emission of light is reversible. Consequently, the ratio γ_+/γ_- is given by N_+/N_-. In thermal equilibrium, the population ratio of states with energy difference $\hbar\omega$ is equal to the Boltzmann factor $\exp(-\beta)$ in Eq. (6.31), which explains our result.

Having established the stationary state of light after complete absorption, we express the evolving state in a form that is going to suggest a simple physical picture. We use our previous result that the Wigner function of partially absorbed light is the scaled and re-parameterized quasiprobability distribution (6.26) with parameter

(6.25) and $s = 0$. Such a quasiprobability distribution is a smoothed Wigner function (4.74) where we express the Gaussian factor as the thermal Wigner function (4.86) with average photon number (6.30). We get

$$W(q,p,t) = \int_{-\infty}^{+\infty} \int_{-\infty}^{+\infty} W\left(q',p',t_0\right) W_{th}\left(q_0',p_0'\right) \frac{dq'dp'}{1-\eta},$$
$$q_0' = \frac{q - q'\sqrt{\eta}}{\sqrt{1-\eta}}, \quad p_0' = \frac{p - p'\sqrt{\eta}}{\sqrt{1-\eta}},$$

(6.32)

which we write in terms of new integration variables q_0 and p_0 that we choose such that

$$q_0' = q\sqrt{1-\eta} + q_0\sqrt{\eta}, \quad p_0' = p\sqrt{1-\eta} + p_0\sqrt{\eta},$$

(6.33)

because in this case

$$q' = q\sqrt{\eta} - q_0\sqrt{1-\eta}, \quad p' = p\sqrt{\eta} - p_0\sqrt{1-\eta}.$$

(6.34)

We obtain for the Wigner function of partially absorbed light

$$W(q,p,t) = \int_{-\infty}^{+\infty} \int_{-\infty}^{+\infty} W\left(q',p',t_0\right) W_{th}\left(q_0',p_0'\right) dq_0 dp_0.$$

(6.35)

Our results (6.35) and (6.34) look exactly like Eqs. (5.34) and (5.35) integrated over one mode, describing the effect of a beam splitter on the transmitted light. All the individual acts of scattering and absorption are subsumed in a single beam splitter with transmissivity $\sqrt{\eta}$ as shown in Fig. 6.1. One of the outgoing modes represents the absorbed part of the light. We integrate over the corresponding variables q_0 and p_0 to indicate that this part is lost – we average over all possible measurement

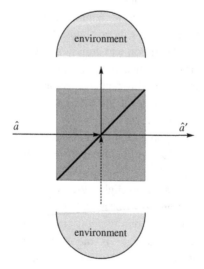

Fig. 6.1. Absorption of light. An absorber can be modelled by an effective beam splitter where the absorption reservoir, the environment, corresponds to both the second input and output of the beam splitter. The absorbed light is reflected into the environment. Simultaneously, fluctuations coming from the environment enter the outgoing light through the second input port of the beam splitter. The picture visualizes Eqs. (6.36) and (6.37).

results of the lost light. The ingoing mode at the second port of the beam splitter assumes the role of the absorption noise. The mode is in a thermal state or, if the temperature of the absorber is negligible, in the vacuum state. So both as input and output the second mode of the beam splitter represents the entire environment the mode is interacting with. In the Heisenberg picture, the beam splitter model of absorption appears as

$$\hat{a}(t) = \hat{a}(t_0)\sqrt{\eta} + \hat{a}_0\sqrt{1 - \eta}. \tag{6.36}$$

Here the first term describes the partially absorbed light that has been reduced in amplitude by the factor $\sqrt{\eta}$. The second term $\hat{a}_0\sqrt{1 - \eta}$, representing the absorption noise, is necessary to preserve the Bose commutation relations of the light mode, because without it the commutator $[\hat{a}(t), \hat{a}^\dagger(t)]$ would be reduced to η. The second mode of the beam splitter model, representing the absorbed light, appears in the Heisenberg picture as

$$\hat{a}_0(t) = \hat{a}_0\sqrt{\eta} - \hat{a}(t_0)\sqrt{1 - \eta}. \tag{6.37}$$

Our simple model allows us to draw interesting conclusions without doing calculations, just by drawing pictures. For instance, the model shows why we normally do not observe macroscopic quantum superpositions of light, the Schrödinger cat states (4.45). They turn out to be extremely fragile to dissipative processes and so disappear as quickly as the Cheshire Cat in Alice in Wonderland. Figure 6.2 shows why. Here we plot the momentum distribution (4.55) of the Schrödinger cat state multiplied by the momentum distribution of the noise the "cat" is interacting with. The momentum distribution carries the typical interference fringes of the superposition state (4.45). The more macroscopic the state is, the more rapidly the momentum distribution oscillates. A beam splitter rotates the momentum wave function, and so it also rotates its modulus squared, the momentum distribution of "cat" and noise. What remains after partial absorption is the average over the noise. Figure 6.2 shows that already a slight rotation is sufficient to wash out the interference fringes. Contact with a dissipative environment "kills the Schrödinger cat".

6.2.3 Amplifier

Having found a simple, intuitive model for absorption, let us consider the case of amplification where the gain outweighs the loss,

$$\gamma_+ > \gamma_-, \tag{6.38}$$

and η exceeds unity. In this case we can no longer employ the beam splitter model (6.36), because the commutator $[\hat{a}(t), \hat{a}^\dagger(t)]$ would give

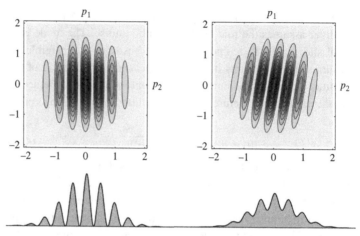

Fig. 6.2. Cat and noise. Left: total momentum quadrature probability distribution of a Schrödinger cat state and the vacuum state of the absorption noise. Right: total quadrature distribution of the outgoing light. Below: projections of the total momentum distribution on the propagating light mode, on the left for the incident light and on the right for the outgoing light. As the total distribution is rotated at the beam splitter modelling the absorber, the quantum interference pattern of the Schrödinger cat state is easily washed out. Contact with a dissipative environment "kills the Schrödinger cat".

$2\eta - 1$ instead of unity, but we can replace it by a similar and equally simple model:

$$\hat{a}(t) = \hat{a}(t_0)\sqrt{\eta} + \hat{a}_0^\dagger \sqrt{\eta - 1}. \tag{6.39}$$

Calculating the commutator,

$$\left[\hat{a}(t), \hat{a}^\dagger(t)\right] = \left[\hat{a}(t_0), \hat{a}^\dagger(t_0)\right]\eta + \left[\hat{a}_0^{\,\dagger}, \hat{a}_0\right](\eta - 1) = 1, \tag{6.40}$$

we see that the model (6.39) is perfectly consistent and physically allowed. The $\hat{a}(t_0)\sqrt{\eta}$ term describes the linearly amplified amplitude and the $\hat{a}_0^\dagger \sqrt{\eta - 1}$ term the amplification noise, for example the spontaneous emission noise of a laser amplifier. Fittingly, the creation operator $\hat{a}_0^\dagger$ represents the noise – amplification is based on the creation of light. We may describe the backaction on the amplifier as

$$\hat{a}_0(t) = \hat{a}_0\sqrt{\eta} + \hat{a}^\dagger(t_0)\sqrt{\eta - 1}, \tag{6.41}$$

in analogy to the backaction on the absorber, the dissipation of the absorbed light (6.37). For example, for a laser amplifier $\hat{a}_0(t)$ accounts for the atomic polarizations affected by the stimulated emission of the amplified light. One verifies that the transformation (6.41) preserves the Bose commutation relations and commutes with $\hat{a}(t)$. Hence it represents a consistent supplement to the amplifier model (6.39).

Our model is not only plausible for linear amplification, it turns out to be exactly equivalent to the solution (6.26) of the Fokker–Planck equation (6.20). In order to prove this statement, we proceed in a similar way as in the case of absorption, apart from a few subtleties. Like there, we express the evolving Wigner function as

$$W(q,p,t) = \int_{-\infty}^{+\infty} \int_{-\infty}^{+\infty} W(q',p',t_0)\, W_{th}(q_0',p_0')\, \frac{dq'\,dp'}{\eta - 1},$$

$$q_0' = \frac{q - q'\sqrt{\eta}}{\sqrt{\eta - 1}}, \quad p_0' = \frac{p - p'\sqrt{\eta}}{\sqrt{\eta - 1}}, \tag{6.42}$$

where we made some subtle but important sign changes: we employ the thermal Wigner function (4.86) with average photon number $\bar{n}$ given by

$$2\bar{n} + 1 = \frac{\gamma_+ + \gamma_-}{\gamma_+ - \gamma_-} \tag{6.43}$$

instead of the photon number (6.30), because otherwise the Gaussian W_{th} would diverge. We also deliberately changed the sign of p_0', because the model (6.39) transforms position and momentum quadratures differently,

$$\hat{q}(t) = \hat{q}(t_0)\sqrt{\eta} + \hat{q}_0\sqrt{\eta - 1}, \quad \hat{p}(t) = \hat{p}(t_0)\sqrt{\eta} - \hat{p}_0\sqrt{\eta - 1}. \tag{6.44}$$

With the chosen sign we obtain

$$q = q'\sqrt{\eta} + q_0'\sqrt{\eta - 1}, \quad p = p'\sqrt{\eta} - p_0'\sqrt{\eta - 1} \tag{6.45}$$

which agrees with the quadrature transformation (6.44). We introduce the new integration variables

$$q_0 = q_0'\sqrt{\eta} + q'\sqrt{\eta - 1}, \quad p_0 = p_0'\sqrt{\eta} - p'\sqrt{\eta - 1} \tag{6.46}$$

that correspond to the quadrature representation of the amplifier's backaction (6.41). Consequently, our model is consistent with the dynamics (6.20) of linear amplification. Moreover, we made our model more specific: the amplification noise is thermal with average photon number (6.30). For the *noise temperature* we obtain from the Planck distribution (3.50)

$$\frac{\gamma_+}{\gamma_-} = \frac{\bar{n} + 1}{\bar{n}} = e^{\beta}. \tag{6.47}$$

Taking some liberty, we express the inverse Boltzmann factor $\exp(\beta)$ in terms of a *negative temperature*,

$$\beta = -\frac{\hbar\omega}{k_B T}, \tag{6.48}$$

to make formula (6.47) appear like the equivalent expression (6.31) for absorption. As in that case, the emission and absorption rates γ_+ and γ_- are proportional to the atomic population in the excited and ground states. Gain outweighs loss if the excited states are more populated than the ground states: linear amplification requires population inversion. A Boltzmann factor with negative temperature describes the population inversion; for $T = -0$ all atoms are excited, for $T = -\infty$ half of the atoms are in excited states and half in their ground states. At infinity the negative temperature scale meets the ordinary positive scale, and at $T = +0$ all atoms are in their ground states. Thermal equilibrium does not create population inversion, so the linear amplifier operates in a far-from-equilibrium regime.

Note that the linear amplifier is not a fully operational laser yet. A laser reaches a steady state, typically at high light intensity, where a pump process maintains the population inversion that is depleted by stimulated emission at the same rate as it is replenished. The stimulated emission depends on the intensity; so the laser stabilizes the light amplitude at the dynamic equilibrium between emission and pumping. The laser is self-organized. In contrast, the exponential growth (6.23) in linear amplification relies on an undepleted reservoir of population inversion. The laser generates a quantum state of well-defined intensity, in the ideal case an ensemble of coherent states with random phase, because the phase is free to drift in the laser. In linear amplification, the outgoing light amplitude is proportional to the amplitude of the incident light, while some amplification noise is added. If the laser amplifier runs idle, with only a vacuum state as input, it emits thermal light, as we see from our solution (6.22) of the characteristic function. We obtain for a vacuum input with $\widetilde{W}(u, v, t_0) = \exp(-(u^2 + v^2)/4)$ the characteristic function (6.29) of a thermal state with average photon number (6.30) given by the expression

$$\bar{n}_{idle} = (\bar{n} + 1)(\eta - 1). \tag{6.49}$$

6.2.4 Eavesdropper

We have discussed absorption $\gamma_- > \gamma_+$, and amplification, $\gamma_+ > \gamma_-$, but not yet the intermediate case where the gain exactly balances the loss,

$$\gamma_- = \gamma_+ = \gamma. \tag{6.50}$$

Such a case may represent a simple eavesdropping attack on a quantum communication line between two partners, often called Alice and Bob. Eve, the notorious eavesdropper, extracts light from the communication channel, analyses it and replenishes the light by linear amplification.

Alice and Bob, however, can detect Eve's presence and disregard the
sent information, because her manipulations have modified the quantum
state of light. Eve has added extra quantum noise that spoils the purity
of the state. We express the purity $\mathrm{tr}\{\hat{\rho}^2\}$ in terms (4.28) of the integral
over the squared Wigner function and obtain from the Fokker–Planck
equation (6.20) in the eavesdropping regime (6.50)

$$
\begin{aligned}
\frac{d\,\mathrm{tr}\{\hat{\rho}^2\}}{dt} &= 4\pi \int_{-\infty}^{+\infty} \int_{-\infty}^{+\infty} W \frac{\partial W}{\partial t}\, dq\, dp, \\
&= -4\pi\gamma \int_{-\infty}^{+\infty} \int_{-\infty}^{+\infty} \left(\left(\frac{\partial W}{\partial q}\right)^2 + \left(\frac{\partial W}{\partial p}\right)^2 \right) dq\, dp \\
&< 0.
\end{aligned}
\tag{6.51}
$$

The purity is positively reduced. Furthermore, eavesdropping changes
the overall quantum state of the communication channel. One verifies
that

$$
\widetilde{W}(u, v, t) = \widetilde{W}(u, v, t_0) \exp\left(-\gamma t \left(u^2 + v^2\right)\right)
\tag{6.52}
$$

is the solution of the Fourier transformed Fokker–Planck equation
(6.21). Consequently, the Wigner function of the channel turns into the
s parameterized quasiprobability distribution

$$
W(q, p, t) = W(q, p, t_0; s), \quad s = 4\gamma t.
\tag{6.53}
$$

Although the eavesdropping attempt does not affect the amplitude, it
introduces extra noise that appears as a linearly growing s parameter.

In *quantum cryptography*, based on *quantum key distribution* by
quantum communication, strategies for secure communication have
been developed against significantly more sophisticated attacks than the
simple optical tap we discussed. Quantum communication has become
a practically appliable research area of quantum information science.

6.3 Continuous quantum measurements

Quantum measurements are genuine irreversible processes. In the textbook case of a quantum measurement, known as the *von Neumann
measurement*, the state $|\psi\rangle$ is turned into an eigenstate $|a\rangle$ depending
on which eigenvalue a has been measured. One cannot reverse this process on an individual quantum object, because all the other components
of the state vector $|\psi\rangle$ are lost after the transition to $|a\rangle$. In quantum state
tomography, we can reconstruct the quantum state as it has been prior
to measurement, but for this we need to observe an entire ensemble of
equally prepared states. We also need to separately measure mutually
incompatible observables, for example the quadratures $\hat{q}\cos\theta + \hat{p}\sin\theta$

at various angles θ. For individual quantum objects, measurements are irreversible.

In quantum optics, the textbook von Neumann measurement is rather difficult to implement in practice, for a simple reason: measurements on light are typically based on photodetection where light is converted into electric signals, but not turned into eigenstates of light itself. Quantum optical von Neumann measurements rely on indirect methods where a light mode interacts with a second system, the quantum meter, which could be another light mode or an atom, for example. After their interaction, the two systems should have become strongly correlated such that measurements on the meter prepare the mode into an eigenstate. In quantum optics, such indirect von Neumann measurements are called *quantum nondemolition measurements*. The quantum state is not completely demolished, but is surely changed by the measurement unless it has been in an eigenstate of the measured observable.

Figure 6.3 shows the diagram of a Gedanken experiment for the quantum nondemolition measurement of quadratures. A beam splitter taps a minute fraction ϵ of the incident light and sends it to a homodyne detector. The tapped light is replenished by a strongly squeezed vacuum through the second port of the beam splitter. We describe the state of the light mode by the quadrature wave function $\psi(q)$ and the squeezed vacuum by the wave function (3.84) with squeezing parameter ζ. The beam splitter rotates the wave function by the angle $\epsilon \ll 1$ and so the homodyne detector measures the quadrature q_0 with probability distribution

$$\mathrm{pr}(q_0) = \int_{-\infty}^{+\infty} |\psi(q + \epsilon q_0)|^2 \, |\psi_s(q_0 - \epsilon q)|^2 dq. \tag{6.54}$$

We introduce the new integration variable $q' = q + \epsilon q_0$, replace $q_0 - \epsilon q$ by $q_0 - \epsilon q'$ in the limit of small ϵ, and get

quadrature measurement

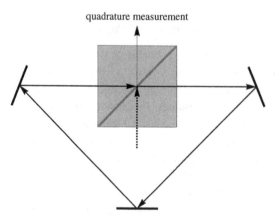

Fig. 6.3. Continuous quadrature measurement. This scheme would perform ideal quantum nondemolition measurements of quadratures. The light is confined in a cavity where an infinitesimally small part is extracted and sent to a homodyne detector for quadrature measurement. The extracted light is replenished by an infinitely squeezed vacuum.

$$\mathrm{pr}(q_0) = \frac{1}{\epsilon} \int_{-\infty}^{+\infty} |\psi(q')|^2 \frac{r}{\sqrt{\pi}} \exp\left(-r^2(q_0/\epsilon - q')^2\right) dq' \tag{6.55}$$

taking the limit $\epsilon \to 0$ and $\zeta \to \infty$ with finite

$$r = \epsilon\, e^\zeta. \tag{6.56}$$

The measured quadrature amplitudes are reduced by ϵ, because only an ϵth of the light is tapped. This reduction of scale is of no consequence, because the scale of quadrature measurements is not naturally defined but calibrated with respect to a vacuum input. We simply recalibrate the measurement with respect to the vacuum state of the untapped light instead of the vacuum immediately in front of the homodyne detector. The parameter r is more important here, because it characterizes the resolution of the quadrature nondemolition measurement. Its resolution cannot be infinite in practice, because then it would prepare a quadrature eigenstate of infinite energy.

How does a quadrature nondemolition measurement affect the quantum state of the measured light? It is understood that we can only predict the quantum backaction of the measurement on average, for many runs of the experiment, because each individual measurement is unpredictable. Let us use Wigner functions for the calculation. The Wigner function W' of the light after the measurement is the joint initial Wigner function of light and squeezed vacuum, rotated by ϵ and integrated over the output heading for the homodyne detector,

$$W'(q,p) = \int_{-\infty}^{+\infty} \int_{-\infty}^{+\infty} W\left(q',p'\right) W_s\left(q'_0,p'_0\right) dq_0\, dp_0. \tag{6.57}$$

We substitute the Wigner function (4.39) of the squeezed vacuum and take q'_0 and p'_0 as integration variables. We obtain

$$\begin{aligned}
W'(q,p) &= \int_{-\infty}^{+\infty} \int_{-\infty}^{+\infty} W\left(\frac{q + q'_0 \sin\epsilon}{\cos\epsilon}, \frac{p + p'_0 \sin\epsilon}{\cos\epsilon}\right) \\
&\quad \times \frac{1}{\pi} \exp\left(-e^{2\zeta} q_0'^2 - e^{-2\zeta} p_0'^2\right) \frac{dq'_0\, dp'_0}{\cos^2\epsilon} \\
&\sim \frac{1}{\cos^2\epsilon} W\left(\frac{q}{\cos\epsilon}, \frac{p}{\cos\epsilon}\right) + \frac{\epsilon^2}{4}\left(e^{-\zeta} \frac{\partial^2 W}{\partial q^2} + e^\zeta \frac{\partial^2 W}{\partial p^2}\right)
\end{aligned} \tag{6.58}$$

by Taylor expansion in $q'_0 \epsilon$ and $p'_0 \epsilon$ to quadratic order in ϵ and calculation of the Gaussian integrals over q'_0 and p'_0. In the limit $\epsilon \to 0$ and $\zeta \to \infty$ with constant (6.56) the only contribution linear in ϵ is

$$W'(q,p) - W(q,p) = \frac{r\epsilon}{4} \frac{\partial^2 W}{\partial p^2}. \tag{6.59}$$

Suppose we continuously monitor the quadratures. For example, imagine the light mode is contained in a cavity, but at each moment of time dt a fraction of the light is tapped, quadrature measured and replenished by the squeezed vacuum. Writing, for infinitesimally small ϵ,

$$\frac{r\epsilon}{8} = \gamma \, dt \qquad (6.60)$$

we read off the backaction as

$$\frac{\partial W}{\partial t} = 2\gamma \frac{\partial^2 W}{\partial p^2} \qquad (6.61)$$

that we immediately translate, via the correspondence rule (6.19), into the quantum master equation

$$\frac{d\hat{\rho}}{dt} = -2\gamma \left[\hat{q}, \left[\hat{q}, \hat{\rho} \right] \right] . \qquad (6.62)$$

In 1986 Ghirardi et al. (1986) formulated such an equation for the localization process of a point particle with position operator $\hat{q}$ (not a light mode with quadrature $\hat{q}$). For example, the equation could describe how an elementary particle materializes in a bubble chamber from a probability wave, drawing a classical trajectory as a string of bubbles. In this case the bubble size sets the resolution of the continuous position measurement. One could, as Ghirardi et al. did, regard the quantum master equation (6.62) as the fundamental equation of the measurement process, arguing that all quantum measurements could be ultimately traced back to spontaneous localizations of probability waves into point particles. Some mysterious part of Nature could materialize probability amplitudes as particles with well-defined position, a process described, on average, by the quantum master equation (6.62). But their argument does not resolve the mystery of quantum mechanics, because it does not explain what this part of Nature is, what really happens behind the scenes of reality, nor does the equation (6.62) govern the fate of individual particles, but rather the average destiny of an ensemble.

Let us return to our case, the quadrature nondemolition measurement. Suppose we vary the reference phase θ of the quadrature, the phase of the local oscillator at the homodyne detector, such that we measure the rotated quadratures $\hat{q}_\theta = \hat{q} \cos\theta + \hat{p} \sin\theta$. Applying optical homodyne tomography we reconstruct the quantum state, so observing *in vivo* the quantum state of light. But in doing so we are changing the state. In order to predict how the state changes we replace $\hat{q}$ in the master equation (6.62) by the rotating quadrature $\hat{q}_\theta$ and average over θ. The average of $\cos^2\theta$ and $\sin^2\theta$ gives $1/2$ while $\cos\theta \sin\theta$ vanishes on average. What remains after averaging is the quantum master equation (6.16) of absorption and amplification in the intermediate regime (6.50) when loss and gain are balanced, the master equation of eavesdropping. Monitoring the quantum state surely amounts to eavesdropping, an offence punished by quantum backaction.

6.4 Questions

6.1 Use the Schrödinger equation with the Hamiltonian $\hat{H}$ (and $\hbar$ set to unity) to deduce the evolution equation of the density matrix

$$\frac{d\hat{\rho}}{dt} = i\left(\hat{\rho}\hat{H}^{\dagger} - \hat{H}\hat{\rho}\right).$$

6.2 Why is it a good idea to describe irreversible processes by the effective Hamiltonian $\hat{H}_{eff} = \hat{H}_0 - i\hat{V}$?

6.3 Why is $\hat{V}$ a positive Hermitian operator? Why is such an operator always represented as $\hat{V} = \sum_a \gamma_a \hat{T}_a^{\dagger} \hat{T}_a$?

6.4 On the other hand, why is the Hamiltonian $\hat{H}_{eff}$ not sufficient to describe the quantum dynamics? Which important feature of the density matrix would be violated?

6.5 What are quantum jumps and how are they described?

6.6 Why is $\hat{T}_a = \hat{T}_a$ and $\Gamma_a = 2\gamma_a$?

6.7 Deduce the quantum master equation from these considerations.

6.8 Describe the absorption and amplification of light in the Lindblad form.

6.9 Deduce the master equation in terms of quadratures,

$$\frac{d\hat{\rho}}{dt} = (\gamma_- - \gamma_+)(\hat{\rho} + i\hat{p}\hat{\rho}\hat{q} - i\hat{q}\hat{\rho}\hat{p}) - \frac{\gamma_- + \gamma_+}{2}\left([\hat{q},[\hat{q},\hat{\rho}]] + [\hat{p},[\hat{p},\hat{\rho}]]\right).$$

6.10 Prove the correspondence rules between the operators $\hat{q}\hat{F}$, $\hat{F}\hat{q}$, $\hat{p}\hat{F}$ and $\hat{F}\hat{p}$ and the Wigner function $W_F(q,p)$. Give an interpretation of these rules.

6.11 Deduce the Fokker–Planck equation for the Wigner function. Give interpretations for the terms in this equation.

6.12 Fourier transform the Fokker–Planck equation. Verify by differentiation that

$$\tilde{W}(u,v,t) = \tilde{W}(u\sqrt{\eta}, v\sqrt{\eta}, t_0)\exp\left(\frac{\gamma_- + \gamma_+}{\gamma_- - \gamma_+}(\eta - 1)\frac{u^2 + v^2}{4}\right)$$

with $\eta = \exp[2(\gamma_+ - \gamma_-)(t - t_0)]$ is indeed the solution of the Fourier transformed Fokker–Planck equation.

6.13 Why is the Wigner function after absorption/amplification an s parameterized quasiprobability distribution of the initial state? Deduce the parameter s for the general case.

6.14 Why does this solution describe linear absorption or amplification?

6.15 Discuss pure absorption, $\gamma_+ = 0$, and pure amplification, $\gamma_- = 0$, using the solution in terms of an s parameterized quasiprobability distribution.

6.16 Consider an absorber where the absorption rate γ_- exceeds the amplification rate γ_+. Show that the stationary state is a thermal state with the temperature T given by $\exp(-\hbar\omega/k_B T) = \gamma_+/\gamma_-$.

6.17 Why do the absorption and amplification rates depend on the temperature?

6.18 Deduce the beam splitter model of absorption.

6.19 Visualize the fragility of Schrödinger cat state states using the beam splitter model of absorption.

6.20 Use the quadrature rotation at the beam splitter to calculate the momentum distribution of a Schrödinger cat state that has been partially absorbed.

6.21 Consider an amplifier. Motivate why the output mode is described as $\hat{a}\sqrt{\eta} + \hat{a}_0^\dagger\sqrt{\eta - 1}$ of the input mode $\hat{a}$. What is the role of $\hat{a}_0^\dagger$?

6.22 Prove that this model describes the solution of the Fokker–Planck equation for the amplifier.

6.23 How does eavesdropping reduce the purity of the quantum state?

6.24 Consider the quantum nondemolition measurements of quadratures by tapping light and replenishing it with a squeezed vacuum. Calculate the quadrature probability distribution of the extracted mode.

6.25 Calculate the Wigner function after a part of the initial signal was extracted and replaced by squeezed vacuum (the quantum backaction).

6.26 Deduce the Fokker–Planck equation of the backaction by continuous quadrature measurements.

6.27 Derive the corresponding quantum master equation.

6.28 Show that the continuous monitoring of the quantum state amounts to the eavesdropping discussed above.

6.5 Homework problem

Lindblad's theorem establishes an elegant description of irreversible processes in general, not only of the absorption and amplification of light. In particular, it can be applied to understand the optical properties of atoms.

(a) Consider two levels of an atom, the ground state $|g\rangle$ and the excited state $|e\rangle$ with energy difference $\hbar\omega_0$. The atom interacts with a classical light field of frequency ω. In the absence of spontaneous emission the field would excite and de-excite the atom with the Rabi frequency (Loudon, 2000)

$$\Omega = \frac{d}{\hbar}E$$

where d is the dipole moment and E the electric field strength. The atomic dynamics is described by the Hamiltonian

$$\hat{H} = \omega_0 |e\rangle\langle e| + \frac{\Omega}{2} e^{-i\omega t}\hat{T}^\dagger + \frac{\Omega}{2} e^{i\omega t}\hat{T}, \quad \hat{T} = |g\rangle\langle e|.$$

For simplicity, the Hamiltonian is considered in units of $\hbar$. The operator $\hat{T}$ describes the coherent transition from the excited to the ground state and $\hat{T}^\dagger = |e\rangle\langle g|$ the reverse transition. The operator $\hat{T}$ also describes the spontaneous emission from $|e\rangle$ to $|g\rangle$, an incoherent, irreversible process that appears in Lindblad's master equation (6.13) with the spontaneous emission rate γ. Represent $\hat{\rho}$ as

$$\hat{\rho} = e^{-i\omega|e\rangle\langle e|t} \hat{\rho}_0 e^{i\omega|e\rangle\langle e|t}$$

and show that from the master equation (6.13) follows

$$\frac{d\hat{\rho}_0}{dt} = i[\hat{\rho}, \hat{H}_0] - \gamma(\hat{T}^\dagger\hat{T}\hat{\rho}_0 - 2\hat{T}\hat{\rho}_0\hat{T}^\dagger + \hat{\rho}_0\hat{T}^\dagger\hat{T}),$$

$$\hat{H}_0 = \Delta|e\rangle\langle e| + \frac{\Omega}{2}(\hat{T}^\dagger + \hat{T}^\dagger), \quad \Delta = \omega_0 - \omega.$$

Here $\hat{H}_0$ appears as an effective Hamiltonian and the transition operators are unchanged in the master equation.

(b) We can always write the density matrix of the two-level system as

$$\hat{\rho}_0 = \begin{pmatrix} 1-n & \chi \\ \chi^* & n \end{pmatrix}.$$

Insert this explicit expression in the effective master equation of part (a) and consider the stationary solution where $d\hat{\rho}_0/dt = 0$. From the diagonal elements deduce a relationship between Im χ and n that, together with the real part of the off-diagonal elements gives a relationship between χ and n. Finally, obtain the complete solution for n and χ from the imaginary part of the off-diagonal elements.

(c) Discuss the solution derived in part (b).

(d) Consider a three-level system with degenerate ground states $|g_1\rangle$ and $|g_2\rangle$ (states with equal energy). The ground states are excited by two polarization modes of light with Rabi frequencies Ω_1 and Ω_2. The effective Hamiltonian for $\hat{\rho}_0$ is

$$\hat{H}_0 = \Delta|e\rangle\langle e| + \frac{\Omega_1}{2}(\hat{T}_1 + \hat{T}_1^\dagger) + \frac{\Omega_2}{2}(\hat{T}_2 + \hat{T}_2^\dagger), \quad \hat{T}_k = |g_k\rangle\langle e|.$$

The rates for the spontaneous emission from the excited state to the two ground states are the same,

$$\gamma_1 = \gamma_2 = \gamma.$$

Show that the stationary solution of the corresponding master equation (6.13) is

$$\hat{\rho} = |\psi\rangle\langle\psi|, \quad |\psi\rangle = \frac{\Omega_2|g_1\rangle - \Omega_1|g_2\rangle}{\sqrt{\Omega_1^2 + \Omega_2^2}}.$$

(e) Give an interpretation for the stationary state derived in part (d).

6.6 Further reading

The theory of irreversible processes is a very large research area. In connection with quantum optics, we recommend as additional literature the books by Carmichael (1993, 2003, 2007), Gardiner (1991), Gardiner and Zoller (2004), and Risken's classic monograph on the Fokker–Planck equation (1996). For quantum trajectories see Carmichael (1987, 2003) and Plenio and Knight (1998).

The beam splitter model of losses was developed in its most general form in Leonhardt (1993) and the related model for amplification in Leonhardt (1994). General consequences of the extra noise in linear amplification have been discussed by Caves (1982).

Quantum cryptography was pioneered by Bennett and Brassard (1984); see also Bennett (1992). Ekert (1991) suggested to apply the nonlocality of quantum mechanics for cryptographic purposes. (Nonlocality is discussed in Section 7.2.3.) For reviews on quantum cryptography, see Gisin et al. (2002) and Dusek et al. (2006).

Continuous quantum measurements are analysed in the books by Braginsky et al. (1995) and Mensky (1993).

Chapter 7
Entanglement

7.1 Parametric amplifier

In 1935 Erwin Schrödinger (Schrödinger, 1935a) coined the scientific term *entanglement*:

> "When two systems, of which we know the states by their respective representatives, enter into temporary physical interaction due to known forces between them, and when after a time of mutual influence the systems separate again, then they can no longer be described in the same way as before, viz. by endowing each of them with a representative of its own. I would not call that one but rather the characteristic trait of quantum mechanics, the one that enforces its entire departure from classical lines of thought. By the interaction the two representatives [the quantum states] have become entangled."

In a related paper (Schrödinger, 1935b) where Schrödinger also mentioned the famous cat, written in his native German, Schrödinger used the word *Verschränkung* for entanglement, which is probably better. Verschränkung is a joiner's term for dovetailing pieces of furniture, Schränke in German. The word entanglement conjures up images of things entangled in nets, whereas entangled quantum states rather resemble neat bridges between different quantum objects, possibly at different places, dovetails across space. Entangled states have turned from posing a theoretical problem into becoming an experimental tool, first for testing the nonlocality of quantum mechanics and later for applying nonlocality and the related massive parallelism of the quantum world in quantum information science. The workhorse of most quantum optical applications of entanglement is the parametric amplifier.

7.1.1 Heisenberg picture

Parametric amplification is a phenomenon of non-linear optics (Shen, 1984; Boyd, 1992) any child can understand, at least the main idea of it, because it is related to the physics of the playground swing. So let us begin with a brief excursion to the playground. One makes the following observation there: after having pushed the swing, each turn takes exactly the same time, regardless of how strongly we pushed the swing, how high or low it goes. In physics terms, the frequency of the swing does not depend on the amplitude. But the child can swing without somebody else pushing it. How is this possible? Sitting on the swing the child rocks it, instinctively, at exactly twice its frequency. The rocking enhances the swinging, because when the swing is up it is pushed downwards with enhanced force, when it is up again on the other side, half a period later, it is pushed downwards as well, etc. The swinging gets stronger each time, because the fundamental frequency of the swing remains the same, despite the growing amplitude. Such a process is called *parametric amplification* (Landau and Lifshitz, Vol. I, 1982) (because the child changes a parameter of the swing, its centre of mass). But how does the swing start? From amplifying minute initial movements occurring at the right phase of the rocking. If the swing stood perfectly still, no amount of rocking could get it going.

In *optical parametric amplification* (see Fig. 7.1), a beam of intense laser light is focused on a crystal, typically β-barium borate (BBO) or potassium titanyl phosphate (KTP). The beam, called the pump, induces electric dipoles in the atoms of the crystal that oscillate with the pump frequency ω_p; the light is rocking the atoms like the child is rocking the swing. The rocking amplifies other dipole oscillations that act as antennae, emitting electromagnetic radiation – light. Energy conservation implies that the energy of the emitted photons, $\hbar\omega_k$, must add up to the energy $\hbar\omega_p$ of a pump photon. The simplest non-trivial case is pair

Fig. 7.1. Diagram of a parametric amplifier. The pump beam, shown in grey, amplifies two incident light modes. As photons are created, the parametric amplifier combines creation and annihilation operators. In particular, the device performs the transformation (7.1).

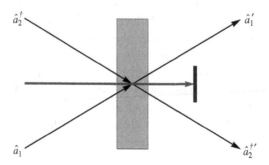

production, where the crystal emits pairs of modes that, for historical reasons, are often called signal and idler. Since for each pair the signal and idler photons originate from the same pump photon they must be strongly correlated. In contrast to the playground swing, photon pairs may start from nothing, from the vacuum state, to be precise. Here the vacuum noise plays the role of the minute movements the swing needs to get going, but the vacuum noise never stops. The optical parametric amplifier may generate a range of signal and idler modes, subject to energy and momentum conservation (called phase matching conditions) and possible constraints of the device, for example a cavity for enhancing one particular pair of modes by resonance. In any case, the mode pairs are independent of each other. So we can consider only one pair to understand all of them. In *degenerate parametric amplification* the two modes coincide. In this case, the pump frequency must be exactly twice the mode frequency, as for the playground swing.

Let us consider the non-degenerate parametric amplification of two modes that we simply call modes 1 and 2. The crystal converts one pump photon into two new photons. So the Hamiltonian of the process must contain a term proportional to $\hat{a}_p\hat{a}_1^\dagger\hat{a}_2^\dagger$. But the Hamiltonian also contains the Hermitian conjugate $\hat{a}_p^\dagger\hat{a}_1\hat{a}_2$ – the reverse process must be allowed as well in principle even if it might be improbable in practice. We assume that the pump is sufficiently strong and coherent such that we can replace $\hat{a}_p$ and $\hat{a}_p^\dagger$ by classical amplitudes α_p and α_p^*. Consequently, the Heisenberg equations of motion (1.32) of optical parametric amplification couple annihilation and creation operators. The solution of the equation will be a linear combination of $\hat{a}_k$ and $\hat{a}_k^\dagger$ that we write as

$$\begin{pmatrix} \hat{a}_1' \\ \hat{a}_2'^\dagger \end{pmatrix} = A \begin{pmatrix} \hat{a}_1 \\ \hat{a}_2^\dagger \end{pmatrix} \tag{7.1}$$

in terms of the amplifier matrix

$$A = \begin{pmatrix} A_{11} & A_{12} \\ A_{21} & A_{22} \end{pmatrix}. \tag{7.2}$$

In the following we simply treat the parametric amplifier as a black box solely characterized by the matrix A. In condensed matter physics, the Bose operator transformation (7.1) is called *Bogoliubov transformation*. Our model describes other optical processes as well, for example *four wave mixing* (Shen, 1984; Boyd, 1992) where two pump modes act together. One may also regard the device (7.1) as a *phase conjugating mirror*, because if we read $\hat{a}_2$ as the input mode for $\hat{a}_1'$ the device creates the phase conjugated image $\hat{a}_2^\dagger$ in the output mode $\hat{a}_1'$. Furthermore, our model (6.39) and (6.41) for the linear amplifier belongs to the same category, with the only difference that we can never observe the noise mode $\hat{a}_2$ and so must average over it.

In the case of the linear amplifier (6.39) and (6.41) we see that the coefficients of the amplification matrix are not independent from each other, which is not a coincidence. Similar to the beam splitter transformation (5.3), the coefficients are constrained to conserve the commutator relation (5.4) between the mode operators. The coefficients must obey

$$|A_{11}|^2 - |A_{12}|^2 = |A_{21}|^2 - |A_{22}|^2 = 1, \quad A_{11}^* A_{21} - A_{12}^* A_{22} = 0. \quad (7.3)$$

The matrix A is said to be quasi-unitary (Cornwell, 1984). It follows that the photon number difference between the two modes is a conserved quantity,

$$\hat{a}_1'^\dagger \hat{a}_1' - \hat{a}_2'^\dagger \hat{a}_2' = \hat{a}_1^\dagger \hat{a}_1 - \hat{a}_2^\dagger \hat{a}_2. \quad (7.4)$$

In parametric amplification, photons are created (or annihilated) in pairs – the photon number difference stays the same. Note that the total photon number $\hat{a}_1^\dagger \hat{a}_1 + \hat{a}_2^\dagger \hat{a}_2$ is not conserved, because energy is exchanged with the pump beam.

The general structure of a matrix with the conditions (7.3) is

$A =$

$$e^{i\Lambda/2} \begin{pmatrix} e^{i\Phi/2} & 0 \\ 0 & e^{-i\Phi/2} \end{pmatrix} \begin{pmatrix} \cosh(\Theta/2) & \sinh(\Theta/2) \\ \sinh(\Theta/2) & \cosh(\Theta/2) \end{pmatrix} \begin{pmatrix} e^{i\Psi/2} & 0 \\ 0 & e^{-i\Psi/2} \end{pmatrix}.$$

$$(7.5)$$

The parametric amplifier can be thought of as effectively acting in three steps: a phase shift, followed by purely real amplification and finally by another phase shift. As in the case of the beam splitter, we can incorporate the phases in the definitions of the modes, but the real amplification would remain. Instead of an ordinary rotation, the real amplification matrix describes a hyperbolic rotation. We discuss in Section 8.1.1 that hyperbolic transformations in space–time are Lorentz transformations. Here, of course, the modes have nothing to do with relativistic space–time coordinates. In the case of real mode transformations we obtain for the quadratures

$$\begin{pmatrix} \hat{q}_1' \\ \hat{q}_2' \end{pmatrix} = \begin{pmatrix} \cosh(\Theta/2) & \sinh(\Theta/2) \\ \sinh(\Theta/2) & \cosh(\Theta/2) \end{pmatrix} \begin{pmatrix} \hat{q}_1 \\ \hat{q}_2 \end{pmatrix},$$

$$\begin{pmatrix} \hat{p}_1' \\ \hat{p}_2' \end{pmatrix} = \begin{pmatrix} \cosh(\Theta/2) & -\sinh(\Theta/2) \\ -\sinh(\Theta/2) & \cosh(\Theta/2) \end{pmatrix} \begin{pmatrix} \hat{p}_1 \\ \hat{p}_2 \end{pmatrix}.$$

$$(7.6)$$

As we have already seen in the case of linear amplification discussed in Section 6.2.3, position and momentum operators are transformed differently. Let us write the hyperbolic rotation matrix as

$$\begin{pmatrix} \cosh(\Theta/2) & \sinh(\Theta/2) \\ \sinh(\Theta/2) & \cosh(\Theta/2) \end{pmatrix} = R^{-1} \begin{pmatrix} \exp(\Theta/2) & 0 \\ 0 & \exp(-\Theta/2) \end{pmatrix} R, \quad (7.7)$$

$$R = \frac{1}{\sqrt{2}} \begin{pmatrix} 1 & 1 \\ -1 & 1 \end{pmatrix}. \quad (7.8)$$

The matrix R describes a rotation by $\pi/4$ (that is 45°). The hyperbolic rotation, or Lorentz transformation, thus amounts to a re-scaling along the rotated axis and inverse re-scaling along the other, by stretching and squeezing along the diagonals, see Fig. 7.3. We define the rotated modes

$$\begin{pmatrix} \hat{a}_{1r} \\ \hat{a}_{2r} \end{pmatrix} = R \begin{pmatrix} \hat{a}_1 \\ \hat{a}_2 \end{pmatrix}, \quad \begin{pmatrix} \hat{a}'_{1r} \\ \hat{a}'_{2r} \end{pmatrix} = R \begin{pmatrix} \hat{a}'_1 \\ \hat{a}'_2 \end{pmatrix} \quad (7.9)$$

and obtain from the quadrature transformations (7.6)

$$\begin{aligned} \hat{q}'_{1r} &= \hat{q}_{1r}e^{+\Theta/2}, & \hat{q}'_{2r} &= \hat{q}_{2r}e^{-\Theta/2}, \\ \hat{p}'_{1r} &= \hat{p}_{1r}e^{-\Theta/2}, & \hat{p}'_{2r} &= \hat{p}_{2r}e^{+\Theta/2}. \end{aligned} \quad (7.10)$$

The non-degenerate parametric amplifier thus behaves like two degenerate amplifiers acting in unison, like a pair of squeezers (3.87). One squeezer reduces the quadrature of the rotated mode, including its quantum noise, while the other amplifies the quadrature. In practice (Takei et al., 2006), high-quality non-degenerate parametric amplifiers are actually made of two squeezers. In the limit of strong amplification, $\Theta \to \infty$,

$$\hat{q}_1 - \hat{q}_2 \to 0, \quad \hat{p}_1 + \hat{p}_2 \to 0, \quad (7.11)$$

the quadratures are strongly correlated. If one measures the q quadrature on one beam, the q quadrature of the other must be exactly the same, if one measures p on one beam, the momentum quadrature of the other is exactly $-p$.

Einstein, Poldolski and Rosen (Einstein et al., 1935) perceived such correlations as posing a paradox in quantum mechanics. Imagine, as they did, two point particles that are as correlated as our quadratures (7.11). Now, if the position of a quantum-mechanical particle is certain its momentum is completely undetermined and vice versa. Position and momentum are mutually exclusive elements of reality. Suppose that an observer decides to measure the position of particle 1 and so forces particle 2 to materialize in a position state. On the other hand, if the observer decides to measure the momentum of particle 1, the second particle must reveal itself in a momentum state. The two states are incompatible as elements of reality, but are brought into existence at the pleasure of the observer by some "spooky action at a distance", in Einstein's words. One may disagree with seeing the Einstein–Podolski–Rosen paradox as a paradox, but it has inspired the discovery of profoundly paradoxical aspects of quantum mechanics, for example the Bell inequalities (Bell, 1964, 1987) that we address in Section 7.2. Schrödinger wrote that Einstein, Poldolski and Rosen's paper (Einstein et al., 1935) directly inspired the article in which he coined the term entanglement.

7.1.2 Schrödinger picture

In the Heisenberg picture, the mode operators are transformed, whereas in the Schrödinger picture we transform the quantum state. To find the Schrödinger-picture transformation for the parametric amplifier, we proceed in the same way as in the case of the beam splitter: we construct the evolution operator, here denoted by $\hat{U}_A$. The evolution operator should transform the mode operator as

$$\begin{pmatrix} \hat{a}_1' \\ \hat{a}_2'^{\dagger} \end{pmatrix} = \hat{U}_A \begin{pmatrix} \hat{a}_1 \\ \hat{a}_2^{\dagger} \end{pmatrix} \hat{U}_A^{\dagger}. \tag{7.12}$$

Similar to the Jordan–Schwinger representation (5.14) we write down the operators

$$\begin{aligned}
\hat{K}_t &= \frac{1}{2} \left(\hat{a}_1^{\dagger}, \hat{a}_2 \right) \mathbb{1}_2 \begin{pmatrix} \hat{a}_1 \\ \hat{a}_2^{\dagger} \end{pmatrix} = \frac{1}{2} \left(\hat{a}_1^{\dagger} \hat{a}_1 + \hat{a}_2 \hat{a}_2^{\dagger} \right), \\
\hat{K}_x &= \frac{1}{2} \left(\hat{a}_1^{\dagger}, \hat{a}_2 \right) \sigma_x \begin{pmatrix} \hat{a}_1 \\ \hat{a}_2^{\dagger} \end{pmatrix} = \frac{1}{2} \left(\hat{a}_1 \hat{a}_2 + \hat{a}_1^{\dagger} \hat{a}_2^{\dagger} \right), \\
\hat{K}_y &= \frac{1}{2} \left(\hat{a}_1^{\dagger}, \hat{a}_2 \right) \sigma_y \begin{pmatrix} \hat{a}_1 \\ \hat{a}_2^{\dagger} \end{pmatrix} = \frac{i}{2} \left(\hat{a}_1 \hat{a}_2 - \hat{a}_1^{\dagger} \hat{a}_2^{\dagger} \right), \\
\hat{K}_z &= \frac{1}{2} \left(\hat{a}_1^{\dagger}, \hat{a}_2 \right) \sigma_z \begin{pmatrix} \hat{a}_1 \\ \hat{a}_2^{\dagger} \end{pmatrix} = \frac{1}{2} \left(\hat{a}_1^{\dagger} \hat{a}_1 - \hat{a}_2 \hat{a}_2^{\dagger} \right)
\end{aligned} \tag{7.13}$$

in terms of the Pauli matrices (5.15). All the $\hat{K}$ operators are Hermitian. We see that $\hat{K}_x$ and $\hat{K}_y$ correspond to the Hamiltonian of parametric amplification for two cases of the pump phase. For $\hat{K}_x$ the pump mode amplitude is real and for $\hat{K}_y$ it is purely imaginary. The $\hat{K}_t$ and $\hat{K}_z$ correspond to the Jordan–Schwinger operators $\hat{L}_t$ and $\hat{L}_z$, apart from unimportant additional terms of $1/2$. They describe, essentially, the photon sum and difference of the two modes. One verifies the commutation relations

$$\left[\hat{K}_x, \hat{K}_y \right] = -i\hat{K}_t, \quad \left[\hat{K}_y, \hat{K}_t \right] = i\hat{K}_x, \quad \left[\hat{K}_t, \hat{K}_x \right] = i\hat{K}_y, \tag{7.14}$$

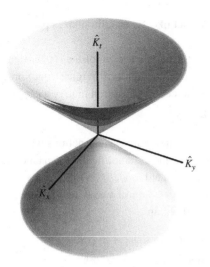

Fig. 7.2. Quantum hyperboloid. The parametric amplifier operates on the surfaces of the hyperboloid sheets given by relation (7.15). The picture shows the first four such hyperboloids.

while $\hat{K}_z$ commutes with the other operators. Hamiltonians that contain the $\hat{K}$ operators would then keep $\hat{K}_z$ invariant, they would conserve the photon number difference (7.4). One also finds

$$\hat{K}_t^2 - \hat{K}_x^2 - \hat{K}_y^2 = \hat{K}_z\left(\hat{K}_z + 1\right), \tag{7.15}$$

the equation of a hyperboloid that plays the role of the Poincaré or Bloch sphere for visualizing the mode transformation in parametric amplification, see Fig. 7.2. The hyperboloid represents a 2+1 dimensional Minkowski space–time with "time" $\hat{K}_t$ and "space" coordinates $\hat{K}_x$ and $\hat{K}_y$. We have seen already that the mode transformations by the parametric amplifier are formal Lorentz transformations: the $\hat{K}_t, \hat{K}_x, \hat{K}_y$ span their Minkowski space. One shows by differentiation and the application of the Bose commutation relations (5.4) that

$$\exp\left(-\mathrm{i}\Theta\hat{K}_y\right)\begin{pmatrix}\hat{a}_1 \\ \hat{a}_2^\dagger\end{pmatrix}\exp\left(\mathrm{i}\Theta\hat{K}_y\right) = \begin{pmatrix}\cosh(\Theta/2) & \sinh(\Theta/2) \\ \sinh(\Theta/2) & \cosh(\Theta/2)\end{pmatrix}\begin{pmatrix}\hat{a}_1 \\ \hat{a}_2^\dagger\end{pmatrix},$$

$$\exp\left(-\mathrm{i}\Lambda\hat{K}_z\right)\begin{pmatrix}\hat{a}_1 \\ \hat{a}_2^\dagger\end{pmatrix}\exp\left(\mathrm{i}\Lambda\hat{K}_z\right) = \exp\left(\mathrm{i}\Lambda/2\right)\begin{pmatrix}\hat{a}_1 \\ \hat{a}_2^\dagger\end{pmatrix}, \tag{7.16}$$

$$\exp\left(-\mathrm{i}\Phi\hat{K}_t\right)\begin{pmatrix}\hat{a}_1 \\ \hat{a}_2^\dagger\end{pmatrix}\exp\left(\mathrm{i}\Phi\hat{K}_t\right) = \begin{pmatrix}\mathrm{e}^{\mathrm{i}\Phi/2} & 0 \\ 0 & \mathrm{e}^{-\mathrm{i}\Phi/2}\end{pmatrix}\begin{pmatrix}\hat{a}_1 \\ \hat{a}_2^\dagger\end{pmatrix}.$$

Consequently, we can express the mode transformation (7.1) with the structure (7.5) in terms (7.12) of the evolution operator

$$\hat{U}_A = \exp\left(-\mathrm{i}\Lambda\hat{K}_z\right)\exp\left(-\mathrm{i}\Phi\hat{K}_t\right)\exp\left(-\mathrm{i}\Theta\hat{K}_y\right)\exp\left(-\mathrm{i}\Psi\hat{K}_t\right). \tag{7.17}$$

We see that the $\hat{K}$ operators indeed act like the Hamiltonian in parametric amplification.

The Schrödinger picture of the parametric amplifier is particularly simple for quadrature wave functions and for Wigner functions. Consider the characteristic case of a real amplification matrix with

$$\Theta = 2\zeta. \tag{7.18}$$

To derive the transformation rules of quadrature wave functions we simply use the representation (7.10) of parametric amplification as a tandem of squeezers acting in unison on the rotated modes (7.9). We apply our previous results for the sqeezing and rotation of wave functions, Eqs. (3.84), (5.29) and (3.85), (5.30), and obtain

$$
\begin{aligned}
\psi'(q_1, q_2) &= \psi(q_1 \cosh\zeta - q_2 \sinh\zeta, -q_1 \sinh\zeta + q_2 \cosh\zeta), \\
\tilde{\psi}'(p_1, p_2) &= \tilde{\psi}(p_1 \cosh\zeta + p_2 \sinh\zeta, p_1 \sinh\zeta + p_2 \cosh\zeta).
\end{aligned}
\tag{7.19}
$$

The quadrature wave functions thus follow the quadrature transformation (7.6). One easily verifies that the Wigner function of the two modes is transformed in the same way. Figure 7.3 shows how the hyperbolic rotation (7.19) of the wave functions creates quadrature correlations. Our simple picture also explains why amplification is not a good idea for making macroscopic quantum superpositions, Schrödinger cat states (4.45) from microscopic superpositions with small amplitude, Schrödinger kittens. Linear amplification is equivalent to parametric amplification where we focus on one of the modes, say mode 1, regard mode 2 as the amplification noise and average over it. Figure 7.4 shows the result: the characteristic quantum interference fringes in the momentum distribution of the Schrödinger cat state are rapidly washed out. Amplification entangles the quantum superposition state with the

Fig. 7.3. Creation of quadrature correlations. The parametric amplifier performs a hyperbolic rotation of quadrature wave functions shown in the position representation. An incident vacuum (left) is turned into the Einstein–Poldolski–Rosen state (right), the two-mode squeezed vacuum (5.33) also shown in Fig. 5.5 (in the momentum representation and with opposite squeezing parameter).

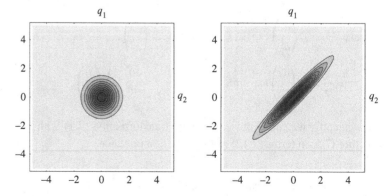

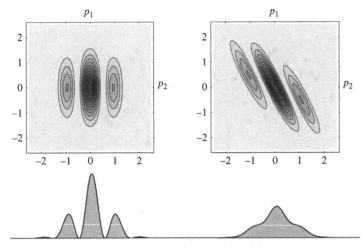

Fig. 7.4. Attempt to amplify a Schrödinger cat. Is it a good idea to amplify a microscopic Schrödinger cat state for turning it into a macroscopic quantum superposition? Left: total momentum quadrature probability distribution of a Schrödinger cat state and the vacuum state of the amplification noise. Right: total momentum distribution of the outgoing light. Below: projections of the momentum distribution on the propagating light mode, on the left for the incident light and on the right for the outgoing light. The figure shows that the hyperbolic rotation of the amplifier erases the tale-telling interference pattern in the momentum distribution of the Schrödinger cat state.

environment: an amplified Schrödinger cat is quantum-mechanically "dead".

7.1.3 Einstein–Podolski–Rosen state

In quantum optics, parametric amplification can start from, literally, nothing. The noise of the quantum vacuum gets the amplifier going. Which quantum state does it make? We know that the parametric amplifier acts like a tandem of squeezers on the rotated modes (7.9). The rotation of two vacuum modes just produces vacuum that is squeezed and anti-squeezed in the rotated modes, but then the interference of the two oppositely squeezed vacuum states generates the *two-mode squeezed vacuum state* (5.33). We see that this quantum state exhibits strong correlations. The quadratures are correlated, because for strong squeezing the wave function vanishes unless $q_1 \sim q_2$. The wave function (5.33) describes the position correlations in the Einstein–Podolski–Rosen paradox (Reid, 1988; Reid and Drummond, 1988; Ou et al., 1992; Takei et al., 2006). We call the corresponding quantum state of light,

$$|\psi\rangle = \exp\left(2\mathrm{i}\zeta \hat{K}_y\right)|0,0\rangle, \tag{7.20}$$

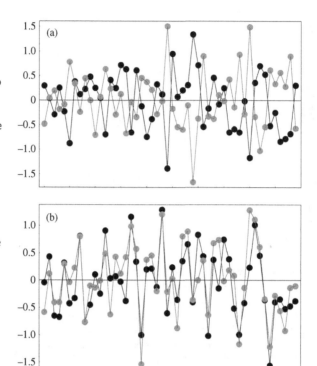

Fig. 7.5. Observed Einstein–Podolski–Rosen correlations (Takei et al., 2006). The figures show quadratures measured on two light modes that are in the Einstein–Podolski–Rosen state (7.20). The quadratures of the two modes are distinguished by black and grey points and are measured one after another in real time on stationary light. For each individual mode, the measurement results are random, but the results of the two modes are related. Part (a) clearly displays the anti-correlations typical for momentum quadratures, whereas part (b) shows the correlations of the position quadratures in good agreement with relation (7.11). (Data: Akira Furusawa and Hidehiro Yonezawa.)

the *Einstein–Podolski–Rosen state*. Figure 7.5 shows the correlations and anti-correlations of measured quadratures for an Einstein–Podolski–Rosen state (Takei et al., 2006) in agreement with Eq. (7.11); Fig. 7.6 displays the sampled quadrature values. The distribution of these values approaches the theoretical quadrature distribution, the modulus squared of the wave function (5.33).

In addition to the quadrature correlations, we expect that the photons of the two beams are strongly correlated. In particular, since initially the modes have been in their vacuum states, after pair production each photon in mode 1 must be accompanied by a partner photon in mode 2, and vice versa. We prove that the state (7.20) has the Fock representation

$$|\psi\rangle = \frac{1}{\cosh \zeta} \sum_{n=0}^{\infty} (\tanh \zeta)^n |n, n\rangle. \tag{7.21}$$

For our proof it is sufficient to show that the Fock expansion (7.21) obeys the same first-order differential equation with respect to ζ as the Einstein–Podolski–Rosen state (7.20), because both representations already agree for the initial value $\zeta = 0$. We perform the following calculation:

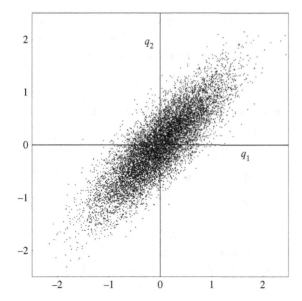

Fig. 7.6. Sampled Einstein–Podolski–Rosen correlations (Takei et al., 2006). The figure displays the data points of quadrature measurements on two light modes in the Einstein–Podolski–Rosen state (7.20), clearly showing correlations similar to Fig. 7.3. (Data: Akira Furusawa and Hidehiro Yonezawa.)

$$2i\hat{K}_y\,|\psi\rangle = (\hat{a}_1^\dagger\hat{a}_2^\dagger - \hat{a}_1\hat{a}_2)\sum_{n=0}^{\infty}\frac{(\tanh\zeta)^n}{\cosh\zeta}\,|n,n\rangle$$

$$= \sum_{n=0}^{\infty}\frac{(\tanh\zeta)^n}{\cosh\zeta}\Big((n+1)|n+1,n+1\rangle - n\,|n-1,n-1\rangle\Big)$$

$$= \sum_{n'=0}^{\infty}\frac{(\tanh\zeta)^{n'-1}}{\cosh\zeta}\Big(n' - (n'+1)\tanh^2\zeta\Big)|n',n'\rangle$$

$$= \frac{\partial}{\partial\zeta}\sum_{n'=0}^{\infty}\frac{(\tanh\zeta)^{n'}}{\cosh\zeta}\,|n',n'\rangle, \tag{7.22}$$

which gives $\partial|\psi\rangle/\partial\zeta$ for the Fock expansion (7.21), the same as for the Einstein–Podolski–Rosen state (7.20) if we apply the operator $2i\hat{K}_y$. Consequently, both formulae describe the same state.

The Einstein–Podolski–Rosen state turns out to be the strongest entangled state of two light modes with given total energy (Barnett and Phoenix, 1989, 1991). But how do we quantify entanglement? How do we measure the entanglement strength? Let us borrow a lemma from linear algebra, the *Schmidt decomposition* (Perez, 1995), a re-statement of the singular value decomposition (Strang, 2005). Consider a pure state $|\psi\rangle$ on the product Hilbert space of the two subsystems, modes 1 and 2 in our case. According to Schmidt, one can always express $|\psi\rangle$ as the series

$$|\psi\rangle = \sum_n c_n\,|u_n\rangle_1|v_n\rangle_2 \tag{7.23}$$

in terms of the real, non-negative coefficients c_n and the orthogonal sets of vectors $\{|u_n\rangle_1\}$ and $\{|v_n\rangle_2\}$. In our case, the Schmidt vectors are the Fock states $|n\rangle_1$ and $|n\rangle_2$, and the Schmidt coefficients are

$$c_n = \frac{(\tanh \zeta)^n}{\cosh \zeta}. \tag{7.24}$$

Consider the quantum states $\hat{\rho}_1$ and $\hat{\rho}_2$ of the two entangled subsystems, the reduced density matrices

$$\hat{\rho}_1 = \mathrm{tr}_2\{|\psi\rangle\langle\psi|\}, \quad \hat{\rho}_2 = \mathrm{tr}_1\{|\psi\rangle\langle\psi|\}. \tag{7.25}$$

We get from the Schmidt decomposition (7.23)

$$\hat{\rho}_1 = \sum_n c_n^2 |u_n\rangle\langle u_n|, \quad \hat{\rho}_2 = \sum_n c_n^2 |v_n\rangle\langle v_n|, \tag{7.26}$$

which suggests a physical interpretation for the Schmidt vectors and coefficients: for each subsystem the Schmidt vectors are the state vectors that occur with probabilities c_n^2 in a statistical ensemble. Moreover, since the Schmidt vectors are orthogonal they describe the eigenvectors of the density matrices $\hat{\rho}_1$ and $\hat{\rho}_2$. The probability for the state $|u_n\rangle$ of mode 1 is exactly the same as the probability for the state $|v_n\rangle$ of mode 2, because the two states are correlated. From this feature it also follows that the entropies (1.21) of the subsystems are the same,

$$S_1 = -k_B \mathrm{tr}\{\hat{\rho}_1 \ln \hat{\rho}_1\} = -k_B \sum_n c_n^2 \ln c_n^2 = -k_B \mathrm{tr}\{\hat{\rho}_2 \ln \hat{\rho}_2\} = S_2. \tag{7.27}$$

The entropy of the total state $|\psi\rangle$ is zero though, because $|\psi\rangle$ is a pure state. The entangled state is pure, but the subsystems are in mixed states. The more disorganized the subsystems are, the stronger are they correlated, the stronger is their entanglement. So we are inclined to adopt the entropy of the two subsystems as a thermodynamic or information-theoretical measure of entanglement. It seems appropriate to use this thermodynamic language, because entanglement serves as a valuable resource like energy or entropy in some "steam engines of the twenty-first century", in quantum information processors.

Coming back to the Einstein–Podolski–Rosen state, we notice that the Schmidt coefficients (7.24) are identical to the photon number probabilities of light in a thermal state,

$$p_n = c_n^2 = (1 - e^{-\beta}) e^{-n\beta} \tag{7.28}$$

with temperature T corresponding to

$$\tanh^2 \zeta = e^{-\beta}, \quad \beta = \frac{\hbar\omega}{k_B T} \tag{7.29}$$

and the average photon number (3.50)

$$\bar{n} = \sinh^2 \zeta. \tag{7.30}$$

We have seen Eq. (7.30) before, in Section 6.2.3, in the discussion of a linear amplifier running idle. Also in this case the amplifier generates thermal light with average photon number (6.49) from vacuum fluctuations. Both cases do agree, because we can model the linear amplifier as an effective parametric amplifier (provided we restrict our attention to only one of the parametrically amplified modes and regard the other one as representing the amplification noise). Now, by definition, the thermal state is the most entropic state for fixed average energy. Consequently, the Einstein–Podolski–Rosen state is the strongest entangled state for a given mean photon number (Barnett and Phoenix, 1989, 1991).

The parametric amplifier creates the series (7.21) of photon pairs from the vacuum state. In the limit of low intensity, called the regime of *spontaneous fluorescence*, only one photon pair is made at a time, because here, approximating $\tanh \zeta$ by ζ and $\cosh \zeta$ by 1, we get

$$|\psi\rangle \approx |0\rangle_1 |0\rangle_2 + \zeta |1\rangle_1 |1\rangle_2. \tag{7.31}$$

If we perform measurements that only give results when one of the beams contains a photon, for example if we detect photons in one of the beams or keep track of the total photon number, we post-select the non-vacuum part of the quantum state (7.31). Effectively, we are dealing with the state

$$|\psi\rangle = |1\rangle_1 |1\rangle_2 \tag{7.32}$$

that occurs with probability ζ^2. We get two photons, one in each mode, a non-entangled pair. Such photon pairs have been applied in a large variety of fundamental experiments where the parametric amplifier has served as the quantum workhorse of choice. Examples are tests of wave–particle duality (Hong et al., 1987; Brendel et al., 1988), quantum erasers (Zou et al., 1991; Kim et al., 2000), discrete quantum teleportation (Bouwmeester et al., 1997; Boschi et al., 1998; Kim et al., 2001) and the tests of the non-locality of quantum mechanics that we discuss in Section 7.2. Photon pairs have also been used to produce single photons (Lvovsky et al., 2001). One of the photons heralds the arrival of the other – when the first photon is detected its partner can be applied elsewhere.

7.1.4 Quantum teleportation

Quantum states comprise the wealth of possibilities in an object. Could we replicate a quantum state, could we clone it? The answer is a clear

No, because cloning would seriously undermine the laws of quantum mechanics. For example, if cloning were allowed we could measure the position of a particle on one clone and the momentum on the other or, by multiple cloning, we could observe mixtures of position and momentum and from them reconstruct the quantum state. If quantum cloning were possible we could observe the quantum state *in vivo*, which is prohibited. Here is a brief formal proof of the *no cloning theorem* (Dieks, 1982; Wootters and Zurek, 1982). Suppose we could clone any quantum state $|\psi\rangle$ by some operator $\hat{X}$. We start from $|\psi\rangle$ and an auxiliary state $|\varphi\rangle$, and get the clones

$$\hat{X}|\psi\rangle|\varphi\rangle = |\psi\rangle|\psi\rangle. \qquad (7.33)$$

Now imagine we replace $|\psi\rangle$ by the quantum superposition

$$|\psi\rangle = c_1|\psi_1\rangle + c_2|\psi_2\rangle. \qquad (7.34)$$

The $|\psi_1\rangle$ and $|\psi_2\rangle$ are supposed to be cloned, and so we obtain

$$\hat{X}|\psi\rangle|\varphi\rangle = c_1|\psi_1\rangle|\psi_1\rangle + c_2|\psi_2\rangle|\psi_2\rangle \neq |\psi\rangle|\psi\rangle. \qquad (7.35)$$

Cloning violates the superposition principle of quantum mechanics. So we cannot clone a quantum state, but can we move the state from one place to another, can we teleport a quantum state (Bennett et al., 1993)?

Figure 7.7 shows the blueprint of a quantum teleportation device, a continuous variable teleporter (Braunstein and Kimble, 1998). The device teleports the quantum state of one light mode to the state of another. Note that such teleportations have been performed in practice (Furusawa et al., 1998). At the heart of the device lies the generation of a strongly entangled state, the Einstein–Podolski–Rosen state (7.20), by parametric amplification. One of the entangled modes "beams up" the quantum state $|\psi\rangle$ while the other is directed to a set-up that closely resembles an eight-port homodyne detector. Here the mode interferes at a 50:50 beam splitter with the state $|\psi\rangle$ that is to be teleported. On one of the interfered modes the quadrature q is measured, on the other p. The measured q and p quadratures are proportional to

$$q = q_{in} - q_2, \quad p = p_{in} + p_2, \qquad (7.36)$$

where q_{in} and p_{in} refer to the quadratures of the quantum state awaiting teleportation, while the q_2 and p_2 quadratures belong to mode 2 of the Einstein–Podolski–Rosen state. In the limit of infinitely strong correlation (infinitely strong parametric amplification) the quadratures of the first mode faithfully follow the quadratures of the second mode, apart from a convenient sign change in the momentum quadrature,

$$q_1 \sim q_2, \quad p_1 \sim -p_2. \qquad (7.37)$$

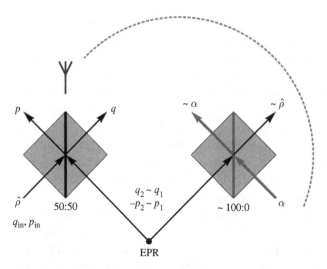

Fig. 7.7. Quantum teleportation device. The diagram shows the blueprint of continuous variable quantum teleportation (Braunstein and Kimble, 1998). The device transports the quantum state $\hat{\rho}$ of a single light mode from the teleport on the left to the teleport on the right. The two teleports are connected by an Einstein–Podolski–Rosen state (7.20) (EPR). The light to be teleported is split by the balanced beam splitter on the left, whereupon the q and p quadratures are measured (as a simultaneous measurement of position and momentum like in Figs. 5.14 and 5.15). One mode of the Einstein–Podolski–Rosen state, however, enters the second input of this beam splitter. The second mode of this state bridges the distance to the second teleport on the right, because the quadratures of the two modes are correlated. At the second teleport, the light interferes with a strong coherent state at a highly transmissive beam splitter, displacing its amplitude in a controlled way. The required displacement is given by the measurement result at the first teleport. This piece of classical information is communicated by a classical channel, indicated by the aerial, to the second teleport. So, in a nutshell, the classical channel communicates the classical amplitude, whereas the quantum channel, the Einstein–Podolski–Rosen state, transfers the quantum noise.

Consequently,

$$q_1 \sim q_{in} - q, \quad p_1 \sim p_{in} - p. \tag{7.38}$$

Apart from a displacement by q and p, the quadratures of mode 1 are exact replicas of the quadratures of the initial state $|\psi\rangle$, the "beamed-up" Wigner function is the displaced Wigner function of the original state. But the displacement is known in every single run of the experiment, because q and p are measured. We could communicate the actual value of q and p to the teleport on the other side. There the displacement is removed by a device we describe below. The original state, of course, is gone. In quantum teleportation, two parts of space are bridged by an entangled state and the state to be teleported interacts with one of the entangled modes. Measurements are made at the first teleport for re-adjusting the quantum state on the second teleport if the measurement results are communicated there. Our example of teleportation via quadrature measurements was inspired by the discrete quantum teleportation of spins or polarization states. Quantum teleportation is a fantastic piece of physics, but it seems limited to small systems. Just imagine we attempt to teleport a macroscopic object of, say, 10^{21} atoms. How many measurements do we need to perform and how much information do we need to send?! It would be unrealistic to assume that the teleportations of all the atoms are perfect. So who would volunteer to be teleported risking certain death and uncertain resurrection?

Let us return to the quantum teleportation device of Fig. 7.7. We still need to discuss how to implement a displacement in the phase space of a light mode, the displacement operator (3.55). The right side of Fig. 7.7 shows a device that does it. Let the light, with quantum state $|\psi\rangle$, interfere at a highly transmitting beam splitter with a strong coherent beam $|-\alpha_0\rangle$. In this case none of the quantum noise of the coherent light gets through to the state $|\psi\rangle$, but only the coherent amplitude – the quantum state is displaced. Consider the beam splitter (5.3) with matrix (5.9) and the coherent state in the limit

$$\tau \to 1, \quad \alpha_0 \varrho = \alpha_0 \sqrt{1 - \tau^2} \to \alpha \tag{7.39}$$

for finite α that we write as $\alpha = (q + ip)/\sqrt{2}$. We show that the state of the transmitted mode is displaced by α. The Wigner function of the transmitted state is given by

$$W'(q_1, p_2) = \int_{-\infty}^{+\infty} \int_{-\infty}^{+\infty} W(q_1', p_1') \, W_{-\alpha_0}(q_2', p_2') \, dq_2 \, dp_2 \tag{7.40}$$

where we integrate over the second outgoing mode in terms of the transformed quadratures (5.35). We take q_1' and p_1' as integration variables with $dq_1' = \varrho \, dq_2$ and $dp_1' = \varrho \, dp_2$, and, similar to Eq. (5.79), we express q_2' and p_2' as

$$q_2' = \frac{1}{\varrho}(-q_1 + \tau q_1'), \quad p_2' = \frac{1}{\varrho}(-p_1 + \tau p_1'). \tag{7.41}$$

We obtain for the Wigner function (4.44) of the coherent state

$$W_{-\alpha_0}(q_2, p_2) = \frac{1}{\pi} \exp\left(-\frac{(\tau q_1' - q_1 - q)^2}{\varrho^2} - \frac{(\tau p_1' - p_1 - p)^2}{\varrho^2}\right)$$
$$\to \varrho^2 \, \delta(q_1 + q - q_1') \, \delta(p_1 + p - p_1'). \tag{7.42}$$

Consequently, the Wigner function of the outgoing light is the displaced Wigner function (4.43). So the device sketched in the right-hand side of

Fig. 7.7 indeed implements the displacement operator. The quantum teleporter is complete.

7.2 Polarization correlations

Parametric amplification correlates the amplitudes of two light modes, most notably in the Einstein–Podolski–Rosen state (7.20). In 1964 John S. Bell published the article "On the Einstein–Podolski–Rosen paradox" (Bell, 1964, 1987). No other paper ever since has given as much insight into the strangeness of quantum mechanics. Experimental tests of Bell's argument employ polarization correlations that the parametric amplifier can generate as well. But before we discuss Bell's correlations let us begin with a word of caution. Quantum mechanics may cause strange correlations, but correlations by themselves are nothing out of the ordinary, the world is full of correlated events. They may be causally connected; for example event A causes event B or events A and B have a common cause C. So if we observe A we can also see B. According to Arthur C. Clarke "somebody once said the laws of chance do not merely permit coincidences, they compel them". So the world also contains a considerable number of strange and wonderful coincidences. Extraordinary quantum correlations are rather subtle and remain elusive without good instruments, but they do exist. Let us discuss how to create them.

7.2.1 Singlet state

Figure 7.8 shows a more complete picture of the optics involved in parametric amplification than the simple black box of Fig. 7.1. In the regime of spontaneous fluorescence the emitted light emerges in two cones for every set of frequencies ω_1 and ω_2 with $\omega_1 + \omega_2 = \omega_p$. On a photographic plate the cones appear as the intersecting rings shown in Fig. 7.9. The two cones of light are orthogonally polarized, one vertically and the other horizontally. The reason for the separation of the polarizations is the birefringence of the crystal where the two polarizations of light experience different refractive indices. The angles and directions of the cones are set by energy and momentum conservation (called phase matching conditions) where the wave momentum, the wave vector, depends on the refractive indices. The opening angles of the cones also depend on the frequencies. The details are not important here; the only important fact is that the polarized cones of light intersect at two lines. At these lines the light belongs to both the vertically and the horizontally polarized cones. So the light must be unpolarized here.

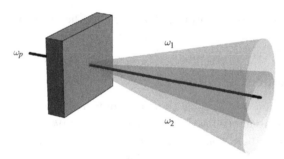

Fig. 7.8. Cones of light in parametric fluorescence. The pump beam with frequency ω_p, indicated in black, creates pairs of cones of light with frequencies ω_1 and ω_2 in the nonlinear optical material of the parametric amplifier. Energy conservation implies that $\omega_1 + \omega_2 = \omega_p$; momentum conservation of the light in the optical material determines the angles of the cones. The two cones of each pair are orthogonally polarized, because the material, a crystal, is birefringent. However, at the lines where the cones intersect the polarization is undecided until measurements are made; here the light is in the unpolarized Einstein–Podolski–Rosen state (7.49).

Fig. 7.9. Rings of entanglement. Negative image of the light emitted in parametric fluorescence. The light propagates in pairs of cones that appear as rings in the detection plate. The angle of the cones depends on the frequencies ω_1 and ω_2 of the emitted light with $\omega_1 + \omega_2 = \omega_p$; for equal frequencies the rings are equal. At the points where the rings of two corresponding frequencies intersect the polarization of light is entangled. (Photo: Paul Kwiat and Michael Reck, ©University of Vienna.)

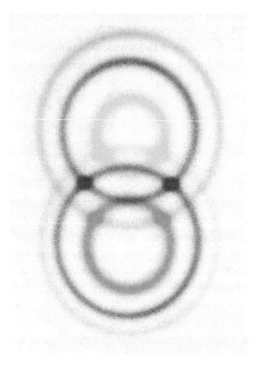

Let us describe the creation of unpolarized light in parametric amplification. We consider only two light beams, the light at the two intersections of the cones, but each light beam has two polarization modes. So four modes are involved that we distinguish by their mode indices 1 and 2 and the polarization $\pm$. Here the label "+" refers to the vertical polarization where, for waves propagating in x direction, the electric field oscillates in y direction and "−" refers to the horizontal polarization in z direction. The Hamiltonian of the parametric amplification should be quadratic in the corresponding annihilation and creation operators $\hat{a}_{k\pm}$ and $\hat{a}_{k\pm}^{\dagger}$. Since the generated light is unpolarized the Hamiltonian must not show preference for any polarization direction. If we change the polarization of both beams the Hamiltonian should remain invariant. To incorporate the polarization invariance, we construct the matrix $\underline{\hat{a}}$ of the mode operators as

$$\underline{\hat{a}} = \begin{pmatrix} \hat{a}_{1+} & \hat{a}_{2+} \\ \hat{a}_{1-} & \hat{a}_{2-} \end{pmatrix}, \tag{7.43}$$

because we can compactly describe polarization changes – unitary transformations (5.3) – for both modes in one matrix multiplication,

$$\underline{\hat{a}}' = B\,\underline{\hat{a}}. \tag{7.44}$$

The matrix structure (5.7) shows that the unitary transformation (5.3) consists of a global phase shift by $\Lambda/2$ and polarization changes. Pure polarization changes correspond to matrices with

$$\det B = 1. \tag{7.45}$$

Which Hamiltonian, which scalar quadratic expression of $\underline{\hat{a}}$, is invariant under polarization changes? The determinant is invariant,

$$\det \underline{\hat{a}} = \hat{a}_{1+}\hat{a}_{2-} - \hat{a}_{1-}\hat{a}_{2+}, \tag{7.46}$$

because $\det \hat{a}' = \det B \det \underline{\hat{a}} = \det \underline{\hat{a}}$. We thus arrive at the Hamiltonian

$$\hat{H} = \mathrm{i}\det \underline{\hat{a}} - \mathrm{i}\det \underline{\hat{a}}^{\dagger} \tag{7.47}$$

where we included the phase factor i for convenience, because in this case we can express $\hat{H}$ as

$$\hat{H} = 2\left(\hat{K}_{y\pm} - \hat{K}_{y\mp}\right) \tag{7.48}$$

where the $\hat{K}_y$ operators (7.13) refer to the modes $\hat{a}_{1+}, \hat{a}_{2-}$ or $\hat{a}_{1-}, \hat{a}_{2+}$, respectively. We could include a different phase factor by phase-shifting the modes, but this would have no physical consequence when the input modes are in their vacuum states. For vacuum input, we obtain the unpolarized Einstein–Podolski–Rosen state

$$|\psi\rangle = \exp\left(-\mathrm{i}\zeta\hat{H}\right)|0\rangle_{1+}|0\rangle_{1-}|0\rangle_{2+}|0\rangle_{2-} = |\psi_+\rangle + |\psi_-\rangle,$$

$$|\psi_\pm\rangle = \frac{1}{\cosh\zeta}\sum_{n=0}^{\infty}(\pm\tanh\zeta)^n |n\rangle_{1\pm}|n\rangle_{2\mp}. \tag{7.49}$$

In the regime of spontaneous fluorescence, for low intensities, the state (7.49) is reduced to

$$|\psi\rangle \approx |0\rangle + \zeta\left(|1\rangle_{1+}|0\rangle_{1-}|0\rangle_{2+}|1\rangle_{2-} - |0\rangle_{1+}|1\rangle_{1-}|1\rangle_{2+}|0\rangle_{2-}\right). \quad (7.50)$$

If we only count the events when photons are present, if we post-select for photons, we are effectively dealing with the state

$$|\psi\rangle = \frac{1}{\sqrt{2}}\left(|1\rangle_{1+}|0\rangle_{1-}|0\rangle_{2+}|1\rangle_{2-} - |0\rangle_{1+}|1\rangle_{1-}|1\rangle_{2+}|0\rangle_{2-}\right)$$

$$= \frac{1}{\sqrt{2}}\left(|+\rangle_1|-\rangle_2 - |-\rangle_1|+\rangle_2\right) \quad (7.51)$$

where we included the normalization factor and abbreviated by $|\pm\rangle_k$ the state when the mode k carries a single photon of polarization $\pm$. The polarization state (7.51) is called the *singlet state*. The singlet state (7.51) clearly is entangled. In fact, it is the most strongly entangled state of two polarizations (or spins or orbits), because the reduced density matrices are maximally disorganized,

$$\hat{\rho}_1 = \frac{1}{2}\mathbb{1}_2 = \hat{\rho}_2, \quad (7.52)$$

describing a mixed state where both polarizations are equally likely. Since, by definition, the singlet state is invariant under polarization changes, we could apply a polarization transformation that would replace the $\pm$ states by new orthogonal polarization states and still get the same singlet state (7.51). For example, we could write down the singlet state in terms of left and right circularly polarized photons in the same form (7.51). Because the singlet state is unpolarized as a whole, the polarizations of the partner photons must be correlated. If we observe a vertically polarized photon in one beam we must get a horizontally polarized photon in the other; if photon 1 is left-circularly polarized, photon 2 must be right-circularly polarized. But before we discuss polarization correlations let us precisely state what we mean by polarization and polarization measurements.

7.2.2 Polarization

It is convenient to use Pauli matrices for describing the polarization of a single photon, because the photon in its two polarizations is completely analogous to a spin. Here the state of "spin up", $(1, 0)$, corresponds to the vertical polarization "+" and "spin down" $(0, 1)$ to the horizontal polarization "−". The density matrix $\hat{\rho}$ of the polarization state is a 2×2 Hermitian matrix. Such a matrix has four independent real components, the two diagonal components and one complex number in the corner of the matrix (the number in the other corner is the complex conjugate). The Pauli matrices and the unity matrix $\mathbb{1}_2$ are linearly independent and Hermitian. Hence we can express any 2×2 Hermitian matrix as a linear combination of $\mathbb{1}_2$ and the Pauli matrices. Density matrices are further constrained though. To begin with, their traces are unity (the total probability for the quantum state). As the traces of the Pauli matrices (5.15) vanish, the component in front of $\mathbb{1}_2$ must be the inverse of the trace of $\mathbb{1}_2$, $1/2$. We write the remaining linear combination of the Pauli matrices as the scalar product of a coefficient vector $\mathbf{s}$ with

$$\boldsymbol{\sigma} = (\sigma_x, \sigma_y, \sigma_z), \tag{7.53}$$

a three-dimensional vector with the Pauli matrices as components, the Pauli vector; we represent the density matrix of a polarization state as

$$\hat{\rho} = \frac{1}{2}\left(\mathbb{1}_2 + \mathbf{s} \cdot \boldsymbol{\sigma}\right). \tag{7.54}$$

The vector $\mathbf{s}$ is called the *Bloch* or *Stokes vector*. To find a physical interpretation for this vector, we use the following mathematical property of the Pauli matrices:[1]

$$\left(\mathbf{a} \cdot \boldsymbol{\sigma}\right)\left(\mathbf{b} \cdot \boldsymbol{\sigma}\right) = \left(\mathbf{a} \cdot \mathbf{b}\right)\mathbb{1}_2 + \mathrm{i}\left(\mathbf{a} \times \mathbf{b}\right) \cdot \boldsymbol{\sigma} \tag{7.55}$$

where $\times$ denotes the vector product. We obtain

$$\mathrm{tr}\left\{\hat{\rho}\left(\mathbf{a} \cdot \boldsymbol{\sigma}\right)\right\} = \mathbf{a} \cdot \mathbf{s}, \tag{7.56}$$

because the trace of $\boldsymbol{\sigma}$ vanishes. So the expectation value $\mathbf{a} \cdot \boldsymbol{\sigma}$ is the projection of $\mathbf{a}$ onto $\mathbf{s}$ and vice versa. From the projection formula (7.56) follows that the expectation values of the Pauli matrices themselves give the components of the Stokes vector; the Stokes vector is the quantum average of the Pauli vector (7.53). The Stokes vector thus expresses the polarization state in terms of observable quantities; it shows how the state can be reconstructed from mutually exclusive measurements. For historical reasons, the *Stokes parameters* (5.18) are the vector components

$$s_0 = \langle\mathbb{1}_2\rangle, \quad s_1 = \langle\sigma_z\rangle, \quad s_2 = \langle\sigma_x\rangle, \quad s_3 = -\langle\sigma_y\rangle. \tag{7.57}$$

We also obtain from the projection property (7.56) that the Stokes vector cannot be arbitrarily large, because we get for the purity

---

[1] This property of the Pauli matrices is related to the multiplication rules of quaternions, hypercomplex numbers with one real and three imaginary parts. One may represent the real part as proportional to $\mathbb{1}_2$ and the imaginary units as $-\boldsymbol{\sigma}$.

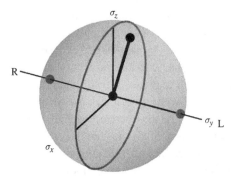

Fig. 7.10. Poincaré sphere (also known as Bloch sphere). The polarization of a single photon is represented by the Stokes vector (7.57), the expectation value of the Pauli vector (7.53), that lies within the unity sphere shown in the picture. The line indicates the propagation direction of light. Pure polarization states correspond to points on the surface of the Poincaré sphere, while partially unpolarized light is represented by points in the interior of the sphere. The centre describes completely unpolarized light. The Stokes vector of linearly polarized light lies on the circle shown in the picture, while circularly polarized light corresponds to the points R and L. The handedness is defined according to the traditional, not the natural nomenclature.

$$\text{tr}\left\{\hat{\rho}^2\right\} = \frac{1}{2}(1 + s^2). \tag{7.58}$$

The Stokes vector must lie within the unity sphere shown in Fig. 7.10, for otherwise the purity would exceed unity. This sphere is called the *Poincaré sphere* in optics (it is called the *Bloch sphere* for spins). Pure polarization states lie on the surface of the Poincaré sphere, mixed states gravitate towards the centre occupied by the completely unpolarized state (7.52). A right circularly polarized photon (Born and Wolf, 1999) $(|+\rangle - i|-\rangle)/\sqrt{2}$ corresponds to the point R in Fig. 7.10, while a left-circularly polarized photon (Born and Wolf, 1999) $(|+\rangle + i|+\rangle)/\sqrt{2}$ appears as the point L on the Poincaré sphere. (For polarization, left and right is defined such that the electric field vector of right polarized light rotates clockwise, while left polarized light rotates counter clockwise. This is right-handed according to the traditional, not the natural nomenclature (Born and Wolf, 1999).) Linearly polarized photons lie on the great circle of the Poincaré sphere shown in Fig. 7.10. Let us represent the Stokes vector as

$$\mathbf{s} = \left(-s\sin\Theta\cos\Phi, \, s\sin\Theta\sin\Phi, \, s\cos\Theta\right). \tag{7.59}$$

One verifies, using the explicit form (5.15) of the Pauli matrices, that the density matrix (7.54) has the eigenvalues

$$\rho_\pm = \frac{1}{2}(1 \pm s) \qquad (7.60)$$

for the eigenvectors

$$
\begin{aligned}
|\varphi_+\rangle &= e^{i\Phi/2}\cos(\Theta/2)\,|+\rangle \;-\; e^{-i\Phi/2}\sin(\Theta/2)\,|-\rangle,\\
|\varphi_-\rangle &= e^{i\Phi/2}\sin(\Theta/2)\,|+\rangle \;+\; e^{-i\Phi/2}\cos(\Theta/2)\,|-\rangle.
\end{aligned}
\qquad (7.61)
$$

Our result (7.60) shows that $\hat\rho$ describes a physically meaningful density matrix, with eigenvalues between 0 and 1, for all Stokes vectors within the Poincaré sphere,

$$s \le 1. \qquad (7.62)$$

Let us give an interpretation for the eigenvectors. We read the $|+\rangle$ state of the single photon (horizontal polarization) as the two-mode Fock state $|1,0\rangle$ and the $|-\rangle$ state (vertical polarization) as $|0,1\rangle$. Comparing the eigenstates (7.61) with the beam splitter transformation (5.43) of single-photon Fock states,

$$|1,0\rangle' = B_{11}|1,0\rangle + B_{21}|0,1\rangle, \quad |0,1\rangle' = B_{12}|1,0\rangle + B_{22}|0,1\rangle, \qquad (7.63)$$

taking into account the matrix structure (5.8), we see that the $|\varphi_\pm\rangle$ states are simply polarization-transformed $|\pm\rangle$ states. The polarization state (7.54) of a single photon thus is a statistical mixture of the pure state $|\varphi_+\rangle$ with Stokes vector (7.59) and the orthogonal polarization state $|\varphi_-\rangle$ (with negative Stokes vector). The effective length s of the Stokes vector for the mixed polarization state (7.54) is the difference between the probabilities (7.60) for the eigenstates $|\varphi_\pm\rangle$.

Consider a polarization measurement. We can model the apparatus as a polarizer, a device that changes the polarization, followed by detectors. For example, we could turn circular polarizations into linear polarizations, separate the two orthogonally polarized components by a polarizing beam splitter and send them to two photodetectors, one for each polarization. When one of the detectors clicks the light has been left-circularly polarized, when the other clicks the polarization has been right-circular. Let us assign the value $+1$ to the first detector and -1 to the second one. If the photon has been in an eigenstate of the measured polarization the expectation value is either $+1$ or -1. We see from the projection formula (7.56) that the operator $\mathbf{a}\cdot\boldsymbol{\sigma}$ has precisely this property for $\mathbf{s} = \pm\mathbf{a}$ and

$$|\mathbf{a}| = 1. \qquad (7.64)$$

Therefore we regard $\mathbf{a}\cdot\boldsymbol{\sigma}$ as the operator of a polarization measurement that we characterize by the Stokes vector $\mathbf{a}$ of the corresponding "+" eigenstate on the Poincaré sphere. For two set-ups with $\pm\mathbf{a}$ the polarization eigenstates are simply interchanged, $-\mathbf{a}$ measures the orthogonal polarization of $+\mathbf{a}$, but the devices are essentially the same. However, for two set-ups with orthogonal Stokes vectors $\mathbf{a}$ and $\mathbf{b}$ the expectation value of $\mathbf{a} \cdot \boldsymbol{\sigma}$ vanishes for the eigenstate of the orthogonal $\mathbf{b}$ polarization and vice versa. This means that the outcomes ± 1 of the $\mathbf{b}$ measurements on the $\mathbf{a}$ states have been completely random. So, for orthogonal Stokes vectors $\mathbf{a}$ and $\mathbf{b}$, an eigenstate of the $\mathbf{a}$ polarization does not contain any information about the $\mathbf{b}$ polarization and vice versa. Or, formulated positively, measuring the $\mathbf{b}$ polarization on an ensemble of equally prepared photons reveals aspects the $\mathbf{a}$ polarization does not contain. Polarization measurements for orthogonal Stokes vectors are complementary.

7.2.3 Bell's theorem

Let us consider an individual single photon. If we make a polarization measurement we ask the photon to reveal itself as either "+" or "−" polarized for a given polarizer setting of our choice. We bluntly enforce a decision, "+" or "−", but the photon may have been ambivalent about its polarization before. Surely we get an answer, but the measurement result does not tell whether the photon really has been $\mathbf{a}$ polarized. It may have been in any polarization state, apart from the eigenstate of the polarization orthogonal to the measured one. The measured polarization of the individual single photon is random. Only after repeated measurements on many identical photons do we notice that the polarization is $\mathbf{a} \cdot \mathbf{s}$ on average (for an ensemble of photons with Stokes vector $\mathbf{s}$). But even in quantum mechanics, some things are absolutely certain. Suppose we create two photons in the singlet state (7.51). If we measure the $\mathbf{a}$ polarization on the first photon with result "$\pm$" then the second photon must be orthogonally polarized, we get the measurement result "$\mp$" with certainty. We could perform any polarization measurements with synchronized settings and always obtain the second photon in the orthogonal polarization of the first one. We could even change the polarizer settings while the photons are still on their way. But these strict polarization correlations are not surprising, because the photons have been created in a state of zero total polarization. The polarizations of the two photons are correlated, because they come from a common source.

Now imagine we turn the second polarizer to a different setting, to the direction $\mathbf{b}$ on the Poincaré sphere, as Fig. 7.11 shows. In this case, the measured polarization of the second photon will not be completely related to the result of the first photon. For example, if the settings are complementary to each other, with orthogonal Stokes vectors, we do not expect to see any correlations at all. Let us quantify the degree of correlation between the results of the two polarization measurements.

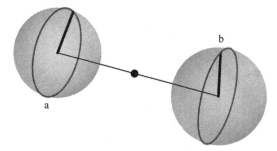

Fig. 7.11. Polarization correlations. The polarization of two photons in the singlet state (7.51) is measured with two polarizer settings characterized by the Stokes vectors *a* and *b* of the corresponding polarization eigenstates. When the two polarizers are in the same settings, the two measured polarizations are perfectly anti-correlated, if they are orthogonal to each other the polarizations are completely uncorrelated. What happens in between?

The operator $\mathbf{a} \cdot \boldsymbol{\sigma}_1$ describes the polarization measurement A on the first photon with eigenvalues ± 1, the measurement results. The operator $\mathbf{b} \cdot \boldsymbol{\sigma}_2$ corresponds to the polarization measurement B on the second photon. The eigenvalues of the product $(\mathbf{a} \cdot \boldsymbol{\sigma}_1)(\mathbf{b} \cdot \boldsymbol{\sigma}_2)$ are obviously $+1$ when both polarization measurements agree and -1 when they disagree. On average, the observable $(\mathbf{a} \cdot \boldsymbol{\sigma}_1)(\mathbf{b} \cdot \boldsymbol{\sigma}_2)$ gives the difference between the number of agreements and disagreements divided by the number of runs, a reasonable quantitative measure for the degree of correlation. Therefore we define the correlation function of the polarization measurement as

$$C(\mathbf{a}, \mathbf{b}) = \langle \psi | (\mathbf{a} \cdot \boldsymbol{\sigma}_1)(\mathbf{b} \cdot \boldsymbol{\sigma}_2) | \psi \rangle. \tag{7.65}$$

We use the explicit form of the Pauli matrices (5.15) and obtain for the singlet state (7.51) the result

$$C(\mathbf{a}, \mathbf{b}) = - \mathbf{a} \cdot \mathbf{b}. \tag{7.66}$$

This formula interpolates between the limiting cases we already know. For equal polarizer settings, the measured polarizations of the two photons are perfectly orthogonal. For complementary settings with orthogonal Stokes vectors, the polarizations are completely uncorrelated. Formula (7.66) seems simple and plausible, but this formula harbours an enigma.

In the polarization measurements the results ± 1 are random for each photon pair. The randomness of the measurement process ought to come from somewhere. "God does not play dice", as Einstein said. Imagine that the fate of the photons is governed by some physics beyond our present understanding, some elusive quantities called *hidden variables*.

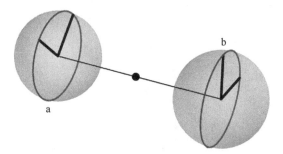

Fig. 7.12. Bell correlations. Polarizer A is set to $\boldsymbol{a} = (0, 1/\sqrt{2}, 1/\sqrt{2})$ and then to the orthogonal setting $\boldsymbol{a}' = (0, -1/\sqrt{2}, 1/\sqrt{2})$ probing the complementary polarization aspects. The Stokes vectors of polarizer B are rotated by $\pi/4$ relative to polarizer A, precisely between perfect anti–correlation and no correlation at all: $\boldsymbol{b} = (0, 0, 1)$ and $\boldsymbol{b}' = (0, 1, 0)$. This setting and similar ones maximally violate Bell's inequality (7.68) for the singlet state (7.51).

We cannot and do not specify what they are, but essentially only assume that they do exist. Now, compare the measurement results A and A' for two settings $\mathbf{a}$ and $\mathbf{a}'$ with the results B and B' for the settings $\mathbf{b}$ and $\mathbf{b}'$ of the second apparatus. For any four numbers with values ± 1 we have

$$(A + A')B + (A - A')B' = \pm 2, \tag{7.67}$$

because either one of $A + A'$ or $A - A'$ vanishes while the other is ± 2. If the randomness of the polarization measurements is caused by the hidden variables the average over many runs of the experiments amounts to the average over the values of the hidden variables. The average of ± 2 is a number between -2 and $+2$, and so we obtain

$$\left| \langle AB \rangle + \langle A'B \rangle + \langle AB' \rangle - \langle A'B' \rangle \right| \le 2. \tag{7.68}$$

On the other hand, an average $\langle AB \rangle$ over many runs is exactly the same as the correlation function $C(\mathbf{a}, \mathbf{b})$. Yet for the polarizer settings shown in Fig. 7.12 we get

$$\left| C(\mathbf{a}, \mathbf{b}) + C(\mathbf{a}', \mathbf{b}) + C(\mathbf{a}, \mathbf{b}') - C(\mathbf{a}', \mathbf{b}') \right| = 2\sqrt{2}. \tag{7.69}$$

How is this possible? In our argument we made in fact two assumptions. One is obvious: hidden variables do exist and cause the apparent randomness in quantum mechanics. The other assumption is more subtle. In asserting that either $(A + A')B$ or $(A - A')B'$ vanishes, we assume that A solely depends on $\mathbf{a}$ and on the hidden variables, but not on the other polarizer settings $\mathbf{b}$ nor on the measurement results B, and the equivalent statement for the other polarization B. In formulae, we regard the correlation as

$$\langle AB \rangle = \int A(\mathbf{a}, \lambda) B(\mathbf{b}, \lambda) \varrho(\lambda) \, d\lambda \tag{7.70}$$

where λ denotes the possibly multi-dimensional hidden variables with the distribution $\varrho(\lambda)$. The two polarizers are assumed to be independent, which seems completely plausible, because they are spatially separated and we could change their setting at will. So unless the detection set-ups act spookily across their distance, there must be a time when a new polarizer setting at one set-up will not have any influence on the other. The hidden variables are said to be *local*, because they act locally, whether they are part of the polarization apparata or are transmitted by the photons. The assumption of local hidden variables as explanation of the randomness of quantum mechanics by some real but elusive physics is called *local realism*. Yet the quantum correlation (7.69) violates the inequality (7.68). Local realism is in conflict with quantum mechanics. This is Bell's theorem (Bell, 1964, 1987). The inequality (7.68), derived by Clauser et al. (1969) is a variation of Bell's original inequality that was slightly less general but equally conclusive.

Experiments are paramount to test whether violations of local realism do indeed occur in reality. After early tests with fixed polarizer settings (Freire Jr, 2006), Alain Aspect lead the first widely convincing experimental tests of Bell inequalities (Aspect et al., 1981, 1982a, 1982b). In particular, Aspect, Dalibard and Roger performed experiments (Aspect et al., 1982b) where the polarizer settings were changed while the photons were still on their way. The pioneering experiments were difficult and heroic (Freire Jr, 2006), partly because parametric amplifiers were not sufficiently well-developed at the time. The experimentalists were forced to use significantly less efficient sources of singlet states of photons. Since then much progress has been made, and the overwhelming experimental evidence points in favour of the prediction (7.69) of quantum mechanics, experimentally refuting local realism. Physics has made a quantitative, empirical contribution on getting to know reality *an sich* (Freire Jr, 2006). Yet what is reality? Is the world completely causal but acting across space, binding everything together? Or is pure chance a fact, not part of our ignorance? Or, for seeing the true picture, are we lacking in imagination?

7.3 Questions

7.1 Describe the parametric amplifier in the Heisenberg picture and explain where this theory comes from.

7.2 Use the Bose commutation relations between the incoming and the outgoing modes in order to show that the amplifier matrix A obeys the relations

$$|A_{11}|^2 - |A_{12}|^2 = |A_{21}|^2 - |A_{22}|^2 = 1, \quad A_{11}^* A_{21} - A_{12}^* A_{22} = 0.$$

7.3 Why is $\hat{a}_1^\dagger \hat{a}_1 - \hat{a}_2^\dagger \hat{a}_2$ a conserved quantity? What is the physical meaning of this conservation law?

7.4 Why is $\hat{a}_1^\dagger \hat{a}_1 + \hat{a}_2^\dagger \hat{a}_2$ not conserved?

7.5 Explain the structure of the amplifier matrix.

7.6 Show that the two-mode parametric amplifier performs a quadrature squeezing and stretching operation in appropriately chosen modes.

7.7 What is the Einstein–Podolski–Rosen paradox?

7.8 Prove the commutation relations for the $\hat{K}_t$, $\hat{K}_x$, $\hat{K}_y$, $\hat{K}_z$ operators.

7.9 Why do the $\hat{K}$ operators lie on hyperboloid sheets?

7.10 How are quadrature wave functions transformed? Why?

7.11 Why are the momentum quadrature wave functions transformed differently than the position quadrature wave functions?

7.12 Use the wave function transformation by the parametric amplifier to show how it generates strong quadrature correlations.

7.13 Use wave functions to visualize the loss of quantum coherence when Schrödinger cat states are amplified.

7.14 What is the Einstein–Podolski–Rosen state?

7.15 Show that, for parametric amplifiers described by a real amplification matrix,

$$\hat{U}_B |0,0\rangle = \frac{1}{\cosh \zeta} \sum_{n=0}^{\infty} (\tanh \zeta)^n |n,n\rangle.$$

7.16 Why is each of the two modes of the Einstein–Podolski–Rosen state in a thermal state (why is the reduced density matrix the density matrix of a thermal state)?

7.17 Explain how to quantify the degree of entanglement of two objects in a pure total state.

7.18 Why is the Einstein–Podolski–Rosen state the most strongly entangled state for a given mean photon number?

7.19 How are photon pairs made? What is heralded downconversion?

7.20 Why is it impossible to clone a quantum state?

7.21 Discuss quantum teleportation.

7.22 Explain how the displacement operator is implemented.

7.23 Deduce the Hamiltonian for unpolarized parametric amplification.

7.24 Show that the unpolarized Einstein–Podolski–Rosen state is

$$|\psi\rangle = |\psi\rangle_+ + |\psi\rangle_-, \quad |\psi\rangle_\pm = \frac{1}{\cosh \zeta} \sum_{n=0}^{\infty} (\pm \tanh \zeta)^n |n\rangle_{1\pm} |n\rangle_{2\mp}.$$

7.25 Consider this state in the limit $\zeta \ll 1$. Why do you get the singlet state $|\psi\rangle = (|+\rangle_1 |-\rangle_2 - |-\rangle_1 |+\rangle_2)/\sqrt{2}$ after post-selection for photons?

7.26 Consider the polarization of a single photon. Why can you express the density matrix of the polarization state as

$$\hat{\rho} = \frac{1}{2}\left(\mathbb{1}_2 + \mathbf{s}\cdot\boldsymbol{\sigma}\right)?$$

7.27 How is the vector $\mathbf{s}$ constrained? Give an interpretation for this vector.

7.28 What are the eigenvalues and eigenstates of the polarization density matrix?

7.29 How can you describe polarization measurements?

7.30 Calculate the polarization correlation $C(\mathbf{a}, \mathbf{b})$ for the singlet state.

7.31 Give an interpretation for the obtained $C(\mathbf{a}, \mathbf{b})$.

7.32 What are local hidden parameters?

7.33 Prove Bell's inequality.

7.34 Show that Bell's inequality is violated for some suitable polarization correlations of the singlet state.

7.4 Homework problem

Linear transformations that combine annihilation and creation operators were introduced in the theory of elementary excitations of Bose–Einstein condensates. They are known as Bogoliubov transformations. Consider the simplest case, a uniform condensate with a constant macroscopic wave function and zero trapping potential,

$$U = 0.$$

In this case, the density $|\psi_0|^2$ is a constant, ρ_0. However, due to the atom–atom interaction, the macroscopic wave function ψ_0 of the condensate oscillates in time with the mean field energy $g\rho_0$,

$$\psi_0 = \sqrt{\rho_0}\,\exp\left(-ig\rho_0 t/\hbar\right),$$

because this expression solves the equation for the condensate of Chapter 2's Problem, part (d). The elementary excitations of the condensate are described by the modes u_k and v_k defined in the Problem of Chapter 2 as well. To find a solution of the mode equations of part (g) we imagine the u_k and v_k as plane waves with wave vector $\mathbf{k}$ and frequency Ω. It is wise to represent Ω as $\omega \pm g\rho_0/\hbar$ for u_k and v_k, respectively, in order to take care of the bulk energy $g\rho_0$ of the condensate. As the modes are to be normalized according to the scalar product (h) of Chapter 2's Problem we write them with the prefactors $\cosh\zeta$ and $\sinh\zeta$ as

$$u_k = \mathcal{A}\cosh\zeta\,\exp\left(i\mathbf{k}\cdot\mathbf{r} - i\omega t - ig\rho_0 t/\hbar\right),$$
$$v_k = \mathcal{A}\sinh\zeta\,\exp\left(i\mathbf{k}\cdot\mathbf{r} - i\omega t + ig\rho_0 t/\hbar\right).$$

(a) Show that from the mode equations (g) of Chapter 2's Problem follows the dispersion relation

$$\omega^2 = c_0^2 k^2\left(1 + \frac{k^2}{4k_0^2}\right), \quad k_0 = \frac{mc_0}{\hbar}, \quad mc_0^2 = g\rho_0.$$

(b) Furthermore, show that

$$\coth(2\zeta) = -\frac{1}{mc_0^2}\left(\frac{\hbar^2 k^2}{2m} + mc_0^2\right).$$

(c) The modes of the condensate describe the oscillations around the mean field ψ_0; they constitute sound waves. Which combination of $\hat{a}_{\mathbf{k}}$ and $\hat{a}_{-\mathbf{k}}$ makes a sound wave in $\hat{\varphi}$ that propagates with wave vector $\mathbf{k}$ and frequency ω?

7.5 Further reading

Entanglement is a key aspect of quantum information science, see, for example (Bouwmeester et al., 2001; Lo et al., 2000; Nielsen and Chuang, 2000). Entanglement measures are reviewed in Bruss (2002) and Plenio and Virmani (2007). On the relation between entanglement and the second law of thermodynamics, see Brandão and Plenio (2008). On Bell's inequalities and other intriguing aspects of quantum mechanics, see Bell's collection of essays and papers (Bell, 1987) and Perez's excellent textbook (1995). For sharper constraints on nonlocal hidden variable theories, see Leggett (2003).

Chapter 8
Horizons

8.1 Minkowski space

At the beginning of the twentieth century, Albert Einstein replaced the aether theory by relativity, but a twenty-first century aether is still puzzling physicists today. This modern aether is the quantum vacuum. The aether was thought to be an all-penetrating mysterious substance that carries light through space like air carries sound. Take away all light, and the aether would still be there, defining a universal frame of reference. Now, according to quantum field theory, the state of absolute darkness, the vacuum state, is still a physical state filling space throughout, similar to the aether. There is an important difference though: one does not notice motion at uniform speed relative to the quantum vacuum, but, as we describe in this chapter, during acceleration the vacuum glows, although slightly, causing friction. Furthermore, as Stephen Hawking predicted in 1974 (Hawking, 1974), the quantum vacuum should also cause black holes to evaporate, because at the event horizon particles are created from nothing, at the expense of the black hole's mass. In this chapter we also describe this creation of radiation at horizons. None of these fascinating phenomema have been observed in astrophysics yet, but they can be demonstrated in laboratory analogues (Philbin et al., 2008a).

The quantum vacuum may also account for the dark energy that, according to astronomical data, constitutes the lion's share of the energy of the Universe (Hogan, 2007). But here theoretical calculations have been contradicting the observational data by more than 10^{100}, an unbelievable number of orders of magnitudes (Brumfiel, 2007). On the other

hand, some aspects of the quantum vacuum have been very success-
ful in solving astronomical conundrums, in particular cosmic inflation
(Abbott and Pi, 1986; Parentani, 2004). Inflation is the accelerated
expansion of the early Universe driven by as yet unknown physics. This
rapid expansion explains why the Universe has become nearly homogeneous (Parentani, 2004). But cosmic inflation also predicts another
feature that is consistent with the fluctuations of the cosmic microwave
background (Parentani, 2004); inflation makes a statement about large-
scale fluctuations. During the inflationary phase the Universe has
amplified the fluctuations of the quantum vacuum, like the child on a
playground swing we discussed in Section 7.1.1 has amplified tiny initial fluctuations of the swing, but with amplification factors on the truly
cosmic scale of 10^{100}. The cosmic fluctuations imprinted their mark on
the matter in the Universe that, in turn, started to form large structures,
galaxies, etc. The matter emitted light that reaches us today as cosmic
microwave radiation. Satellite data have shown that the actual fluctuations of the microwave background are consistent with the theory of
amplified vacuum fluctuations, which supports the idea that the large
structures of the Universe have grown from amplified fluctuations of
the quantum vacuum. What a fantastic insight!

Coming back to Earth, some effects of the quantum vacuum may
even appear in daily life: the Casimir force that we discussed in
Section 2.3 is closely related to the van der Waals force that causes
things to stick. Geckos (Autumn et al., 2002), for example, effortlessly
climb up walls and cling to the ceiling, thanks to the quantum vacuum
force between walls and the micro hairs of their extraordinary feet, see
Fig. 2.5.

We clearly got some aspects of the quantum vacuum right, but
not all of them. Obviously, our understanding is best developed where
experimental evidence has been available, although there are still surprises to be found in seemingly well-established features of the quantum
vacuum (Philbin and Leonhardt, 2009). For further understanding the
twenty-first century aether, experiments are vital, experiments that can
be done in quantum optics, for example using fibre-optical analogues of
the event horizons (Philbin et al., 2008a). Their design and interpretation often depends on the quantum theory of light in moving media.

Let us, in this final chapter, return to our starting point, quantum
electrodynamics in media. To explain the key insights there and working out the most important quantitative results with as little technical
effort as possible, we focus on one spatial dimension, ignoring the
other two. We assume that both light and media move in one direction, say the x axis, and are either irrelevant or uniform in y and z. This
assumption is natural to light in optical fibres (Agrawal, 2001) where

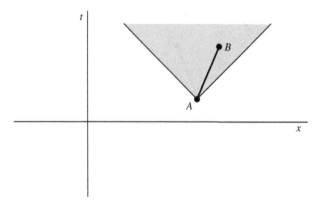

Fig. 8.1. Causality. It takes time for an event A to have an effect on a distant place B. Physical interactions are local; they are mediated by fields and travel at a maximal velocity c that turns out to be the speed of light in vacuum. The grey area in the picture illustrates the causal cone where A can influence other physical events B; it is bounded by the light cone.

demonstrations of Hawking radiation can be performed (Philbin et al., 2008a). It is also generic to the physics of the event horizon, as we explain in Section 8.3.3. But before we discuss accelerating observers and moving media, let us begin with a brief self-consistent section on Einstein's special theory of relativity.

8.1.1 Locality and relativity

At the heart of field theory (Landau and Lifshitz, Vol. II, 1987) lies the idea of locality. Cause and effect are local. Physical interactions[1] are mediated by fields, for example the electromagnetic field. The field responds to an event, for instance the transition of an atom, and transmits its effect through space, here as emitted light that a distant atom may absorb. The field is the messenger, but we may regard it as leading a life of its own, as an independent physical object, the quantum field of light. So it takes time for an event at A to have an effect at a distant place B, as the space–time diagram in Fig. 8.1 illustrates. If all physical interactions are local, they should propagate with a maximal speed that we denote as c and later identify as the speed of light in vacuum.

The second ingredient of special relativity is *Galilei invariance*, the insight that inertial motion does not have any influence on the laws of mechanics. More precisely and more generally, the fundamental laws of physics are the same in each inertial frame. We define an inertial frame as a Cartesian coordinate system that moves uniformly relative to a Cartesian master system that we call a rest frame, see Fig. 8.2. Which inertial frame we consider the rest frame is completely arbitrary, we only assume that such a frame exists. Consequently, if c is the maximal

---

[1] By physical interactions we mean the interactions of quantum objects, not the measurement process.

Fig. 8.2. Moving frame. The dotted line indicates the origin of a moving frame as seen in the space–time diagram of Fig. 8.1, the rest frame. The causal cone should appear to be the same in both the moving and the rest frame. How is this possible? By the Lorentz transformation illustrated in Fig. 8.3.

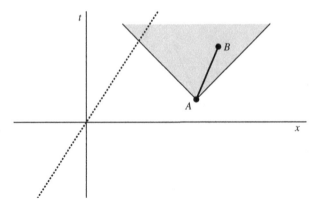

speed of interactions in an inertial frame it ought to be the same in all inertial frames. How is this possible?

We require that the space–time coordinate transformations between inertial frames are linear transformations. Why? A free particle should appear to move freely in all inertial frames. Free motion corresponds to a straight line in a space–time diagram. As linear transformations preserve the linearity of trajectories, a freely moving particle appears to draw a straight trajectory in all inertial frames. Hence we express the coordinates t' and x' in the moving frame by a linear transformation of the rest frame coordinates t and x and vice versa in the form

$$\begin{pmatrix} ct \\ x \end{pmatrix} = \begin{pmatrix} \Lambda^0_0 & \Lambda^0_1 \\ \Lambda^1_0 & \Lambda^1_1 \end{pmatrix} \begin{pmatrix} ct' \\ x' \end{pmatrix} = \begin{pmatrix} \Lambda^0_0 ct' + \Lambda^0_1 x' \\ \Lambda^1_0 ct' + \Lambda^1_1 x' \end{pmatrix} \tag{8.1}$$

where we include c in the time coordinates for later convenience. The $\Lambda^\alpha_{\alpha'}$ must depend on the velocity of the moving frame u that we assume not to exceed c,

$$|u| < c. \tag{8.2}$$

Let us work out the coefficients $\Lambda^\alpha_{\alpha'}$. Figure 8.2 shows the coordinate lines of the moving frame as seen by the rest frame. The coordinate origin, where $x' = 0$, moves at $x = u\,t$. On the other hand, we obtain from the transformation (8.1) that $t = \Lambda^0_0 t'$ and $x = \Lambda^1_0 ct'$ for $x' = 0$. Consequently, $\Lambda^1_0 / \Lambda^0_0 = u/c$. We parametrize Λ^0_0 as $\cosh \zeta$ and Λ^1_0 as $\sinh \zeta$ with

$$\tanh \zeta = \frac{u}{c} \tag{8.3}$$

which is always possible, because $|u| < c$. Now, consider propagation at the universal speed c in both the rest frame and the moving frame, drawing the causal cones illustrated in the space–time diagrams of Fig. 8.3. First, focus on the right-hand side of Fig. 8.3 where the propagation occurs in the forward direction, $x' = ct'$ and $x = ct$. By subtracting the

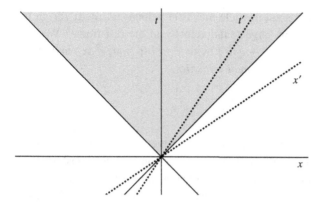

Fig. 8.3. Lorentz transformation. The dotted line labelled by t' shows the origin of the moving frame in the space–time diagram of the rest frame. Here $x' = 0$; so this line represents the time axis of the moving frame. Light cones lie exactly between the time lines measured as ct or ct' and the spatial lines x and x'. The x' axis is chosen such that the light cone in the moving frame is exactly the same as in the rest frame.

first line of the transformation (8.1) from the second line we obtain $\Lambda^1_1 - \Lambda^0_1 = \Lambda^0_0 - \Lambda^1_0 = \cosh\zeta - \sinh\zeta$. Then consider the left-hand side of Fig. 8.3, propagation at velocity $-c$ when $x' = -ct'$ and $x = -ct$. By adding the first and the second line of the transformation (8.1) we find $\Lambda^1_1 + \Lambda^0_1 = \Lambda^0_0 + \Lambda^1_0 = \cosh\zeta + \sinh\zeta$. Consequently,

$$\begin{pmatrix} ct \\ x \end{pmatrix} = \begin{pmatrix} \cosh\zeta & \sinh\zeta \\ \sinh\zeta & \cosh\zeta \end{pmatrix} \begin{pmatrix} ct' \\ x' \end{pmatrix} \tag{8.4}$$

with ζ given by relation (8.3). We discussed such transformations for quadratures in parametric amplification, but there they have nothing to do with space–time coordinate transformations between moving frames. Nevertheless, we see from our matrix representation (7.8) that the transformation (8.4) stretches space–time in one direction of the causal cone and squeezes it in the other, see Fig. 8.3. The causal cones themselves are not changed – c is indeed a universal speed. The transformation (8.4) also shows that both space and time are transformed, not only points at equal time. Motion thus changes the measure of time. Frequently, $\cosh\zeta$ in the transformation (8.4) is abbreviated as γ and expressed in terms of the velocity u of the moving frame. We obtain from relation (8.3)

$$\gamma = \frac{1}{\sqrt{1 - u^2/c^2}} \tag{8.5}$$

and write the space–time coordinate transformation (8.4) as

$$x = \gamma \left(x' + u t' \right), \quad t = \gamma \left(t' + \frac{u}{c^2} x' \right), \tag{8.6}$$

the classic *Lorentz transformation*. The inverse transformation from the rest frame to the moving frame is simply given by inverting the sign of u or, equivalently, of ζ, because relative to the moving frame the rest frame moves at velocity $-u$.

Suppose that some particle moves at velocity $v' = dx'/dt'$ in the moving frame. How large is the velocity in the rest frame? We simply divide $dx = \gamma (dx' + u \, dt')$ by $dt = \gamma (dt' + u/c^2 dx')$ and obtain *Einstein's addition theorem of velocities*,

$$v = \frac{v' + u}{1 + \frac{v' u}{c^2}}. \qquad (8.7)$$

Note that in the three-dimensional case the other components of the velocity vector are also affected, because the Lorentz transformation (8.6) changes the measure of time. An application of Einstein's addition theorem (8.7) is the *dragging of light* in media. Imagine a medium, say water, moving at speed u. Light propagates in the water with speed c divided by the refractive index n, here 1.33 for water. In the laboratory frame the velocity of light is

$$v = \frac{u + \frac{c}{n}}{1 + \frac{u}{cn}} \approx \frac{c}{n} + \left(1 - \frac{1}{n^2}\right)u \qquad (8.8)$$

for $|u| \ll c/n$. Light propagating with the flow advances more rapidly, whereas against the flow it is slowed down. The effect of the moving water is small, but observable in an interference experiment. Suppose a beam is split in two, one propagating with the flow and the other against the flow, as Fig. 8.4 shows. If the beams are brought together again the acquired relative delay appears in their interference. Remarkably, in 1851 Hippolyte Fizeau (Fizeau, 1851) was able to see such

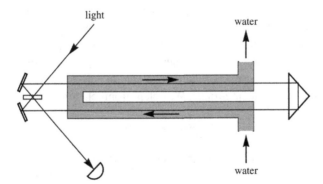

Fig. 8.4. Fizeau experiment. In one arm of the interferometer, light always travels with the flow of water, in the other arm it propagates against the current. The two light beams interfere at the detector. Their relative phase depends on the Fresnel drag of light in the moving water, the relativistic addition (8.7) of the velocity of light in the water, c/n and the flow speed u. Fizeau observed this relativistic effect in 1851, Fresnel predicted it in 1818.

a subtle effect, without using lasers. Even more remarkably, in 1818 Augustine Jean Fresnel (Fresnel, 1818) predicted the correct result (8.8), without knowing relativity, nearly a century before Einstein's 1905 Electrodynamics of Moving Bodies (Einstein, 1905b).

8.1.2 Space–time geometry

The Lorentz transformation (8.6) acts as a hyperbolic rotation (8.4) of the space–time coordinates. As Hermann Minkowski said in 1908 "The views of space and time which I wish to lay before you have sprung from the soil of experimental physics, and therein lies their strength. They are radical. Henceforth space by itself, and time by itself, are doomed to fade away into mere shadows, and only a kind of union of the two will preserve an independent reality." This union of space and time carries Minkowski's name now: it is called *Minkowski space* or *Minkowski space–time*. The spatial and temporal coordinates are combined in the *space–time coordinates* x^α, multiplying the time t by c to give it the physical units of a length,

$$x^\alpha = (ct, x). \tag{8.9}$$

The index α runs from 0 to 1 in our one-dimensional model (and from 0 to 3 in the three-dimensional world). We denote derivatives with respect to the space–time coordinates, *space–time gradients*, as

$$\partial_\alpha = \left(\frac{1}{c} \frac{\partial}{\partial t}, \frac{\partial}{\partial x} \right). \tag{8.10}$$

Although spatial and temporal coordinates are mixed in Lorentz transformations, the causal order of any events is always preserved, as Fig. 8.3 shows: space–time points within the causal cone, events that may influence each other, remain causally connected after Lorentz transformation to a moving frame. Conversely, points that are causally disconnected remain so. Time is not absolute, but causality is. Instead of earlier and later we speak of cause and effect or the absence of causal connections. Causally connected events are said to be *time-like* to each other, while causally disconnected events are *space-like*. Locality simply implies, in its most radical form, that the measure of time may differ from place to place.

What is the measure of time? Imagine a clock like the one illustrated in Fig. 8.5 moving at variable speed v. The clock measures its own time in its own frame, but this frame is moving relative to the rest frame, and at varying velocity in general. To calculate the time counted by the moving clock, relative to a stationary clock in the rest frame, we use

Fig. 8.5. Flying clock. (Artist: Maria Leonhardt, 7 years.)

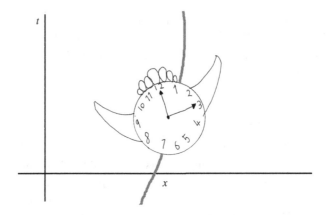

the following convenient mathematical construction: we introduce the *metric tensor* as the matrix[2]

$$g_{\alpha\beta} = \begin{pmatrix} 1 & 0 \\ 0 & -1 \end{pmatrix} \tag{8.11}$$

and define the *space–time metric* of the infinitesimal increments dt and dx as

$$ds^2 = \sum_{\alpha\beta} g_{\alpha\beta}\, dx^\alpha\, dx^\beta = c^2 dt^2 - dx^2. \tag{8.12}$$

The metric establishes a measure of space–time distance, the hyperbolic version of the Pythagorean measure of spatial lengths $dl^2 = dx^2 + dy^2$. The important point is that the hyperbolic rotation (8.4) of the Lorentz transformation preserves the metric in the same way as an ordinary rotation preserves distances in space,

$$ds'^2 = ds^2. \tag{8.13}$$

Suppose we attach co-moving inertial frames to the moving clock. The frames may differ at different laboratory times, the clock is gliding from one frame to the next, but we can always use Lorentz transformations to the rest frame for comparing time, transformations at variable speed, though. In the co-moving frames the moving clock remains at the same position, but time ticks away. Only time increments dt' thus contribute to the metric (8.12) in primed coordinates. We use the invariance (8.13) of the metric to relate dt' to the laboratory frame and get $dt' = ds/c$. The total time τ of the moving clock is called *proper time*. We obtain

$$\tau = \int \frac{ds}{c} = \int \sqrt{c^2 - \frac{dx^2}{dt^2}} \frac{dt}{c} = \int \sqrt{1 - \frac{v^2}{c^2}}\, dt. \tag{8.14}$$

---

[2] We adopt the Landau–Lifshitz convention (Landau and Lifshitz, Vol. II, 1987) for the sign of the metric.

As the factor $\sqrt{1 - v^2/c^2}$ is smaller than unity, the moving clock ticks slower than the clock at rest. Or, if we replace the clocks by the biological clocks of two identical twins, one travelling and the other staying at home, the travelling twin ages slower than the one at rest. On return of the traveller, the twins may recognize with amazement their different ages, a feature of relativity known as the *twin paradox*. The twin paradox has been experimentally tested and verified, although not with twins, but with highly accurate atomic clocks, because the time differences are usually tiny. We have seen that Minkowski space has a natural measure of space–time distance that we identified as the time increment experienced by moving objects (multiplied by c). Minkowski space thus establishes a geometry measured by time.

8.1.3 Light

We mentioned in Section 8.1.1 that the speed of causal interactions, c, is the speed of light in vacuum (meaning here empty space filled with light, not the vacuum state). Why is this the case? The causal speed c is the same in all inertial frames. So if we show that the propagation velocity of light fields has this defining property of the causal speed, we have shown that c is indeed the speed of light in vacuum. How is light transformed from one inertial frame to another? For simplicity, throughout this chapter we assume that the light propagates in the x direction. Consequently, it must be polarized in the y, z plane, and in empty space the polarization remains uniform. Any uniform polarization can be thought of as an optical superposition of two linear polarizations with fixed coefficients. So it is sufficient to consider one linear polarization that we describe by the vector potential component $\hat{A}(x, t)$. Imagine an inertial frame moving in the x direction. As the light is polarized orthogonal to x it is reasonable to assume that the polarization state remains the same in the moving frame; so the component $\hat{A}$ is not transformed,[3] but of course the argument of $\hat{A}(x, t)$, the space–time coordinates x and t are Lorentz-transformed. In formulae,

$$\hat{A}'(x', t') = \hat{A}(x, t).\tag{8.15}$$

From relation (2.3) we obtain the electric field component $\hat{E}$ and the component $\hat{B}$ of the magnetic induction,

$$\hat{E} = -\frac{\partial \hat{A}}{\partial t}, \quad \hat{B} = \frac{\partial \hat{A}}{\partial x}.\tag{8.16}$$

---

[3] In three-dimensional space, $\hat{A}_\alpha$ is a four-vector (Landau and Lifshitz, Vol. II, 1987) that is transformed like a space–time gradient. One can then use a gauge transformation (Jackson, 1998) to put $\hat{A}_0$ to zero. After this procedure, the transversal components are not transformed, as we have also seen in our simple one-dimensional model.

These fields are transformed in the moving frame,

$$\hat{E}' = -\frac{\partial \hat{A}}{\partial t'}, \quad \hat{B}' = \frac{\partial \hat{A}}{\partial x'}, \tag{8.17}$$

because the derivatives are transformed. We obtain from the chain rule of differentiation, $\partial/\partial t' = (\partial t/\partial t')\,\partial/\partial t + (\partial x/\partial t')\,\partial/\partial x$, etc, and the Lorentz transformation (8.6)

$$\frac{\partial}{\partial t'} = \gamma \left(\frac{\partial}{\partial t} + u\frac{\partial}{\partial x} \right), \quad \frac{\partial}{\partial x'} = \gamma \left(\frac{\partial}{\partial x} + \frac{u}{c^2}\frac{\partial}{\partial t} \right). \tag{8.18}$$

Consequently,

$$\hat{E}' = \gamma \left(\hat{E} - u\hat{B} \right), \quad \hat{B}' = \gamma \left(\hat{B} - \frac{u}{c^2}\hat{E} \right). \tag{8.19}$$

The laws of physics, including the laws of quantum electromagnetism, should be the same in all inertial frames. The Coulomb gauge (2.4) is always satisfied when the vector potential points orthogonally to the spatial direction where it evolves. The second equation of quantum electromagnetism in the Coulomb gauge, the wave equation (2.5), remains the same if we require that

$$\hat{D} = \gamma \left(\hat{D}' + \frac{u}{c^2}\hat{H}' \right), \quad \hat{H} = \gamma \left(\hat{H}' + u\hat{D}' \right), \tag{8.20}$$

because, for one-dimensional propagation, the wave equation (2.5) reduces to the Maxwell equation

$$\frac{\partial \hat{H}}{\partial x} + \frac{\partial \hat{D}}{\partial t} = 0 \tag{8.21}$$

that turns into

$$\frac{\partial \hat{H}'}{\partial x'} + \frac{\partial \hat{D}'}{\partial t'} = 0 \tag{8.22}$$

for the transformations (8.18) and (8.20) of the derivatives and the fields. As our simple one-dimensional model indicates, the laws of electromagnetism are the same in all inertial frames if the fields are appropriately transformed. Electric fields partially appear as magnetic fields and vice versa.[4] In empty space, where $\varepsilon = \mu = 1$, we obtain from the Maxwell equation (8.21), the constitutive equations (2.2) and the expressions (8.17) for the electromagnetic field

$$0 = \frac{1}{c^2}\frac{\partial^2 \hat{A}}{\partial t^2} - \frac{\partial^2 \hat{A}}{\partial x^2} = \frac{1}{c^2}\frac{\partial^2 \hat{A}}{\partial t'^2} - \frac{\partial^2 \hat{A}}{\partial x'^2}. \tag{8.23}$$

---

[4] In three spatial dimensions, the electromagnetic fields are components of four-dimensional antisymmetric tensors of rank two (Landau and Lifshitz, Vol. II, 1987).

We introduce the new variables

$$t_\pm = t \mp \frac{x}{c} \tag{8.24}$$

where the d'Alembertian appears in a particularly instructive form,

$$\frac{\partial^2}{\partial t_+ \partial t_-} = \frac{1}{c^2}\frac{\partial^2}{\partial t^2} - \frac{\partial^2}{\partial x^2}. \tag{8.25}$$

So any solution of the wave equation (8.23) is the t_- derivative of a constant with respect to t_+. A constant in t_+ is a function of t_-. Integrating this function with respect to t_- gives another function of t_- plus an integration constant. The integration constant is a constant for t_-, but a function of the other variable, t_+. In this way we obtain d'Alembert's classic solution of the wave equation (8.23)

$$\hat{A}(x,t) = \hat{A}_+(t_+) + \hat{A}_-(t_-), \tag{8.26}$$

and the same logic applies in the transformed coordinates. Equation (8.26) shows that, in one dimension, light is an optical superposition of left moving ($-$) and right moving ($+$) wavepackets that propagate according to Eq. (8.24) at the speed c without changing their shapes.[5] In any other inertial frame the situation is exactly the same. The causal speed c is indeed the speed of light. The constancy of the speed of light has been used to define the measure of length. The SI metre is specified by the time of light travel and no longer by what is stored in the "Bureau des Longitudes". Geometry literally is measured by light and time.

8.2 Accelerated observers

As we have seen, the light field $\hat{A}$ remains the same in a uniformly moving frame. Is the quantum vacuum changed? Let us return to the fundamentals. According to definition (3.23), the vacuum state is the eigenstate of the annihilation operators with eigenvalue zero. The annihilation operators are the quantum coefficients of orthogonal modes with positive norm (2.24) in the mode expansion (2.16). Monochromatic modes (2.31) with positive frequencies have positive norms (2.32). Therefore, as long as positive-frequency waves are transformed into positive-frequency waves, annihilation operators are turned into annihilation operators (2.30). Their zero eigenstate, the quantum vacuum, is not affected. In the rest frame, the monochromatic modes are

---

[5] In three-dimensional empty space, light is diffracted (Born and Wolf, 1999), but still propagates with c (Jackson, 1998).

plane waves $\mathcal{A}\exp(-i\omega t_\pm)$ that in the moving frame appear as plane waves $\mathcal{A}\exp(-i\omega' t'_\pm)$, as we obtain from the Lorentz transformation (8.6) of the space–time coordinates in the times (8.24). Note that the frequency is changed, $\omega' = \gamma(1 \pm u/c)\omega$, by the Doppler effect. However, positive frequencies in the rest frame still correspond to positive frequencies in the moving frame, provided of course the frame does not move at superluminal speed. So the vacuum state remains the same in a uniformly moving frame; one can race through the quantum vacuum without noticing it. But now imagine an accelerated observer. With increasing velocity the observer would notice increasing Doppler shifts, yet positive frequencies should remain positive. We would expect that the quantum vacuum remains a vacuum, which, however, turns out to be wrong.

8.2.1 Rindler coordinates

How can we characterize the field seen by the accelerated observer? Let us introduce the following coordinates and argue that they define a frame of uniformly accelerated observers. Suppose we describe Minkowski space in a hyperbolic version of polar coordinates,

$$x = \xi \cosh \eta, \quad ct = \xi \sinh \eta, \tag{8.27}$$

known as *Rindler coordinates* named after Wolfgang Rindler. Both ξ and η run from $-\infty$ to $+\infty$, but ξ has the dimension of a length while η is dimensionless. Figure 8.6 shows a space–time chart of Rindler coordinates. The coordinate lines for fixed ξ are hyperbolae and the coordinate lines for fixed η are straight rays in a space–time diagram. Both are always space-like relative to the origin at $x = ct = 0$, because for the parameterization (8.27) x^2 exceeds $c^2 t^2$. The Rindler coordinates do not completely cover space–time, only the space-like part on either the left or the right of the origin. We could introduce another set of coordinates for the time-like part, but the two sets would be disconnected. For space-like events (ct, x) the Rindler coordinates are

$$\xi = \pm\sqrt{x^2 - c^2 t^2}, \quad \eta = \frac{1}{2}\ln\left(\frac{x+ct}{x-ct}\right). \tag{8.28}$$

In the Rindler frame, the Minkowski metric (8.12) turns into

$$ds^2 = c^2 dt^2 - dx^2 = \xi^2 d\eta^2 - d\xi^2, \tag{8.29}$$

as one easily verifies by differentiation where $dt = (\partial t/\partial \xi)d\xi + (\partial t/\partial \eta)d\eta$, etc. We can already draw an important conclusion from the

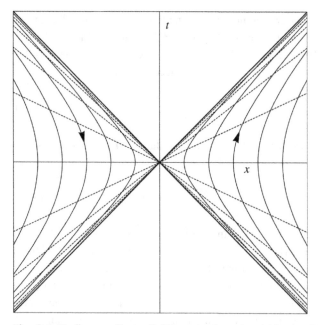

Fig. 8.6. Rindler coordinates (8.27) as seen in an inertial frame of time t and position x. Solid lines: η coordinates, dashed lines: ξ coordinates. The η coordinate lines with $\xi = $ const describe the space–time trajectories of accelerated observers. The acceleration of each observer is constant, but observers on different trajectories experience different accelerations (8.31) that are inversely proportional to ξ. The proper time (8.30) measured by an accelerated observer is proportional to η. Rindler coordinates do not cover the entire Minkowski space–time; they end at horizons where accelerated observers would instantly reach the speed of light. On the left side of this Rindler wedge the coordinate time must run backwards such that the proper time of accelerated observers runs forwards.

Rindler metric (8.29): for coordinate lines with fixed $\xi = \xi_0$ (and $d\xi = 0$ of course) the proper time (8.14) is

$$\tau = \frac{\xi_0\,\eta}{c}. \qquad (8.30)$$

So, for a physical object on a space–time trajectory with fixed Rindler parameter ξ, the Rindler η is proportional to the proper time. Note that on the left side of the space–time diagram, where $\xi < 0$, the parameter η must run backwards such that proper time τ runs forwards.

The object on the Rindler trajectory turns out to be uniformly accelerated. Why? One can imagine acceleration as continuous Lorentz boosting with infinitesimal velocities du. The Lorentz transformations correspond to the hyperbolic rotations (8.4) of Minkowski space. In Rindler coordinates (8.27) the "angle" η would simply "rotate" to become $\eta + \zeta$ after Lorentz transformation, so $d\eta = d\zeta$. For small Lorentz boosts the parameter ζ of Eq. (8.3) reduces to u/c and from relation (8.30) follows $du/d\tau = c^2/\xi_0$. The object on the Rindler trajectory thus experiences the uniform acceleration

$$a = \frac{c^2}{\xi_0}. \tag{8.31}$$

We also get this result by explicit calculation. The velocity of the object is

$$v = \frac{dx}{dt} = \frac{dx/d\eta}{dt/d\eta} = c \tanh \eta, \tag{8.32}$$

increasing, asymptotically, from $-c$ to $+c$. However, what counts in relativistic dynamics (Landau and Lifshitz, Vol. II, 1987) is not the velocity with respect to coordinate time t but the velocity with respect to proper time, $dx/d\tau$. According to formula (8.14) for τ and the definition (8.5) the two velocities deviate by the relativistic factor γ (which accounts for the difference between rest mass and dynamic mass (Landau and Lifshitz, Vol. II, 1987))

$$\frac{dx}{d\tau} = \gamma \frac{dx}{dt}. \tag{8.33}$$

In our case, we obtain

$$\frac{dx}{d\tau} = \frac{c}{\xi_0} \frac{dx}{d\eta} = c \sinh \eta = \frac{c^2}{\xi_0} t \tag{8.34}$$

from which follows

$$a = \frac{d}{dt} \frac{dx}{d\tau} = \frac{c^2}{\xi_0}. \tag{8.35}$$

The Rindler coordinates (8.27) thus establish the space–time coordinates of uniformly accelerated observers. However, the acceleration (8.31) differs for observers on different space–time trajectories (for different ξ_0). No global coordinate system can describe uniformly accelerated observers that have started at different places x at the same coordinate time t – they would follow displaced hyperbolae that intersect, their space–time paths would cross – something that is not allowed for coordinate lines. But the Rindler coordinates (8.27) are completely sufficient to describe any single uniformly accelerated observer. We apply them as useful mathematical constructions, as a coordinate system that, extending from the accelerated observer, captures the field around the observer.

8.2.2 Accelerated modes

How do light waves appear in Rindler coordinates? The d'Alembertian in the wave equation (8.23) is as similar to the Laplacian $\partial^2/\partial x^2 + \partial^2/\partial y^2$ as the Rindler coordinates (8.27) are similar to polar coordinates. So we obtain, by analogy,

$$\frac{1}{c^2}\frac{\partial^2}{\partial t^2} - \frac{\partial^2}{\partial x^2} = -\frac{\partial^2}{\partial \xi^2} - \frac{1}{\xi}\frac{\partial}{\partial \xi} + \frac{1}{\xi^2}\frac{\partial^2}{\partial \eta^2}, \tag{8.36}$$

a result one can of course verify by calculation. The wave equation (8.23) for light modes thus appears as

$$\frac{\partial^2 A}{\partial \eta^2} = \xi \frac{\partial}{\partial \xi} \xi \frac{\partial A}{\partial \xi} \tag{8.37}$$

with the solutions

$$A = \mathcal{A}\,\xi^{\pm i\nu}e^{-i\nu\eta} \tag{8.38}$$

for any real constants ν. In Rindler coordinates, the waves (8.38) play the role of monochromatic plane waves. However, as the Rindler coordinates (8.27) only cover space-like events, the waves (8.38) should be confined to either the left or the right side of the Rindler wedge shown in Fig. 8.6,

$$A_R = \mathcal{A}\,\Theta(\xi)\,\xi^{\pm i\nu}e^{-i\nu\eta} \;=\; \mathcal{A}\,\Theta(x \mp ct)\,(x \mp ct)^{i\nu}, \tag{8.39}$$
$$A_L = \mathcal{A}\,\Theta(-\xi)(-\xi)^{\pm i\nu}e^{-i\nu\eta} \;=\; \mathcal{A}\,\Theta(ct \mp x)\,(ct \mp x)^{i\nu} \tag{8.40}$$

where we have applied the inverse coordinate transformation (8.28) and use the Θ function $\Theta(x)$ that is 0 for $x < 0$ and 1 for $x > 0$. The Rindler waves appear as left-moving $(-)$ or right-moving $(+)$ waves in Minkowski space, in agreement with d'Alembert's solution (8.26) of the wave equation (8.23). Moreover, the Rindler waves (8.39) and (8.40) are either left or right moving. Let us focus, without loss of generality, on the right moving waves.

First we find a simple physical interpretation for ν. The phase of the wave evolving in Rindler coordinates, $\nu\eta$, must be the same as the phase $\omega\tau$ perceived by the accelerated observer with frequency ω. Consequently, we obtain from Eqs. (8.30) and (8.31)

$$\nu = \frac{\xi_0\,\omega}{c} = \frac{c\,\omega}{a}. \tag{8.41}$$

As η runs backwards on the left side of the Rindler wedge, we should use negative frequencies there. This important qualification also arises in the normalization of modes.

Modes are normalized with respect to a scalar product. For example, in empty Minkowski space the scalar product is given by Eq. (2.17) with $\varepsilon = 1$,

$$(A_1, A_2) = \frac{i\varepsilon_0}{\hbar} \int_{-\infty}^{+\infty} \left(A_1^* \frac{\partial A_2}{\partial t} - A_2 \frac{\partial A_1^*}{\partial t} \right) dx. \qquad (8.42)$$

The scalar product must be a conserved quantity with respect to the relevant time coordinate. The prefactor is chosen such that the Bose commutation relations (2.28) of the normal mode operators follow from the fundamental commutator (2.14). In Rindler coordinates, we propose

$$(A_1, A_2) = \frac{i\varepsilon_0 c}{\hbar} \int_{-\infty}^{+\infty} \left(A_1^* \frac{\partial A_2}{\partial \eta} - A_2 \frac{\partial A_1^*}{\partial \eta} \right) \frac{d\xi}{\xi}. \qquad (8.43)$$

We obtain from the wave equation (8.37) that the scalar product (8.43) is indeed conserved,

$$\frac{d(A_1, A_2)}{d\eta} = 0. \qquad (8.44)$$

Having established the scalar product of modes in Rindler coordinates, we try it out on modes (8.39) and (8.40),

$$
\begin{aligned}
(A_{R1}, A_{R2}) &= \frac{2\varepsilon_0 c\omega}{\hbar} \mathcal{A}^2 \int_0^{+\infty} \xi^{ic(\omega_2 - \omega_1)/a} \frac{d\xi}{\xi}, \quad \zeta = \ln\xi, \\
&= \frac{2\varepsilon_0 c\omega}{\hbar} \mathcal{A}^2 \int_{-\infty}^{+\infty} \exp\left(i\frac{c}{a}(\omega_2 - \omega_1)\zeta \right) d\zeta \\
&= \frac{4\pi\varepsilon_0 a\omega}{\hbar} \mathcal{A}^2 \, \delta(\omega_2 - \omega_1), \\
(A_{L1}, A_{L2}) &= \frac{2\varepsilon_0 c\omega}{\hbar} \mathcal{A}^2 \int_{-\infty}^0 (-\xi)^{ic(\omega_2 - \omega_1)/a} \frac{d\xi}{\xi} \\
&= -\frac{2\varepsilon_0 c\omega}{\hbar} \mathcal{A}^2 \int_0^{+\infty} \xi^{ic(\omega_2 - \omega_1)/a} \frac{d\xi}{\xi} \\
&= -\frac{4\pi\varepsilon_0 a\omega}{\hbar} \mathcal{A}^2 \, \delta(\omega_2 - \omega_1),
\end{aligned}
\qquad (8.45)
$$

in the limit $\omega_1 \to \omega_2 \equiv \omega$ where the scalar product does not vanish. This calculation shows that one should use negative frequencies to describe positive norm modes on the left side of the Rindler wedge, because here the Rindle time parameter η is running backwards.[6] Richard P. Feynman regarded particles running backwards in time as

---

[6] The conversion of positive frequency waves into waves with negative frequencies has been observed on water waves (Rousseaux et al., 2008) proving that negative-frequency waves are meaningful physical concepts.

antiparticles. If the anti-waves on the left side are somehow brought in contact with the waves on the right side, we expect that particles and antiparticles are created, or, as the anti-photons are just photons, photon pairs.

8.2.3 Unruh effect

Here we show that the creation of radiation from, literally, nothing is not only a possibility in uniform acceleration, but a necessity. In empty space, the quantum vacuum is the vacuum in the Minkowski frame. The vacuum is defined with respect to normal modes that are plane waves with positive frequencies or superpositions (2.29) of them, Fourier integrals over positive frequencies,

$$A = \int_0^\infty \tilde{A}(\omega) \exp\left(i \frac{\omega}{c} x_+\right) d\omega, \quad x_+ = x - ct, \tag{8.46}$$

for right-moving waves. The Minkowski modes (8.46) turn out to bridge both sides of the Rindler wedge, connecting waves running forwards in proper time with waves running backwards in time, as we show with an elegant trick from complex analysis (Ablowitz and Fokas, 1997). Suppose we analytically continue (Ablowitz and Fokas, 1997) the wave (8.46) to the complex x_+ plane. As the function (8.46) is a linear combination of the analytic functions[7] $\exp(i\omega x_+/c)$, it is also analytic, unless the integral (8.46) does not converge. For positive frequencies, $\exp(i\omega x_+/c)$ is exponentially decreasing on the upper half-plane, where $\mathrm{Im}\, x_+ > 0$. Consequently, if $\tilde{A}(\omega)$ is chosen such that the mode (8.46) is well-behaved in real space, no singularity can be created on the upper half x_+ plane: Minkowski modes (8.46) are analytic there. Analytic functions cannot vanish on either the positive or the negative real axis, because if they are zero in any interval they must vanish everywhere (Ablowitz and Fokas, 1997). So the Minkowski modes indeed bridge both sides of the Rindler wedge.

However, we can easily combine pairs of Rindler modes (8.39) and (8.40) to establish analytic functions and, in doing so, Minkowski modes. The Rindler waves depend on $x_+^{i\nu} = \exp(i\nu \ln x_+)$. The logarithm $\ln z$, being the integral of z^{-1}, gains $i\pi$ when the integration contour goes from the positive to the negative real axis on the upper side around $z = 0$. So $x_+^{i\nu}$ differs by $\exp(-\nu\pi)$ on the positive and the negative real axis. If we combine $\Theta(x_+) x_+^{i\nu}$ with $\Theta(-x_+)(-x_+)^{i\nu}$

---

[7] By analytic functions we mean functions $f(z)$ obeying the Cauchy–Riemann differential equations (Ablowitz and Fokas, 1997).

multiplied by $\exp(-\nu\pi)$ the result is a function on the real axis that is analytic on the upper half complex plane,

$$\Theta(x_+)x_+^{i\nu} + \Theta(-x_+)(-x_+)^{i\nu}e^{-\nu\pi} = x_+^{i\nu} \quad \text{analytic}, \tag{8.47}$$

and, exchanging ν by $-\nu$ and multiplying by $\exp(-\nu\pi)$,

$$\Theta(-x_+)(-x_+)^{-i\nu} + \Theta(x_+)x_+^{-i\nu}e^{-\nu\pi} = x_+^{-i\nu} \quad \text{analytic}. \tag{8.48}$$

Consequently, a linear combination of Rindler waves with relative amplitude $\exp(-\nu\pi)$ corresponds to the analytic Minkowski modes. However, on the left side of the Rindler wedge the frequency ω is negative and so is ν in Eqs. (8.40) and (8.41). We must therefore combine Rindler waves with their complex conjugate partners, waves with antiwaves. Such mode mixing ought to convert orthonormal modes into a new set of orthonormal modes. As the scalar product (8.43) shares features (2.20) of the commutator, the mode combination has the same structure as the combination of creation and annihilation operators in parametric amplification, Eqs. (7.1) and (7.5). We thus arrive at the *Bogoliubov transformation* of modes

$$A_1 = A_R \cosh\zeta + A_L^* \sinh\zeta, \quad A_2 = A_L \cosh\zeta + A_R^* \sinh\zeta \tag{8.49}$$

where the relative amplitude of the conjugate partners is given by

$$\tanh\zeta = e^{-\nu\pi}. \tag{8.50}$$

Here ζ does not denote the parameter of a Lorentz transformation, but rather a squeezing parameter. One verifies that the mode combination (8.49) preserves the orthonormality conditions (2.24), as required.

All we have done is a mode transformation; the quantum field of light remains the same, but the mode expansion (2.16) is changed. Specifically, for the transformed pairs of modes and their conjugates, we have

$$A_1\hat{a}_1 + A_2\hat{a}_2 + A_1^*\hat{a}_1^\dagger + A_2^*\hat{a}_2^\dagger = A_R\hat{a}_1' + A_L\hat{a}_2' + A_R^*\hat{a}_1'^\dagger + A_L^*\hat{a}_2'^\dagger. \tag{8.51}$$

Consequently, we obtain for the operators

$$\hat{a}_1' = \hat{a}_1 \cosh\zeta + \hat{a}_2^\dagger \sinh\zeta, \quad \hat{a}_2' = \hat{a}_2 \cosh\zeta + \hat{a}_1^\dagger \sinh\zeta. \tag{8.52}$$

The mode operators are transformed in the same way as in parametric amplification, although acceleration has nothing to do with parametric processes. All pair-creation processes involve linear combinations of annihilation and creation operators. So they must all be of the structure

(7.1) of parametric amplification, possibly with additional beam splitter type transformations (5.3). Our black box models of single optical instruments are fairly general.

As we discussed in Section 7.1.3, the mode transformation (8.52) creates Einstein–Podolski–Rosen states (7.20) from the vacuum state. A single accelerated observer has only access to one of the entangled modes, the right side Rindler modes (8.39) in the case of positive acceleration assumed here. (For negative acceleration a we can represent $-x$ instead of x in Rindler coordinates (8.27) and obtain the same results, replacing a by $|a|$ of course.) The reduced Einstein–Podolski–Rosen state is the thermal state (3.49) with temperature T corresponding to the squeezing parameter (7.29). We obtain from the squeezing parameter (8.50) of uniform acceleration and relation (8.41) the temperature

$$k_B T = \frac{\hbar a}{2\pi c}. \tag{8.53}$$

Remarkably, the temperature (8.53) is the same throughout the entire electromagnetic spectrum, in contrast to parametric amplification where each amplified mode may have a temperature of its own. The temperature (8.53) is universal, depending only on the acceleration and on natural constants. We focused on right-moving waves in our analysis, but the same algebra carries through to left-moving waves.[8] In all regards, the quantum vacuum of empty space appears to a single accelerated observer as thermal radiation with temperature (8.53).

The thermal transmutation of the vacuum by acceleration is called the *Unruh effect* (Unruh, 1976) after William G. Unruh who theoretically predicted the effect studying an accelerated detector (using a different method than explained here). The temperature of acceleration is incredibly small. Putting in the numbers in Unruh's formula (8.53) shows that as much as 10^{23} m s^{-2} are required – 10^{22} times the gravitational acceleration g on Earth – to reach room temperature. Unruh radiation has not been observed yet. Nevertheless, the Unruh effect is of fundamental importance, because of its role in elucidating the nature of the quantum vacuum.

The energy required for turning the quantum vacuum into thermal radiation is extracted from the accelerated object (Parentani, 1995). How? Consider the accelerated observer in the Minkowski frame. Here the Bogoliubov transformation (8.52) is reversed, but this reversed transformation also generates Einstein–Podolski–Rosen states. So, seen in an inertial frame of empty Minkowski space, the accelerated observer appears to spontaneously emit photon pairs. The two photons of each pair have opposite momenta in the Minkowski frame, but the momenta

---

[8] In three dimensions, the accelerated observer sees thermal radiation in all directions.

have different magnitudes in a frame co-moving with the observer at the time of emission, due to the Doppler effect. This momentum imbalance leads to friction that extracts energy from the accelerated observer. The random kicks play the role of the corresponding fluctuations, according to the fluctuation dissipation theorem (Callen and Welton, 1951; Landau and Lifshitz, Vol. V, 1996) we briefly discussed in Section 6.1.3. In this way, the vacuum is causing friction and Brownian motion, but in contrast to ordinary viscosity (Landau and Lifshitz, Vol. VI, 1987), the quantum friction depends on the acceleration, but not on the velocity. The quantum vacuum is a strange fluid!

The quantum vacuum appears to be particularly paradoxical if the accelerated observer is accompanied by a partner that is accelerated in the opposite direction. Imagine the two observers are on two conjugate Rindler trajectories, one on a space–time hyperbola with fixed positive ξ (on the right side of the Rindler wedge) and the second observer on a trajectory with a negative ξ (on the left side of the Rindler wedge), like the arrows in Fig. 8.6 but moving forward in coordinate time. Both observers are part of the same space–time coordinate system, the Rindler chart (8.27). Therefore we can use the Rindler modes (8.39) and (8.40) to describe the electromagnetic field experienced by them. Their mode operators are the Bogoliubov-transformed annihilation and creation operators (8.52). Instead of the Minkowski vacuum, the two accelerated observers witness the Einstein–Podolski–Rosen state (7.20) (Reznik, 2003; Reznik et al., 2005; Massar and Spindel, 2006). Both accelerated observers individually perceive the Minkowski vacuum as thermal radiation, but if they compared their records after the acceleration stage they would notice that the thermal photons were always correlated. The recoil kicks that slow them down have been synchronized! So, apparently random events caused by the quantum vacuum have been correlated across space; they have been entangled. On the other hand, in order to witness such correlations the two observers must move on fine tuned space–time trajectories, because they must be part of one and the same Rindler frame, or sufficiently close to it. If their paths are less correlated their correlations are less visible (Massar and Spindel, 2006). But it still seems utterly strange that two distant observers, by choosing the right acceleration and initial position, suffer friction correlated across space.

The two partners in observation are on the two sides of the Rindler wedge, they are separated by a *horizon*. The horizon is the space–time line[9] where the acceleration is so strong that an observer would instantly

---

[9] In three-dimensional space the horizon is an area moving through time.

reach the speed of light. One might object that such a horizon merely is an artefact of the Rindler coordinates, but this coordinate horizon turns out to share the essential properties of the *event horizon* of the black hole, as we show in the next section.

8.3 Moving media

Black holes are fascinating, perhaps for three rather different reasons. They are awe inspiring, their gravity is so strong that nothing, not even light, can escape. Many people have heard about black holes as the ultimate destroyers, as dangerous things that may swallow the Earth if a black hole comes close by or is accidentally created by malicious scientists. One might doubt whether black holes deserve such a sinister reputation. Yes, they do swallow everything in their way, but it is not entirely easy to cross their paths, because they tend to be rather small.

For physicists, black holes are interesting, because they are quite simple. Black holes are ideal objects completely characterized by their mass and angular momentum (and total charge if they have any). And yet in these simple objects quantum mechanics and gravity come together, revealing tantalizing glimpses into possible connections between the quantum microcosmos and the astrophysical macrocosmos that are not yet fully understood. In short, the black hole makes a good research subject.

For philosophical minds the black hole is astonishing as well, because it is something we can never experience and return to tell the tale, but we can make reasonably accurate predictions about it.[10] Black holes are not only awe-inspiring manifestations of the powers of Nature, but also show the powers of the human mind.

Black holes are, in fact, easy to understand. In 1972 William G. Unruh (Unruh, 2008) was asked to give a colloquium at Oxford where he illustrated the inner workings of black holes by drawing an analogy from daily life. The space around a black hole is flowing like a river towards the central singularity. Figure 8.7 shows a slight modification of Unruh's original idea (Unruh, 2008), a river populated by fish. Suppose the fish cannot swim faster than a certain maximal speed, say c'. If the speed of the river, u, exceeds the speed of the fish they are doomed. The fish are not able to swim upstream any more, but gradually drift towards the waterfall, Singularity Falls. The fish would not

---

[10] There is indirect astronomical evidence that black holes exist. For example, the core of each galaxy seems to harbour a supermassive black hole of about a million Solar masses.

Fig. 8.7. Aquatic analogue of
the event horizon.
(Reproduced with friendly
permission of Yan
Nascimbene.)

The "Point of no return" is the analogue of the event
horizon for fish in a river. To the left of it, the water
flows faster than a fish can swim. So, if a fish happens
to drift beyond this line, it can never get back upstream; it is doomed to be crushed in Singularity Falls.

notice anything unusual at the place where the water overtakes them,
the horizon. But beyond the horizon their fate is sealed.

Unruh's aquatic analogy is elevated from being an illustration of
black holes to a precise scientific analogue if the fish are replaced by
waves. While preparing a course on fluid mechanics, Unruh noticed and
published in 1981 (Unruh, 1981) that the equation of waves in moving
media is exactly the same as the equation of waves in the space–time
geometry of black holes. Both cases are mathematically identical. Grig-
ory E. Volovik (Volovik, 2003) discovered such an analogy during his
undergraduate studies in the 1960s, but was discouraged from pursuing
and publishing his idea. Richard White (1973) and Vincent Moncrief
(1980) noticed that sound perceives a medium as a space–time geometry. Matt Visser (1998) and also Paul Piwnicki and I (Leonhardt and
Piwnicki, 1999, 2000) independently rediscovered the analogy between
moving media and gravitation.[11]

Horizons for waves are easy to make (and do not infringe the rights
of animals as horizons for fish certainly would). In fact, horizons can
be made in a simple kitchen experiment. Just let water flow out of the
tap onto a metal surface. Figure 8.8 shows what happens there. Inside
a certain ring the water surface is very smooth, but outside waves are
appearing. Where the stream from the tap hits the metal, the water flows
faster than the wave velocity. Then the water flows outwards and gets
slower. Waves cannot enter the region where the water exceeds their
velocity, but they are formed at the critical radius where the water has
reached the wave velocity. In fluid mechanics (Landau and Lifshitz,

[11] Sir Isaac Newton toyed with the idea that gravitation is caused by a medium. Andrei D.
Sakharov, inventor of the Soviet hydrogen bomb and the tokamak fusion reactor, dissi-
dent and Nobel Peace Prize winner in absentia, made Newton's idea precise (Sakharov,
1968).

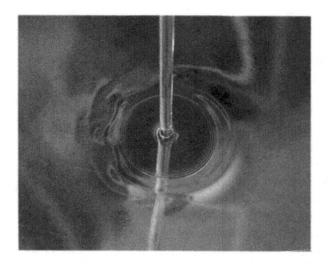

Fig. 8.8. White holes in the kitchen. Tap water establishes a white hole horizon for water waves, the ring shown in the picture. (Photo: Piotr Pieranski.)

Vol. VI, 1987), this phenomenon is known as the *hydraulic jump*. Seen from an astrophysics perspective, the hydraulic jump resembles a *white hole*, an object that nothing can enter. White holes are time-reversed black holes – if we run a movie of the kitchen experiment backwards in time and turn it upside down, the water seems to flow towards a waterfall, just like in Unruh's analogy. White hole horizons also form when rivers flow into the sea, causing the blocking of waves coming from the sea. Horizons are ubiquitous.

8.3.1 Motivation

Analogues of the event horizon occur in daily life in various disguises – so why are they worth studying? How far does the analogy go and where are the differences to astrophysical black holes? The true event horizon of the black hole is universal – nothing can escape it, because here space itself seems to flow at the speed of light in vacuum, c. All objects are drawn into the black hole, not only specific waves (or unfortunate fish). Also the physics involved in establishing the horizon is very different. Black holes are formed by masses curving the geometry of space and time sufficiently strongly (Landau and Lifshitz, Vol. II, 1987; Misner et al., 1999). They are the remnants of the gravitational collapse of stars.[12] In black hole analogues, moving media directly establish effective space–time geometries for waves.

[12] J. Robert Oppenheimer, father of the atomic bomb, wrote the first paper on gravitational collapse (Oppenheimer and Snyder, 1939) together with his then graduate student Hartland S. Snyder in 1939, before Oppenheimer joined the Manhattan Project as its leader. John A. Wheeler coined the term "black hole" in 1969.

Any medium appears to change the geometry of space for the waves it carries. For example, according to Fermat's principle (Born and Wolf, 1999; Leonhardt and Philbin, 2009), light in static media follows paths of the shortest or longest optical length where length is measured with respect to the refractive index. The refractive index thus establishes a spatial geometry where light rays are no longer straight, but bent, in general. The bending of light in materials causes many optical illusions and it may cause the ultimate illusion, invisibility (Leonhardt, 2006a, 2006b; Pendry et al., 2006; Schurig et al., 2006; Leonhardt and Tyc, 2009). Moving media not only appear to change the geometry of space, but also of space and time, as we are going to discuss in Section 8.3.4. But why are horizon analogues more than mere illustrations of black holes, why are they worth the time and money to investigate them in quantum optics?

In 1974 Stephen Hawking published a theoretical prediction (Hawking, 1974) about the quantum physics of black holes that has been one of the intellectually most influential results of theoretical astrophysics, but that most probably can only be observed in analogues of the event horizon. Hawking predicted that black holes are not black after all, but radiate. Similar to the radiation perceived by accelerated observers, the black hole appears to emit thermal radiation known as *Hawking radiation*. Furthermore, Hawking radiation consists of Einstein–Podolski–Rosen correlated particle pairs. In each pair, one particle is created outside of the horizon and the other inside, one is escaping into space and the other is falling into the singularity. As the Hawking partners are separated by the horizon, their correlations can never be observed, though.

One can apply simple Newtonian physics to "deduce" Hawking's result (Hawking, 1974) from the temperature (8.53) of acceleration. Let us follow the ideas of John Michell who suggested the possible existence of dark stars – black holes – in 1783 (Schaffer, 1979). Imagine a spherically symmetric star of mass M. The horizon is the area at the radius r where the escape velocity is the speed of light c, where a particle of mass m would need the kinetic energy $mc^2/2$ in order to withstand the gravitational potential GmM/r (with G being Newton's gravitational constant). The radius r is known as the *Schwarzschild radius* (Landau and Lifshitz, Vol. II, 1987; Misner et al., 1999)

$$r = \frac{2GM}{c^2}. \tag{8.54}$$

Karl Schwarzschild (Schwarzschild, 1916) and Roy P. Kerr (Kerr, 1963) obtained the same result (8.54) from their exact solutions of Einstein's equations in general relativity (Landau and Lifshitz, Vol. II, 1987; Misner et al., 1999) for non-rotating and rotating black holes, respectively. Let us continue with Newtonian physics. We obtain from Newton's inverse square law the acceleration at the horizon,

$$a = \frac{GM}{r^2} = \frac{c^2}{2r}.$$

(8.55)

If we regard this acceleration as the one responsible for particle creation we obtain from Unruh's formula (8.53)

$$k_B T = \frac{\hbar c}{4\pi r}$$

(8.56)

and from the Schwarzschild radius (8.54) Hawking's result (Hawking, 1974)

$$k_B T = \frac{\hbar c^3}{8\pi GM}.$$

(8.57)

The natural constants in Hawking's formula (8.57) belong to three distinct areas of physics: G and c represent gravitation and relativity – Einstein's general relativity, $\hbar$ stands for quantum mechanics and the Boltzmann constant k_B belongs to thermodynamics and statistical physics. The natural constants already indicate that Hawking radiation connects three seemingly disconnected areas of physics. The physics of the very large, general relativity, meets the physics of the quantum world. Hawking's result (8.57) also supports Jacob D. Bekenstein's idea (Bekenstein, 1973, 2001) that black holes have a thermodynamic entropy that is proportional to the area of the horizon. Candidates for the quantum theory of gravity, string theory (Green et al., 1987) or loop quantum gravity (Rovelli, 1998) have been at pains to reproduce the Bekenstein–Hawking entropy. Hawking radiation has been the acid test for quantum gravity.

Is it possible to gain empirical evidence for Hawking radiation? Let us put numbers in Hawking's formula (8.57). A star needs to be sufficiently massive in order to collapse into a black hole. Otherwise it turns into a neutron star where the Fermi pressure (Landau and Lifshitz, Vol. IX, 1980) of the neutrons balances gravity. Subrahmanyan Chandrasekhar calculated, during his sea voyage from India to England in 1930, that the minimal mass is 1.44 Solar masses. In this case, the Schwarzschild radius is about 4.5 km and the Hawking temperature (8.56) is as low as $4 \cdot 10^{-8}$ K, nearly eight orders of magnitude below the 3 K temperature of the cosmic microwave background. More massive black holes are even colder. So the testbed of some of the most advanced theories of physics seems destined to remain theory itself, unless Hawking radiation is observed in analogues of the event horizon (Volovik, 2003; Unruh and Schützhold, 2007; Philbin et al., 2008a).

8.3.2 Trans-Planckian problem

What can we learn from artificial black holes? In our "derivation" of Hawking's result (8.57) we combined eighteenth century physics with some aspects of the quantum vacuum, the twenty-first century aether. That our result is correct comes close to a miracle. Now, if Hawking radiation is such a robust phenomenon that it almost follows from Newtonian gravity, how can it seriously discriminate between the various contenders for the quantum theory of gravity? What is the validity range of Hawking's formula (8.57)? In the next sections we present a "proper" derivation of Hawking radiation where we are going to encounter a problem known as the *trans-Planckian problem*. What is it? Consider the analogue of a white hole horizon. Imagine waves propagating against the current. As the flow speed increases the waves begin to pile up in front of the horizon where they come to a standstill and have no more room to oscillate further – their wavelengths dramatically decrease to zero. Since black holes are time-reversed white holes, the same wave phenomenon occurs there, but in reverse order: the waves emerging just in front of the horizon originate from waves with very small wavelengths. In astrophysics, Hawking radiation would originate from waves with wavelengths at extremely short scales, shorter than the *Planck length*

$$l_P = \sqrt{\frac{\hbar G}{c^3}}, \tag{8.58}$$

about $1.6 \cdot 10^{-35}$ m, the only length scale obtainable from $\hbar$, G and c that thus may play a role in quantum gravity. Some as-yet-unknown mechanism ought to regularize the waves at the horizon. The event horizon seems to probe a piece of unknown trans-Planckian physics. So, this being the case, can we trust Hawking's prediction (Hawking, 1974)? This issue is called the trans-Planckian problem.

In horizon analogues, the equivalent of the trans-Planckian regularization mechanism is very simple: it is dispersion. When waves change their wavelengths or frequencies in media their velocities change. For example, the speed of light in materials depends on the frequency. Dispersion limits the extreme frequency shifting or wavelength reduction at horizons. For instance, in the kitchen experiment shown in Fig. 8.8, the capillary water waves (Landau and Lifshitz, Vol. VI, 1987) increase their velocities with decreasing wavelength. Hence they can partly penetrate the horizon, as Fig. 8.8 also shows. The concept of the horizon becomes fuzzy. What precisely is the wave velocity that matters in Hawking radiation, the group or the phase velocity? The *group velocity* is the speed at which wavepackets move, whereas the *phase velocity* is the speed at which the wave oscillates; both differ in the case of dispersion (Born and Wolf, 1999).

In astrophysics, dispersion could become important near the Planck scale (8.58). For instance, if space were discrete at a fundamental level instead of continuous, dispersion would herald the discreteness of space. Consider for example the following simple toy model: a one-dimensional chain of coupled harmonic oscillators with amplitudes $\hat{A}_k$ that represent the electromagnetic potential at discrete locations at Planck's length away from each other. Each oscillator interacts with its nearest neighbours according to the equation of motion

$$\frac{\partial^2 \hat{A}_k}{\partial t^2} = \kappa \left(\hat{A}_{k+1} - \hat{A}_k \right) + \kappa \left(\hat{A}_{k-1} - \hat{A}_k \right). \tag{8.59}$$

In the continuum limit, we regard $\hat{A}_k(t)$ as $\hat{A}(x_k, t)$ and obtain, by Taylor expansion of $\hat{A}(x, t)$ to fourth order in $x_{k+1} - x_k = l_P$ and identifying $\kappa\, l_P^2$ with c^2,

$$\frac{\partial^2 \hat{A}}{\partial t^2} = c^2 \left(\frac{\partial^2 \hat{A}}{\partial x^2} + \frac{l_P^2}{12} \frac{\partial^4 \hat{A}}{\partial x^4} \right), \tag{8.60}$$

a modification of the wave equation (8.23). In this case, plane waves oscillate as $\mathcal{A} \exp(ikx - i\omega t)$ with the dispersion relation

$$\frac{\omega^2}{c^2} = k^2 \left(1 - \frac{l_P^2 k^2}{12} \right). \tag{8.61}$$

The phase velocity of light, ω/k, would depend on the wave number k: discreteness appears as dispersion.

In laboratory analogues of the event horizon the dispersion is certainly much stronger than near astrophysical horizons and it also depends on material properties that can be tailored and tuned. One can study a large range of radiation phenomena that resemble Hawking's effect and find the validity range of Hawking's formula (8.57), testing it to the limit. Experimental tests are both a sobering and an exhilarating experience. They expose the limits of our understanding, revealing blunders and errors, but, occasionally, they also show how things work, sometimes far beyond expectation. In the absence of empirical evidence science becomes sterile, but with the right clues from experimental data the natural sciences are "unreasonably effective", as Wigner put it (Wigner, 1960). Then he writes "The miracle of the appropriateness of the language of mathematics for the formulation of the laws of physics is a wonderful gift which we neither understand nor deserve."

Observing Hawking radiation in the laboratory requires the sophisticated tools of modern experimental physics. Here we focus on optical analogues of the event horizon.[13] At first glance, making optical black holes appears to be a hopeless task. According to Unruh's analogy we need media that move at the speed of light in the material. This is possible in principle, because optical media typically reduce the speed of light in vacuum by the refractive index, but seems very difficult in practice. One could slow down light such that it matches normal velocities (Leonhardt and Piwnicki, 2000) or is brought to a complete standstill (Leonhardt, 2002). However, it turns out (Philbin et al., 2008a) that moving an optical medium at the speed of light is easy. In fact, it happens all the time in optical telecommunication. In fibre-optical communication, the information carriers are light pulses confined to the core of optical fibres. Each pulse adds a slight contribution to the refractive index of the fibre (Agrawal, 2001) (a contribution proportional to the intensity profile of the pulse), as if the pulse were adding an extra piece of glass. This is called the Kerr effect (Agrawal, 2001). The contribution to the refractive index moves with the pulse. The pulse thus establishes a moving medium (Philbin et al., 2008a), although nothing material is moving, only light. In a frame co-moving with the pulse, the material of the fibre appears to race in the opposite direction. The pulse stands still and imprints, by the Kerr effect (Agrawal, 2001), the additional contribution δn to the refraction index n_0 of the fibre. In the simplest case, the pulse does not change its shape and so the total refractive index $n = n_0 + \delta n$ depends only on the spatial coordinate x,

$$n = n_0 + \delta n, \quad \delta n = \delta n(x). \tag{8.62}$$

Pulses in optical fibres appear as media moving at uniform speed but having a slightly non-uniform refractive index profile.[14] The effective medium naturally moves at the speed of light, because it is made by light itself. In materials, the velocity of light depends on the frequency and sometimes on the polarization as well; different colours move at different speeds. This works to our advantage, because we can launch probe light that follows the pulse at a faster velocity, but is slowed down due to the additional contribution to the refractive index, experiencing horizons, see Fig. 8.9.

---

[13] Hawking radiation can also be observed in Bose–Einstein condensates (Garay et al., 2000; Pitaevskii and Stringari, 2003) and liquid helium 3 (Volovik, 2003).

[14] This simple model describes the essence of fibre-optical horizons (Philbin et al., 2008a); the reality is more complex, though (Agrawal, 2001; Philbin et al., 2008b).

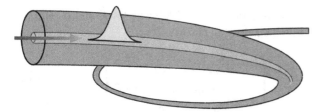

Fig. 8.9. Fibre-optical analogue of the event horizon. A light pulse in an optical fibre slows down a continuous wave of light attempting to overtake it. The pulse acts like a medium that moves at the group velocity of the pulse. This medium establishes the optical analogues of horizons at the places where the effective speed of light matches the group velocity of the pulse, a white hole horizon at the trailing end and a black hole horizon at the front (Philbin et al, 2008a).

8.3.3 Light in moving media

Let us focus on the simplest model for a horizon in optics, made by a one-dimensional moving medium. This case closely resembles the fibre-optical analogue of the event horizon (Philbin et al., 2008a). Our model also describes the essentials of astrophysical horizons, because near the event horizon the wavelengths become so small that the lateral extension of the horizon is irrelevant. Horizon physics is one-dimensional in space, or rather, as we will see, it is 1+1–dimensional in space–time (Bekenstein and Mayo, 2001).

Consider a medium moving with the velocity profile $u(x)$ on the x axis. As in the previous sections of this chapter, the light is assumed to propagate along the x axis with fixed polarization. We describe the quantum field of light by the vector-potential component $\hat{A}(x, t)$ that generates the electromagnetic field strengths (8.16). The moving medium appears in the constitutive equations. We find them by imagining a locally co-moving inertial frame in each point in space. In each co-moving frame the medium is locally at rest and so it obeys the constitutive equations (2.2) here, but for the transformed field components denoted with primes,

$$\hat{D}' = \varepsilon_0 \varepsilon \hat{E}', \quad \hat{B}' = \mu_0 \mu \hat{H}', \quad \varepsilon_0 \mu_0 = c^{-2}. \tag{8.63}$$

From the transformations (8.19) and (8.20) we obtain the *constitutive equations in the laboratory frame*,

$$\hat{D} = \varepsilon_0 \gamma^2 \left(\varepsilon(\hat{E} - u\hat{B}) + \frac{u}{\mu}\left(\hat{B} - \frac{u}{c^2}\hat{E}\right) \right),$$

$$\hat{H} = \varepsilon_0 \gamma^2 \left(\frac{c^2}{\mu}\left(\hat{B} - \frac{u}{c^2}\hat{E}\right) + u\,\varepsilon(\hat{E} - u\hat{B}) \right) \tag{8.64}$$

where γ is the relativistic factor (8.5). We relate the $\hat{D}$ and $\hat{H}$ fields to the potential $\hat{A}$ with the help of the matrix

$$g^{\alpha\beta} = \begin{pmatrix} \dfrac{\varepsilon c^2 - \mu^{-1}u^2}{c^2 - u^2} & \dfrac{(\varepsilon - \mu^{-1})uc}{c^2 - u^2} \\[3mm] \dfrac{(\varepsilon - \mu^{-1})uc}{c^2 - u^2} & \dfrac{\varepsilon u^2 - \mu^{-1}c^2}{c^2 - u^2} \end{pmatrix}. \qquad (8.65)$$

We substitute the relations (8.16) in the constitutive equations (8.64) and obtain, adopting the notation (8.10) for the space–time derivatives,

$$\hat{D} = -\varepsilon_0 c \sum_\beta g^{0\beta}\, \partial_\beta \hat{A}, \quad \hat{H} = -\varepsilon_0 c^2 \sum_\beta g^{1\beta}\, \partial_\beta \hat{A}. \qquad (8.66)$$

Maxwell's equation (8.21) is valid in both the co-moving frames and the laboratory frame. We obtain from Eqs. (8.21) and (8.66) the *wave equation of light in moving media*

$$\sum_{\alpha\beta} \partial_\alpha\, g^{\alpha\beta}\, \partial_\beta \hat{A} = 0. \qquad (8.67)$$

For the quantum field theory of light in moving media we need, in addition to the wave equation (8.67), the correct scalar product between orthonormal modes (2.24) in order to determine the creation and annihilation operators (2.25). Let us try the scalar product (2.17) but with the constitutive equations (8.66) in the moving medium,

$$(A_1, A_2) = \frac{1}{i\hbar} \int \left(A_1^* D_2 - A_2 D_1^* \right) \mathrm{d}x. \qquad (8.68)$$

As this scalar product agrees with the one in the case of non-moving media, with the right prefactor, all we need to do is prove that (A_1, A_2) is a conserved quantity,

$$\frac{\mathrm{d}(A_1, A_2)}{\mathrm{d}t} = 0. \qquad (8.69)$$

We obtain from Maxwell's equation (8.21) and the relationships (8.16)

$$\frac{\partial(AD)}{\partial t} = \frac{\partial A}{\partial t}D - A\frac{\partial H}{\partial x} = -ED + BH - \frac{\partial(AH)}{\partial x}. \qquad (8.70)$$

The quantity $ED - BH$ is invariant in Lorentz transformations,[15] as we see from the transformations (8.20) and (8.19) or rather the inverse of the latter,

$$ED - BH = \gamma^2\left(E' + uB'\right)\left(D' + \frac{u}{c^2}H'\right) - \gamma^2\left(B' + \frac{u}{c^2}E'\right)\left(H' + uD'\right)$$
$$= E'D' - B'H'. \qquad (8.71)$$

---

[15] The quantity $\frac{1}{2}(ED - BH)$ is the Lagrangian density (Landau and Lifshitz, Vol. II, 1987) of the electromagnetic field in moving media, a Lorentz invariant.

When we substitute the expressions (8.70) and (8.71) in the integral of the time derivation of the scalar product (8.68) the spatial derivative $\partial(AH)/\partial x$ produces vanishing boundary terms and the $E'D' - B'H'$ cancel due to the constitutive equations (8.63). This argument proves the crucial condition (8.69). Our quantum field theory of light in moving media is complete.

8.3.4 Geometry of light

The appearance of the matrix $g^{\alpha\beta}$ in the wave equation (8.67) is not a coincidence – in the following we show that moving media establish effective space–time geometries. Assume, for simplicity, that the medium is *impedance matched* (Jackson, 1998) to the vacuum, meaning that the electric permittivity ε and the magnetic permeability μ are equal, but may vary in space,

$$\varepsilon = \mu = n(x). \tag{8.72}$$

The function $n(x)$ describes the refractive index profile of the medium. In practice, the impedance matching condition is rarely satisfied, but our model still captures the essence of horizon physics (Philbin et al., 2008b).

Similar to the wave propagation in empty space considered in Section 8.1.3 we define the d'Alembertian variables

$$t_\pm = t - \int \frac{dx}{v_\pm}, \tag{8.73}$$

but here with respect to the velocity-added speed of light in the medium according to Einstein's addition theorem (8.7)

$$v_\pm = \frac{u \pm \dfrac{c}{n}}{1 \pm \dfrac{u}{cn}}. \tag{8.74}$$

The velocities may vary in space – hence the integral in the variables (8.73). We expect that the $t_\pm$ describe right- and left-moving waves. To prove this idea, we need to write the wave equation in new variables. We obtain from the chain rule of differentiation $\partial/\partial t = \partial/\partial t_+ + \partial/\partial t_-$ and $\partial/\partial x = -v_+^{-1}\partial/\partial t_+ - v_-^{-1}\partial/\partial t_-$. By solving for $\partial/\partial t_\pm$ we get

$$\begin{aligned}
\frac{\partial}{\partial t_+} &= \frac{v_+}{v_+ - v_-}\left(\frac{\partial}{\partial t} + v_-\frac{\partial}{\partial x}\right), \\
\frac{\partial}{\partial t_-} &= \frac{v_-}{v_- - v_+}\left(\frac{\partial}{\partial t} + v_+\frac{\partial}{\partial x}\right).
\end{aligned} \tag{8.75}$$

From Einstein's addition theorem (8.74), defining the velocities $v_\pm$, follows

$$\frac{1}{v_+ - v_-} = \frac{c^2 n^2 - u^2}{2nc(c^2 - u^2)}, \quad \frac{v_+ + v_-}{v_+ - v_-} = \frac{(n^2 - 1)uc}{n(c^2 - u^2)}. \tag{8.76}$$

We obtain for the product of the derivatives (8.75), by comparing the expressions (8.76) with the coefficients (8.65) of the $g^{\alpha\beta}$ matrix,

$$4 \frac{\partial^2}{\partial t_+ \, \partial t_-} = \frac{c^2}{\Lambda} \sum_{\alpha\beta} \partial_\alpha \, g^{\alpha\beta} \, \partial_\beta \tag{8.77}$$

where Λ denotes the dimensionless factor

$$\Lambda = \frac{c}{2} \frac{v_- - v_+}{v_+ \, v_-} = \frac{n(c^2 - u^2)}{c^2 - n^2 u^2}. \tag{8.78}$$

Consequently, the wave equation (8.67) has the simple solution (8.26) à la d'Alembert. In impedance-matched moving media (8.72), light waves thus propagate either to the right or to the left; they do not mix – light is not reflected in impedance-matched one-dimensional media.[16] The scalar product of modes (8.68) also turns out to be separated into right-moving and left-moving parts. We obtain from the geometric formula (8.66) for the dielectric displacement and the expansions (8.76)

$$(A_1, A_2) = \frac{i\varepsilon_0 c}{\hbar} \int (A_1^* \, \partial^0 A_2 - A_2 \, \partial^0 A_1^*) dx \qquad \text{with}$$

$$\partial^0 = \sum_\beta g^{0\beta} \partial_\beta \qquad \text{(note the upper index)}$$

$$= \frac{2}{v_+ - v_-} \frac{\partial}{\partial t} + \frac{v_+ + v_-}{v_+ - v_-} \frac{\partial}{\partial x}$$

$$= \frac{1}{v_+} \frac{\partial}{\partial t_+} - \frac{1}{v_-} \frac{\partial}{\partial t_-}, \tag{8.79}$$

and by transforming the integral to t_+ and t_- variables, where $dt_\pm = -v_\pm^{-1} dx$ by definition (8.73),

$$(A_1, A_2) = -\frac{i\varepsilon_0 c}{\hbar} \int \left(A_1^* \frac{\partial A_2}{\partial t_+} - A_2 \frac{\partial A_1^*}{\partial t_+} \right) dt_+$$

$$+ \frac{i\varepsilon_0 c}{\hbar} \int \left(A_1^* \frac{\partial A_2}{\partial t_-} - A_2 \frac{\partial A_1^*}{\partial t_-} \right) dt_-. \tag{8.80}$$

As the wave propagation is exactly the same as in empty space – only written in different coordinates – we express the d'Alembertian variables in terms of transformed Cartesian coordinates,

$$t_\pm = t' \mp \frac{x'}{c}, \tag{8.81}$$

---

[16] In three dimensions, impedance matching is not sufficient to completely suppress scattering and reflection (Leonhardt and Philbin, 2009).

where space seems to be empty. The impedance-matched moving medium appears to perform a mere coordinate transformation from a virtual empty space (x', t') to physical space (x, t). The medium is a transformation medium (Leonhardt and Philbin, 2006; 2009).

What is the geometry of the space–time coordinate transformation (8.73) and (8.81)? We simply write the metric (8.12) of the empty virtual space–time in terms of the physical coordinates,

$$ds'^2 = c^2 dt'^2 - dx'^2 = c^2 dt_+ dt_- = \Lambda \sum_{\alpha\beta} g_{\alpha\beta} \, dx^\alpha dx^\beta \qquad (8.82)$$

with the prefactor (8.78) and the matrix

$$g_{\alpha\beta} = \begin{pmatrix} \dfrac{\varepsilon^{-1}c^2 - \mu u^2}{c^2 - u^2} & \dfrac{(\mu - \varepsilon^{-1})uc}{c^2 - u^2} \\[3ex] \dfrac{(\mu - \varepsilon^{-1})uc}{c^2 - u^2} & \dfrac{\varepsilon^{-1}u^2 - \mu c^2}{c^2 - u^2} \end{pmatrix}$$

for $\varepsilon = \mu = n$. The matrix (8.83) is the inverse of the matrix (8.65) that occurs in the wave equation. For light

$$ds' = 0, \qquad (8.83)$$

so the prefactor Λ is of little consequence in the space–time geometry (8.82). It is called a *conformal factor*. The condition (8.83) describes light cones in virtual space–time, so the matrix $g_{\alpha\beta}$ determines the deformation of light cones in physical space–time. The matrix (8.83) is the metric tensor (Landau and Lifshitz, Vol. II, 1987; Misner et al., 1999). A moving medium makes a space–time geometry.

Walter Gordon published the connection between moving media and geometries in a well-forgotten 1923 paper (Gordon, 1923), an insight that was independently rediscovered later on (Quan, 1957, 1957/58; Leonhardt and Piwnicki, 1999). This geometric interpretation of media is not restricted to one–dimensional propagation, but is valid in three-dimensional space as well. According to Einstein's general theory of relativity (Landau and Lifshitz, Vol. II, 1987; Misner et al., 1999) gravitation is mediated by the geometry of space–time. Gravity thus appears like a medium and media appear like gravity.

8.3.5 Hawking radiation

We argued in Section 8.3.2 that astrophysical event horizons are essentially one-dimensional in space and 1+1-dimensional in space–time (Bekenstein and Mayo, 2001), because the near-Planckian decrease in wavelength makes all other dimensions obsolete at the horizon. On the other hand, we have seen in the previous section that the one-dimensional wave propagation in space–time geometries of the type (8.83) or, equivalently, in moving media reduces to a mere coordinate transformation of waves in empty space. Horizons are transformation media. How do they transform the quantum vacuum?

We focus on a single horizon and assume that the medium moves in negative x direction such that u is negative. The horizon is formed where the velocity of the medium matches the speed of the light in the medium,

$$u = -\frac{c}{n}. \tag{8.84}$$

We define our coordinate system such that the horizon is located at $x = 0$. Light propagating against the current is brought to a standstill here – the right-moving velocity v_+ vanishes and the integral (8.73) in the t_+ variable contains a pole. To investigate the behaviour of t_+ near the horizon, we expand v_+ to linear order in the distance x from the horizon,

$$v_+ = \alpha x, \quad \alpha \equiv \left.\frac{dv_+}{dx}\right|_0. \tag{8.85}$$

We obtain from Eqs. (8.74) and (8.84)

$$\alpha = \frac{1}{1 - n^{-2}} \left.\left(\frac{du}{dx} - \frac{c}{n^2}\frac{dn}{dx}\right)\right|_0. \tag{8.86}$$

A black hole is an object where nothing can escape beyond the horizon. In our model – the moving medium – the modulus of the velocity u must exceed c/n for $x < 0$. As u is negative, it must increase around $x = 0$ to reach a smaller modulus, and so α is positive for black hole horizons. Conversely, a white hole is an object nothing can enter, where $|u|$ exceeds c/n for $x > 0$ and hence α is negative. Let us focus on the black holes, because, as white holes are time-reversed black holes, we can deduce their quantum properties once we understand the black holes.

Near the horizon, we get from the coordinate transformation (8.73) and the linearization (8.85)

$$t_+ = t - \frac{\ln|x|}{\alpha}. \tag{8.87}$$

So it would take an exponential decrease in spatial distance x to advance linearly in the time variable t_+ of the wave $A(t_+)$. Right-moving waves are exponentially compressed near the horizon. Formula (8.87) is thus consistent with our picture of the extreme reduction in wavelength and, moreover, it casts this picture in quantitative terms. For fixed t, the time t_+ in Eq. (8.87) and, more generally, in Eq. (8.73) runs from $+\infty$ to $-\infty$ for $x > 0$, where waves farther away come from times t_+ further in the past, and from $-\infty$ to $+\infty$ for $x < 0$. Consequently, a wave $A(t_+)$ with $-\infty < t_+ < +\infty$ may spend its entire propagation time on either side of the horizon: the black hole separates space–time into two separate regions. In one region waves do escape, in the other they are drifting away (and would fall into the singularity for astrophysical black holes). On the left side of the horizon t_+ advances forward into the future instead of reaching into the past. Similar to the Rindler waves we discussed in Section 8.2.3, we expect that the norms (8.68) of positive frequency waves are negative there. Indeed, we obtain for stationary waves $A = \mathcal{A}\exp(-\mathrm{i}\omega t_+)$ from the expression (8.80) of the scalar product

$$(A_1, A_2) = -\frac{2\varepsilon_0 c\omega}{\hbar}\mathcal{A}^2 \int \exp\left(-\mathrm{i}(\omega_2 - \omega_1)t_+\right) \mathrm{d}t_+ \qquad (8.88)$$

in the limit $\omega_1 \to \omega_2 \equiv \omega$. The scalar product is negative when t_+ runs from $-\infty$ to $+\infty$. We obtain from the logarithmic time (8.87) in complete analogy to the Rindler modes (8.39) and (8.40)

$$A_R = \mathcal{A}\,\Theta(x)\,e^{-\mathrm{i}\omega t_+} \ \sim \ \mathcal{A}\,\Theta(x)\,x^{\mathrm{i}\omega/\alpha}\,e^{-\mathrm{i}\omega t}, \qquad (8.89)$$
$$A_L = \mathcal{A}\,\Theta(-x)\,e^{\mathrm{i}\omega t_+} \ \sim \ \mathcal{A}\,\Theta(-x)\,(-x)^{\mathrm{i}\omega/\alpha}\,e^{\mathrm{i}\omega t}, \qquad (8.90)$$

near the horizon $x \sim 0$. As the A_R and A_L are confined to either side of the horizon, they describe the outgoing modes, for example light that has managed to escape from the horizon. What are the ingoing modes? At a beam splitter, the light in an ingoing mode is partially reflected and partially transmitted, but it is transmitted in only one direction; no light comes from the other direction of incidence. At a horizon, the ingoing modes continue to propagate in one direction; in our case they are purely right moving. But, for each frequency, we have two right-moving ingoing modes, one partially escaping to the right and one drifting away on the left side of horizon. Suppose that far away from the horizon the medium has a uniform refractive index and moves at uniform speed. In such regions v_+ is constant; v_+ is positive at the right side of the horizon and negative on the left side. Right-moving modes are given by the integrals

$$A_1 = \int_0^\infty \tilde{A}(\omega)\exp\left(+\mathrm{i}\frac{\omega}{v_+}x - \mathrm{i}\omega t\right)\mathrm{d}\omega \quad \text{for } v_+ > 0, \qquad (8.91)$$

$$A_2 = \int_0^\infty \tilde{A}(\omega)\exp\left(-\mathrm{i}\frac{\omega}{v_+}x + \mathrm{i}\omega t\right)\mathrm{d}\omega \quad \text{for } v_+ < 0. \qquad (8.92)$$

Similar to the Minkowski modes seen by the accelerated observer, as discussed in Section 8.2.3, the ingoing modes A_1 and A_2 are analytic in x on the upper half-plane (Ablowitz and Fokas, 1997). Imagine we trace the ingoing waves back in time to their origin near the horizon. The wave propagation (8.26) with the time variable t_+ given by the integral

(8.73) never develops singularities for analytic $v_+(x)$, except at the zeros of $v_+(x)$, at horizons. Apart from isolated points, analytic wavepackets remain analytic. So the ingoing modes must have originated from analytic waves, but the outgoing waves are not analytic, because they are discontinuous (Ablowitz and Fokas, 1997). Some ingoing waves must have tunnelled through the horizon and escaped.

To calculate the tunnelling amplitude we follow the same procedure (Damour and Ruffini, 1976) as in the case of the accelerated observer. As the modes beyond the horizon oscillate with negative frequencies, we employ the Bogoliubov transformation (8.49) to combine the non-analytic modes (8.89) and (8.90) with relative weight (8.50) and

$$v = \frac{\omega}{\alpha}. \tag{8.93}$$

The outgoing mode operators are Bogoliubov transformations of the ingoing ones, mixing annihilation and creation operators. The black hole horizon thus creates the Einstein–Podolski–Rosen states (7.20) from the ingoing quantum vacuum. Seen from the outside of the black hole, the photon pairs constitute a thermal radiation (3.49), *Hawking radiation*, with temperature T related to the squeezing parameter (7.29),

$$k_B T = \frac{\hbar \alpha}{2\pi}. \tag{8.94}$$

The black hole appears as a black body radiator with a universal temperature across the spectrum. In contrast, typical laboratory sources of Einstein–Podolski–Rosen pairs, parametric amplifiers, have frequency-dependent noise temperatures. For astrophysical black holes, one can use suitable space–time coordinates, for example Painlevé–Gullstrand coordinates (Misner et al., 1999), where the radial wave equation reduces to the one–dimensional equation (8.67) near the horizon (Visser, 1998), with

$$\alpha = \frac{c^3}{4GM}, \tag{8.95}$$

thus proving Hawking's formula (8.57). The smaller the astrophysical black hole, the stronger is the gravitational force (8.55) at the horizon and, in turn, the more intense is the Hawking radiation. In black hole analogues, the Hawking temperature (8.94) is set by the gradients of the velocity u and the refractive index n according to relation (8.86). The steeper the gradient is, the higher is the Hawking temperature. As white holes are time-reversed black holes, the ingoing modes become the outgoing modes and vice versa at white holes. The mode transformation is reversed, but the Hawking temperature stays the same.

In practice, it matters a great deal whether a subtle effect such as Hawking radiation can be made sufficiently strong. Let us briefly discuss our realistic example of an artificial black hole, the fibre-optical analogue of the event horizon (Philbin et al., 2008a). As mentioned in Section 8.3.2, an ultrashort, intense light pulse establishes a moving medium in the frame co-moving with the pulse. If the pulse propagates with group velocity v_p the medium appears to move in the opposite direction to the pulse with velocity $u = -v_p$. In the co-moving frame, the refractive index contribution δn due to the Kerr effect of the pulse (Agrawal, 2001) varies in space, but not in time. Hence we can use our procedure for calculating the Hawking temperature (8.94) here. However, Hawking radiation is observed in the laboratory frame. We need to Lorentz-transform our results (8.94) and (8.86) and express it in experimentally accessible quantities normally used in non linear fibre optics (Agrawal, 2001). As we perform a Lorentz transformation from the co-moving inertial frame with coordinates x and t, the laboratory coordinates are the primed ones here. Pulses are described as wavepackets in retarded time $\tau = t' - x'/v_p$ (not to be confused with proper time (8.14) here). We obtain from the Lorentz transformation (8.18) of space–time gradients

$$\frac{dn}{dx} = \frac{d\delta n}{dx} = \gamma \left(\frac{\partial \delta n}{\partial x'} - \frac{u}{c^2} \frac{\partial \delta n}{\partial t'} \right) = \frac{1}{\gamma u} \frac{d\delta n}{d\tau} = -\frac{n}{\gamma c} \frac{d\delta n}{d\tau} \tag{8.96}$$

using the inverse of relation (8.18), the definition (8.5) and, in the last step, the horizon condition (8.84). The Kerr contribution typically is very small, even in fibres with high nonlinear response. So how can the small gradient (8.96) create observable Hawking radiation? We obtain for the α parameter (8.86), expressing $1 - n^{-2}$ as γ^{-2} by virtue of the horizon condition (8.84),

$$\alpha = \frac{\gamma}{n} \frac{d\delta n}{d\tau}, \tag{8.97}$$

but this parameter sets the Hawking temperature in the co-moving frame, not in the laboratory. In the co-moving frame, the frequencies (3.47) in the Planck spectrum (3.50) are significantly reduced by the Doppler effect such that modest temperatures turn out to have a strong effect. To see this, we Lorentz-transform the coordinates (8.6) in the outgoing waves (8.91) far away from the horizon where $n = n_0$. We obtain the transformed – Doppler-shifted – frequency

$$\omega = \gamma \left(1 + n_0 \frac{u}{c} \right) \omega_{lab} = \frac{\gamma}{n} \delta n \Big|_0 \omega_{lab}, \tag{8.98}$$

using the relation (8.62) and the horizon condition (8.84). The effective α parameter in the temperature (8.94) of the Planck spectrum (3.47) and (3.50) is transformed by the same factor. Consequently,

$$\alpha = \frac{1}{\delta n}\frac{\mathrm{d}\delta n}{\mathrm{d}\tau}. \tag{8.99}$$

The Hawking temperature does not depend on the magnitude of δn at all, only on the relative change of δn. Even tiny Kerr contributions δn may produce respectable Hawking temperatures, if δn varies on ultrafast time scales comparable to a single optical cycle (provided a horizon is established in the first place). This little miracle makes the observation of Hawking radiation possible in nonlinear fibre optics (Philbin et al., 2008a).[17]

Let us finally return to the trans-Planckian problem. In our derivation of Hawking's formula (8.57) we did not consider the dispersion emerging due to the near-Planckian physics or its analogue. What would happen? Picture the outgoing wavepackets traced back in time, as the space–time diagrams in Figs. 8.10 and 8.11 show. Near the horizon,

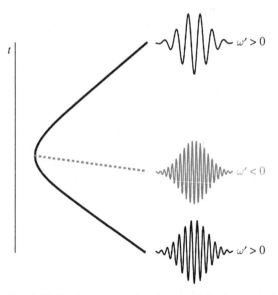

Fig. 8.10. Tracing wave packets backwards in time at the horizon of a black hole. Schematic space–time diagram showing a wave packet escaping into space (top), potentially reaching an observer. This wave packet oscillates at positive frequencies, but it originates from two distinct waves, one with positive and another one with negative frequencies, shown below the escaping wave packet in the space–time diagram (for times in the past). This mixing of positive and negative frequencies is the classical root of the quantum Hawking radiation. Note that the deflection of the incident waves at the horizon depends on the dispersion properties of the "space–time medium". In astrophysics, these properties are unknown, in contrast to laboratory analogues.

[17] Other aspects of nonlinear fibre optics enhance the luminosity of white hole horizons formed at the trailing end of pulses. In particular, optical shocks may create steep relative index gradients δn there (Agrawal, 2001).

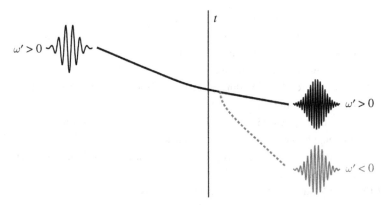

Fig. 8.11. Hawking partner. Schematic space–time diagram of a wave packet propagating against the "space–time flow" on the other side of the horizon, drifting towards the singularity of the black hole. Like the wave illustrated in Fig. 8.10, this wave packet originates from waves with positive and negative frequencies. These waves are mixtures of the escaping waves of Fig. 8.10 traced backwards in time; hence the escaping quanta and the in falling quanta form entangled partners.

the wavelength is dramatically reduced, and due to dispersion the speed of light changes. But the horizon is defined as the place where the speed of light matches the speed of the medium. The horizon seems to dissolve. Depending on whether the speed of light in the material falls or rises with the wavelength, the waves may originate from the slow or the fast side around the place where we suspect the horizon to be. But as long as there is an intermediate regime when the linearization (8.85) is applicable, Hawking's result should still stand (Brout et al., 1995b; Philbin et al., 2008a). Yet what exactly constitutes this regime has not been worked out so far. Furthermore, what happens for strong dispersion? Is the mere thread of a horizon sufficient to induce Hawking radiation? Which velocity matters in establishing the horizon – the phase or the group velocity (Born and Wolf, 1999)? What really is a horizon? Black holes have not lost their mystery or their lure.

8.4 Questions

8.1 Deduce the Lorentz transformations. Give a geometrical picture for them.

8.2 Derive Einstein's addition theorem of velocities.

8.3 What is Fresnel drag?

8.4 What is proper time and how is it related to the velocity of a moving object?

8.5 Why is the vector potential $\hat{A}$ not Lorentz-transformed in one-dimensional wave propagation?

8.6 How are the $\hat{E}$ and $\hat{B}$ fields transformed?

8.7 Explain how to transform the $\hat{D}$ and $\hat{H}$ fields such that the wave equation of light remains the same in all inertial frames.

8.8 Why does light propagate with c?

8.9 How are Rindler coordinates defined? How are they expressed in Minkowski coordinates?

8.10 Write the Minkowski metric in Rindler coordinates and use your result to give an interpretation for the η coordinate.

8.11 Develop two arguments why a particle on a trajectory with fixed Rindler ξ experiences uniform acceleration.

8.12 In the Rindler frame, the acceleration depends on ξ. Why is the acceleration not uniform in an accelerated frame?

8.13 Express the wave equation of light in empty space in terms of Rindler coordinates and find their stationary solutions.

8.14 Why do waves on the left side of the Rindler wedge oscillate at negative frequencies?

8.15 How are Rindler waves normalized?

8.16 Why do positive norm waves in Minkowski space consist of positive and negative norm Rindler waves?

8.17 Explain how to construct Minkowski modes from Rindler modes.

8.18 How does the Minkowski vacuum appear to the accelerated observer?

8.19 How does the accelerated particle experience friction?

8.20 What are analogues of black holes?

8.21 Deduce the temperature of Hawking radiation from Newtonian physics and the Unruh effect.

8.22 What is the trans-Planckian problem?

8.23 Why would a hypothetical discreteness of space appear as dispersion?

8.24 Deduce the constitutive equations of light in moving media.

8.25 Show that the scalar product of light in moving media is a conserved quantity.

8.26 Express the wave equation of light in moving media in terms of suitable d'Alembertian times $t_{\pm}$.

8.27 Why do the $t_{\pm}$ correspond to left- and right-moving wave packets? Why are these wave packets completely independent?

8.28 Why do moving media appear to transform space–time?

8.29 Deduce the metric tensor of light in moving media.

8.30 Discuss Hawking radiation tracing wave packets back in time.

8.31 Why and how are positive and negative frequency modes partially converted into each other at horizons?

8.32 Derive Hawking's result for the radiation temperature from the required mode transformations.

8.33 Deduce the Hawking temperature for fibre-optical black holes.

8.34 What is a horizon?

8.5 Homework problem

Astrophysical black holes are black body radiators with the Hawking temperature (8.57). According to the Stefan–Boltzmann law of Chapter 3's Problem, the net energy flux out of the black hole is

$$\sigma A = \frac{\pi^2 A}{60\,c^2\hbar^3}(k_B T)^4, \quad A = 4\pi r^2$$

where r is the Schwarzschild radius (8.54). This energy must come from the total energy E of the black hole,

$$\frac{dE}{dt} = -\sigma A.$$

(a) Use Einstein's $E = mc^2$ to deduce a differential equation for the energy E.
(b) Show that

$$E = \sqrt[3]{3\beta(t_0 - t)}, \quad \beta = \frac{\hbar c^{10}}{15360\,\pi\,G^2}$$

is the solution of the differential equation for E.
(c) We interpret the t_0 in the solution of part (b) as the life span Δt of a black hole with initial energy E_0. In quantum mechanics, frequency and time are canonically conjugate, see Chapter 4's Problem, with the uncertainty relation

$$\Delta\omega\,\Delta t \geq \frac{1}{2}.$$

As the frequency corresponds to the energy, a black hole with well-defined energy and life time Δt can only form above a certain energy threshold. Use the rough relation

$$E_0\,\Delta t \geq \hbar$$

to find an estimate for the minimal energy E_0 of quantum mechanically metastable black holes. Black holes with smaller energies would not materialize, but decay immediately after their creation.
(d) Use your result of part (c) to show that present particle accelerators cannot create standard black holes.

8.6 Further reading

The Classical Theory of Fields by Landau and Lifshitz (Vol. II, 1987) contains a beautiful introduction to special relativity and also to general relativity. Here we did not use the mathematical machinery of general relativity, because we did not assume any prior knowledge on differential geometry. Some of our derivation would have become significantly more elegant had we applied geometry. The connection between media and geometries is reviewed by Schleich and Scully (1984) and Leonhardt and Philbin (2009).

Note that the Unruh effect is related to a subtle question in classical electromagnetism: do accelerated charges radiate? See Peierls' *Surprises in Theoretical Physics* (1979) and Boulware (1980). Jacobson explained black holes as space–time rivers (Jacobson, 1999). Unruh described the background, motivation and some of the history of the analogy between waves in moving media and in space–time geometries (Unruh, 2008). Quantum field theory in curved space is summarized by Birrell and Davies (1982). Brout et al. (1995a) wrote a primer for black-hole quantum physics that is completely self-contained and very readable.

Appendix A
Stress of the quantum vacuum

Casimir's 1948 calculation (Casimir, 1948) of the vacuum force between two dielectric bodies is restricted to an idealized situation: two infinitely conducting parallel plates of infinite extension. Real dielectrics are neither infinitely conducting nor infinitely extended though. Moreover, in most experimental tests of the Casimir effect (Lamoreaux, 1997, 1999; Bordag et al., 2001; Lamoreaux, 2005; Munday et al., 2009) the force between a plate and a sphere was measured, because it is very difficult in practice to keep microscopic plates exactly parallel, a problem avoided by using a sphere above a plate instead. The most comprehensive general theory of the Casimir effect is known as *Lifshitz theory* (Landau and Lifshitz, Vol. IX, 1980). This theory was pioneered by Evgeny M. Lifshitz in 1955 (Lifshitz, 1955) and further developed by Igor E. Dzyaloshinskii, Lifshitz and Lev P. Pitaevskii in 1961 (Dzyaloshinskii et al., 1961). Here we explain the central concepts of Lifshitz theory. We test the theory on Casimir's case and mention some of the results beyond it. Lifshitz theory avoids the artefacts of Casimir's simple calculation (Casimir, 1948) and it is significantly more flexible and general, but it is also technically complicated and sometimes not very intuitive. Although we try to elucidate Lifshitz theory as much as possible, it still remains a rather heavy theoretical machinery *à la russe*. It seems remarkable that such a complicated theory may give such simple results as Casimir's formula (2.50) and the generalizations we are going to discuss. Maybe some significant simplifications are hidden behind the technicalities; here, at least, we develop some modest but still labour-saving improvements.

Although Lifshitz theory is applicable to dielectrics with dispersion and loss, we derive it here within the general framework of this book, for lossless and dispersionless dielectrics. We also consider only vacuum forces, whereas Lifshitz theory can be easily extended to fields in thermal states (3.49). But the vacuum forces are the conceptually more interesting and practically relevant ones. Our starting point is given by Maxwell's equations in dielectric media (2.1), but in the presence of external charges and currents where

$$\nabla \cdot \hat{\mathbf{D}} = \hat{\rho}, \quad \nabla \times \hat{\mathbf{H}} - \frac{\partial \hat{\mathbf{D}}}{\partial t} = \hat{\mathbf{j}}. \tag{A1}$$

Here $\hat{\rho}$ denotes the operator for the density of external fluctuation charges and $\hat{\mathbf{j}}$ the corresponding current density. In macroscopic electromagnetism (Jackson, 1998) we discriminate between external charges and currents and the induced charges and dielectric-polarization currents in dielectric media. The external charges and currents feel the force of the electromagnetic field, the *Lorentz force*, with density (Jackson, 1998)

$$\hat{\mathbf{f}} = \hat{\rho} \, \hat{\mathbf{E}} + \hat{\mathbf{j}} \times \hat{\mathbf{B}}. \tag{A2}$$

The dielectric polarization charges and currents are subject to additional restoring forces in the atoms and molecules the material is made of. Macroscopic forces like the Casimir force are the external forces (A2) (Pitaevskii, 2006). Note that the order of operators is irrelevant in the Lorentz force (A2), because $\hat{\rho}$ and $\hat{\mathbf{j}}$ belong to different physical systems than the fields and hence their operators must commute.

As the next step, we express the Lorentz force (A2) in terms of *Maxwell's stress tensor* (Jackson, 1998) defined as

$$\hat{\sigma} = \hat{\mathbf{E}} \otimes \hat{\mathbf{D}} + \hat{\mathbf{B}} \otimes \hat{\mathbf{H}} - \frac{1}{2} \left(\hat{\mathbf{E}} \cdot \hat{\mathbf{D}} + \hat{\mathbf{B}} \cdot \hat{\mathbf{H}} \right) \mathbb{1}_3 \tag{A3}$$

where $\mathbb{1}_3$ denotes the three-dimensional unity matrix and $\otimes$ describes the tensor product of two vectors (a matrix $\mathbf{F} \otimes \mathbf{G}$ with components $F_l G_m$). As the fields are related to each other by the linear constitutive relations (2.2), their operator order is arbitrary in the stress tensor. We obtain from the first group of Maxwell's equations (2.1) and the second group (A1)

$$\begin{aligned}
\nabla \cdot (\hat{\mathbf{D}} \otimes \hat{\mathbf{E}}) = \hat{\mathbf{E}}(\nabla \cdot \hat{\mathbf{D}}) + (\hat{\mathbf{D}} \cdot \nabla)\hat{\mathbf{E}} &= \hat{\rho}\hat{\mathbf{E}} + (\hat{\mathbf{D}} \cdot \nabla)\hat{\mathbf{E}} \\
\nabla \cdot (\hat{\mathbf{B}} \otimes \hat{\mathbf{H}}) = \hat{\mathbf{H}}(\nabla \cdot \hat{\mathbf{B}}) + (\hat{\mathbf{B}} \cdot \nabla)\hat{\mathbf{H}} &= (\hat{\mathbf{B}} \cdot \nabla)\hat{\mathbf{H}}.
\end{aligned} \tag{A4}$$

Furthermore, we differentiate $\hat{\mathbf{D}} \times \hat{\mathbf{B}}$ with respect to time and use Maxwell's equations (A1) again

$$\frac{\partial(\hat{\mathbf{D}} \times \hat{\mathbf{B}})}{\partial t} = (\nabla \times \hat{\mathbf{E}}) \times \hat{\mathbf{D}} + (\nabla \times \hat{\mathbf{H}} - \hat{\mathbf{j}}) \times \hat{\mathbf{B}}. \tag{A5}$$

We expand the double vector products according to the formula

$$\mathbf{a} \times (\mathbf{b} \times \mathbf{c}) = \mathbf{b}\,(\mathbf{a} \cdot \mathbf{c}) - \mathbf{c}\,(\mathbf{a} \cdot \mathbf{b}) \qquad (A6)$$

and apply the linear constitutive relations (2.2) in order to obtain

$$\frac{\partial(\hat{\mathbf{D}} \times \hat{\mathbf{B}})}{\partial t} = (\hat{\mathbf{D}} \cdot \nabla)\hat{\mathbf{E}} - \frac{\nabla}{2}(\hat{\mathbf{D}} \cdot \hat{\mathbf{E}}) + (\hat{\mathbf{B}} \cdot \nabla)\hat{\mathbf{H}} - \frac{\nabla}{2}(\hat{\mathbf{B}} \cdot \hat{\mathbf{H}})$$
$$- \hat{\mathbf{j}} \times \hat{\mathbf{B}}. \qquad (A7)$$

From this piece of vector algebra follows

$$\nabla \cdot \hat{\sigma} = \hat{\mathbf{f}} + \frac{\partial(\hat{\mathbf{D}} \times \hat{\mathbf{B}})}{\partial t}. \qquad (A8)$$

The divergence of Maxwell's stress tensor gives the force density $\hat{\mathbf{f}}$ and the time derivative of $\hat{\mathbf{D}} \times \hat{\mathbf{B}}$ that constitutes the *Minkowski momentum density* (Jackson, 1998) of the electromagnetic field in media. The stress tensor thus describes the *momentum transport*. When the field is in the vacuum state or the thermal state (3.49) the expectation values of time derivatives are zero, because these states are stationary. In this case, the divergence of the stress tensor directly gives the force density.

However, the vacuum stress turns out to be as infinite as the zero–point energy (2.37). In order to understand where this infinity comes from and how to remove it, we adopt the following regularization procedure: consider instead of the expectation value of the stress tensor the more general correlation function

$$\sigma\left(\mathbf{r}, \mathbf{r}'\right) = \tau\left(\mathbf{r}, \mathbf{r}'\right) - \frac{1}{2}\,\mathrm{Tr}\,\tau\left(\mathbf{r}, \mathbf{r}'\right)\mathbb{1}_3. \qquad (A9)$$

Here Tr denotes the trace of three-dimensional matrices and τ is defined as

$$\tau\left(\mathbf{r}, \mathbf{r}'\right) = \langle 0|\hat{\mathbf{D}}(\mathbf{r}) \otimes \hat{\mathbf{E}}(\mathbf{r}') + \hat{\mathbf{H}}(\mathbf{r}) \otimes \hat{\mathbf{B}}(\mathbf{r}')|0\rangle \qquad (A10)$$

with the correlation functions

$$\langle 0|\hat{\mathbf{D}}(\mathbf{r}) \otimes \hat{\mathbf{E}}(\mathbf{r}')|0\rangle = \varepsilon_0 \varepsilon(\mathbf{r})\,\langle 0|\hat{\mathbf{E}}(\mathbf{r}) \otimes \hat{\mathbf{E}}(\mathbf{r}')|0\rangle$$
$$\langle 0|\hat{\mathbf{H}}(\mathbf{r}) \otimes \hat{\mathbf{B}}(\mathbf{r}')|0\rangle = \frac{\varepsilon_0 c^2}{\mu(\mathbf{r})}\,\langle 0|\hat{\mathbf{B}}(\mathbf{r}) \otimes \hat{\mathbf{B}}(\mathbf{r}')|0\rangle. \qquad (A11)$$

As we will see, $\sigma(\mathbf{r}, \mathbf{r}')$ is finite for $\mathbf{r}' \neq \mathbf{r}$ and approaches infinity for $\mathbf{r}' \to \mathbf{r}$. So the infinite contribution to the vacuum stress comes from infinitely short distances where we can regard the material as nearly homogeneous. We subtract from $\sigma(\mathbf{r}, \mathbf{r}')$ the stress tensor $\sigma_\infty(\mathbf{r}, \mathbf{r}')$ we would obtain in a homogeneous material near the point $\mathbf{r}$ (Dzyaloshinskii et al., 1961)

$$\sigma(\mathbf{r}) = \lim_{\mathbf{r}' \to \mathbf{r}} \left(\sigma(\mathbf{r}, \mathbf{r}') - \sigma_\infty(\mathbf{r}, \mathbf{r}') \right). \tag{A12}$$

We calculate the required correlation functions $\langle 0|\hat{\mathbf{E}}(\mathbf{r}) \otimes \hat{\mathbf{E}}(\mathbf{r}')|0\rangle$ and $\langle 0|\hat{\mathbf{B}}(\mathbf{r}) \otimes \hat{\mathbf{B}}(\mathbf{r}')|0\rangle$ by substituting in the expressions (2.3) for the electromagnetic field strengths the expansion (2.16) in terms of monochromatic modes (2.31). All expectation values of the type $\langle 0|\hat{a}_k\hat{a}_{k'}|0\rangle$, $\langle 0|\hat{a}_k^\dagger\hat{a}_{k'}|0\rangle$ and $\langle 0|\hat{a}_k^\dagger\hat{a}_{k'}^\dagger|0\rangle$ vanish, because, according to the definition of the vacuum state, $\hat{a}_k|0\rangle = 0$ and $\langle 0|\hat{a}_k^\dagger = 0$. We obtain from the Bose commutation relations (2.28) that $\langle 0|\hat{a}_k\hat{a}_{k'}^\dagger|0\rangle$ gives $\langle 0|\hat{a}_{k'}^\dagger\hat{a}_k|0\rangle + \delta_{kk'} = \delta_{kk'}$. Consequently,

$$\langle 0|\hat{\mathbf{E}}(\mathbf{r}) \otimes \hat{\mathbf{E}}(\mathbf{r}')|0\rangle = \sum_k \omega_k^2 \mathbf{A}_k(\mathbf{r}) \otimes \mathbf{A}_k^*(\mathbf{r}'),$$

$$\langle 0|\hat{\mathbf{B}}(\mathbf{r}) \otimes \hat{\mathbf{B}}(\mathbf{r}')|0\rangle = \sum_k \nabla \times \mathbf{A}_k(\mathbf{r}) \otimes \mathbf{A}_k^*(\mathbf{r}')\times \overleftarrow{\nabla}' \tag{A13}$$

where $\overleftarrow{\nabla}'$ indicates that differentiations are performed on $\mathbf{A}_k(\mathbf{r}')$ from the right; $\mathbf{A}_k^*(\mathbf{r}')\times \overleftarrow{\nabla}'$ means $\nabla' \times \mathbf{A}_k^*(\mathbf{r}')$. In the following, we relate the correlation functions (A13) to the *classical Green's function* of the electromagnetic field in media.

The classical Green's function is proportional to the electric field $\mathbf{E}(\mathbf{r})$ of a single external dipole placed at position $\mathbf{r}'$ where it oscillates with frequency ω. The dipole generates electromagnetic radiation that probes the properties of the medium. This dipolar probe may point in three possible spatial directions; and so the Green's function is a matrix where each column corresponds to any of the directions of the dipole. Mathematically, the Green's function is defined as the solution of the inhomogeneous electromagnetic wave equation

$$\nabla \times \frac{1}{\mu} \nabla \times G - \varepsilon \frac{\omega^2}{c^2} G = \mathbb{1}_3 \, \delta(\mathbf{r} - \mathbf{r}'). \tag{A14}$$

We expand the Green's function in terms of the complete set $\{\mathbf{A}_k(\mathbf{r})\}$ as $G = \sum_k G_k \mathbf{A}_k(\mathbf{r}) \otimes \mathbf{A}_k^*(\mathbf{r}')$. In order to find the coefficients G_k we substitute this expansion in the inhomogeneous wave equation (A14) where $\nabla \times \mu^{-1}\nabla \times \mathbf{A}_k(\mathbf{r})$ gives $\varepsilon(\omega_k^2/c^2)\mathbf{A}_k(\mathbf{r})$, multiply the resulting series by $\mathbf{A}_k(\mathbf{r}')$ from the right, integrate over $\mathbf{r}'$ and use the orthonormality conditions (2.24) with respect to the scalar product (2.32). In this way, we arrive at the expression

$$G = \frac{2\varepsilon_0 c^2}{\hbar} \sum_k \frac{\omega_k}{\omega_k^2 - \omega^2} \mathbf{A}_k(\mathbf{r}) \otimes \mathbf{A}_k^*(\mathbf{r}'). \tag{A15}$$

Alternatively, we express G as a series of $\mathbf{A}_k^*(\mathbf{r}) \otimes \mathbf{A}_k(\mathbf{r}')$ and obtain by a similar procedure

$$G = \frac{2\varepsilon_0 c^2}{\hbar} \sum_k \frac{\omega_k}{\omega_k^2 - \omega^2} \mathbf{A}_k^*(\mathbf{r}) \otimes \mathbf{A}_k(\mathbf{r}'). \tag{A16}$$

From Eqs. (A15) and (A16) follows the *reciprocity relation* (Lanczos, 1998) between the Green's function at position $\mathbf{r}$ for a dipole source at $\mathbf{r}'$ and the transposed matrix of the Green's function at position $\mathbf{r}'$ for a source at $\mathbf{r}$, the relation

$$G(\mathbf{r}, \mathbf{r}', \omega) = G^T(\mathbf{r}', \mathbf{r}, -\omega). \tag{A17}$$

In our case of dispersionless media, G depends on ω^2, because in the wave equation (A14) the dielectric functions ε and μ are independent of frequency, but the reciprocity relation of the form (A17) is a general result that holds also for dispersive materials (Lanczos, 1998).

The Green's function depends on the continuous frequency ω, the spectral parameter, whereas the modes oscillate at their eigenfrequencies ω_k. We analytically continue G to complex frequencies (Ablowitz and Fokas, 1997), by allowing ω to be complex, because then we can apply Cauchy's theorem for analytic functions $f(z)$ (Ablowitz and Fokas, 1997),

$$f(z_0) = \frac{1}{2\pi i} \oint \frac{f(z)\, dz}{z - z_0}, \tag{A18}$$

in order to express the correlation functions (A13) in terms of the Green's function (A15):

$$\langle 0|\hat{\mathbf{E}}(\mathbf{r}) \otimes \hat{\mathbf{E}}(\mathbf{r}')|0\rangle = \frac{\hbar}{2\pi i\varepsilon_0 c^2} \oint G(\mathbf{r}, \mathbf{r}', \omega)\, \omega^2\, e^{i\omega t}\, d\omega \Big|_{t\to+0} \tag{A19}$$

for the integration contour shown in Fig. A1 and $t > 0$. The contribution of the closing contour segment, indicated by the dotted line in Fig. A1, vanishes when we extend this path to infinity on the lower half-plane, because there $\exp(i\omega t)$ vanishes exponentially fast. We use the reciprocity relation (A17) to transform the contour indicated by the dashed line to the contour of the full line with the transposed Green's matrix as integrand. So the only relevant integration contour is the full line shown in Fig. A1 with the integral

$$\langle 0|\hat{\mathbf{E}}(\mathbf{r}) \otimes \hat{\mathbf{E}}(\mathbf{r}')|0\rangle = \frac{\hbar}{\pi i\varepsilon_0 c^2} \int_C G_s(\mathbf{r}, \mathbf{r}', \omega)\, \omega^2\, d\omega \tag{A20}$$

where we took the limit $t \to 0$ and used the symmetrized Green's function G_s defined as

$$G_s(\mathbf{r}, \mathbf{r}', \omega) = \frac{1}{2}\Big(G(\mathbf{r}, \mathbf{r}', \omega) + G^T(\mathbf{r}', \mathbf{r}, \omega)\Big). \tag{A21}$$

We move this contour to the upper imaginary axis where $\omega = i\xi$ for real ξ. Here the Green's function $G(\mathbf{r}, \mathbf{r}', i\xi)$ is real, because its defining

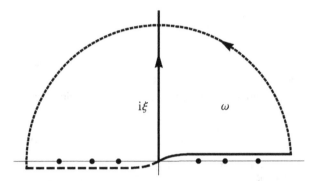

Fig. A1. Green's function as contour integral. The figure shows the contour of the integral (A19) in the complex ω plane. The points indicate the singularities $\pm\omega_k$ of the Green's function (A15). The contribution of the dotted line vanishes when it is extended to infinity. According to the reciprocity relation (A17) the dashed line integral gives the transposed Green's function of the full line integral. For the symmetrized Green's function (A21) the full line contour is deformed to the positive imaginary axis.

equation (A14) is real for purely imaginary frequencies. We obtain, finally,

$$\langle 0|\hat{\mathbf{E}}(\mathbf{r}) \otimes \hat{\mathbf{E}}(\mathbf{r}')|0\rangle = -\frac{\hbar}{\varepsilon_0 c^2 \pi} \int_0^\infty \xi^2\, G_s(\mathbf{r},\mathbf{r}',\mathrm{i}\xi)\, \mathrm{d}\xi. \qquad (A22)$$

In a similar way we obtain

$$\langle 0|\hat{\mathbf{B}}(\mathbf{r}) \otimes \hat{\mathbf{B}}(\mathbf{r}')|0\rangle = \frac{\hbar}{\varepsilon_0 c^2 \pi} \int_0^\infty \nabla \times G_s(\mathbf{r},\mathbf{r}',\mathrm{i}\xi) \times \overleftarrow{\nabla'}\, \mathrm{d}\xi. \qquad (A23)$$

These expressions for the field strength correlation functions have the advantage that the summation over the modes (A13) or their frequencies (2.37) is replaced by an integration over the Green's function, the solution of the inhomogeneous wave equation (A14). More importantly, the expressions (A22) and (A23) remain valid for dispersive and dissipative media (Dzyaloshinskii et al., 1961; Raabe and Welsch, 2005).

Consider a uniform medium where ϵ and μ are constant. In this case, we express the electromagnetic Green's function $G(\mathbf{r},\mathbf{r}',\mathrm{i}\xi)$ in terms of the scalar Green's function g_0 that obeys the inhomogeneous wave equation

$$\left(\nabla^2 - \varepsilon\mu\kappa^2\right) g_0 = \delta(\mathbf{r} - \mathbf{r}') \qquad (A24)$$

for imaginary wave numbers

$$\kappa = \frac{\xi}{c} = -\mathrm{i}\frac{\omega}{c}. \qquad (A25)$$

The electromagnetic Green's function $G(\mathbf{r}, \mathbf{r}', i\xi)$ is given by

$$G_0 = \left(\frac{\nabla \otimes \nabla}{\varepsilon \kappa^2} - \mu \mathbb{1}_3 \right) g_0, \qquad (A26)$$

because G_0 obeys the defining electromagnetic wave equation (A14) for constant ε and μ, as a consequence of the scalar wave equation (A24) and the identity

$$\left(\nabla \times \nabla \times + \kappa'^2 \mathbb{1}_3 \right) \left(\nabla \otimes \nabla - \kappa'^2 \mathbb{1}_3 \right) = \kappa'^2 \mathbb{1}_3 \left(\nabla^2 - \kappa'^2 \right) \qquad (A27)$$

for $\kappa'^2 = \varepsilon \mu \kappa^2$ that follows from the double vector product (A6). The scalar Green's function is given by the expression

$$g_0 = -\frac{1}{4\pi r} \exp\left(-\sqrt{\varepsilon\mu}\,\kappa r\right), \qquad r = |\mathbf{r} - \mathbf{r}'|, \qquad (A28)$$

because this g_0 satisfies the inhomogeneous wave equation (A24) with the volume delta function on the right-hand side. First, it obeys the homogeneous wave equation for $\mathbf{r} \neq \mathbf{r}'$, written in spherical coordinates $\{r, \theta, \phi\}$ with $\mathbf{r}'$ at the origin,

$$\left(\frac{1}{r^2} \frac{\partial}{\partial r} r^2 \frac{\partial}{\partial r} - \varepsilon\mu\kappa^2 \right) g_0 = 0 \quad \text{for} \quad r \neq 0. \qquad (A29)$$

Second, the integral of $\nabla^2 g_0$ over an infinitely small volume around $\mathbf{r}'$ is 1, as we see using Gauss's theorem and regarding g_0 as $-(4\pi r)^{-1}$,

$$\int \nabla^2 g_0 \, dV = \oiint \nabla g_0 \cdot d\mathbf{S} = \int_0^\pi \int_0^{2\pi} \frac{1}{4\pi r^2} r^2 \sin\theta \, d\phi \, d\theta = 1. \qquad (A30)$$

Consequently, g_0 indeed obeys the defining wave equation (A24). The scalar Green's function shows the symmetry

$$g(\mathbf{r}, \mathbf{r}') = g(\mathbf{r} - \mathbf{r}') = g(\mathbf{r}' - \mathbf{r}). \qquad (A31)$$

This symmetry carries through to the correlation function (A9),

$$\sigma(\mathbf{r}, \mathbf{r}') = \sigma(\mathbf{r} - \mathbf{r}') = \sigma(\mathbf{r}' - \mathbf{r}) \qquad (A32)$$

and so $\nabla \cdot \sigma$ equals $-\nabla \cdot \sigma$; in other words the divergence of σ vanishes: uniform media do not feel the force of the quantum vacuum, which is of course what we expect, because otherwise uniform media would disintegrate. On the other hand, we see from the explicit expression (A28) that the Green's function and hence the vacuum stress tends to infinity for $\mathbf{r} \to \mathbf{r}'$; but, as we have seen, without causing physical effects.

The infinity of the Green's function for uniform media characterizes the infinity in the non-uniform case, because, around the singularity of the delta function in the wave equation (A14) we can assume the medium to be uniform; the dominant, diverging contribution to G is given by the uniform Green's function (A26) with ε set to $\varepsilon' = \varepsilon(\mathbf{r}')$ and μ set to $\mu' = \mu(\mathbf{r}')$. On the other hand, we know that this contribution does not generate a force. In this way, we have deduced a simple recipe[1] for removing the infinite but physically insignificant contribution $\sigma_\infty(\mathbf{r}, \mathbf{r}')$ from the correlation function (A9): $\sigma_\infty(\mathbf{r}, \mathbf{r}')$ is constructed from the uniform Green's function (A26) taking the local values of ε and μ at $\mathbf{r}'$ (Dzyaloshinskii et al., 1961). In a non-uniform medium, however, we have to be a bit careful, because the Green's function must obey the reciprocity relation (A17). A suitable expression is

$$G_0 = -\left(\frac{\nabla \otimes \nabla'}{\sqrt{\varepsilon\varepsilon'}\kappa^2} + \sqrt{\mu\mu'}\,\mathbb{1}_3 \right) g_0 \qquad \text{(A33)}$$

with the scalar Green's function

$$g_0 = -\frac{1}{4\pi|\mathbf{r} - \mathbf{r}'|} \exp\left(-\kappa \int_{\mathbf{r}'}^{\mathbf{r}} \sqrt{\varepsilon\mu}\,d\mathbf{r}^2 \right) \quad \text{for} \quad \mathbf{r} \sim \mathbf{r}'. \qquad \text{(A34)}$$

The Green's function (A34) generates the infinite contribution σ_∞ that we need to subtract from σ. Of course, we can also directly subtract the uniform Green's function (A34) from the integrals (A22) and (A23) for removing the physically meaningless infinite contribution to the vacuum stress. Where does σ_∞ come from? We could interpret the Casimir force as the interaction of dipoles induced in the material by vacuum fluctuations. The Green's function describes how these dipoles interact with each other, but for $\mathbf{r} = \mathbf{r}'$ it would describe the interaction of dipoles with themselves, something that is unphysical. A similar problem already occurs in classical electrodynamics (Jackson, 1998) where a charge ought not to interact with itself, although in theory it would. In removing G_0 we remove this spurious self interaction.

In the following we apply our procedure to the simplest non-trivial geometry where vacuum forces occur. Consider planar dielectrics where ε and μ depend on x, but not on y and z, for example the dielectric sandwich shown in Fig. A2. We are going to develop both a simplified and a generalized Lifshitz theory for the vacuum stress in such planar dielectrics. In view of the symmetry in y and z directions it is wise to

---

[1] This regularization procedure does not always completely remove the infinity in the stress tensor; in some cases additional regularizations are needed (Leonhardt and Philbin, 2007).

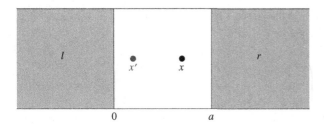

Fig. A2. Dielectric sandwich. Planar dielectric consisting of three uniform layers with constant electric permittivities ε_l, ε, ε_r and magnetic permeabilities μ_l, μ, μ_r. The left and right outer layers are idealized to be infinitely thick, the inner layer extends from 0 to a.

represent the Green's function in terms of its spatial Fourier transform in y and z, as

$$G = \frac{1}{(2\pi)^2} \int_{-\infty}^{+\infty} \int_{-\infty}^{+\infty} \widetilde{G}(x, x', u, v)\, e^{iu(y-y')+iv(z-z')}\, du\, dv. \quad (A35)$$

On such Fourier transforms, the derivatives with respect to the coordinates appear as

$$\nabla = (\partial_x, iu, iv), \quad \nabla' = (\partial_x', -iu, -iv) \quad (A36)$$

where ∂_x and ∂_x' abbreviate $\partial/\partial x$ and $\partial/\partial x'$. The Fourier transform $\widetilde{G}$ obeys the Fourier transformed wave equation (A14)

$$\nabla \times \frac{1}{\mu} \nabla \times \widetilde{G} + \varepsilon\kappa^2 \widetilde{G} = \mathbb{1}_3\, \delta(x - x'). \quad (A37)$$

We obtain from expression (A26) for the Green's function in a uniform medium

$$\widetilde{G}_0 = \frac{1}{\varepsilon\kappa^2} \begin{pmatrix} \partial_x^2 - \varepsilon\mu\kappa^2 & iu\,\partial_x & iv\,\partial_x \\ iu\,\partial_x & -u^2 - \varepsilon\mu\kappa^2 & -uv \\ iv\,\partial_x & -uv & -v^2 - \varepsilon\mu\kappa^2 \end{pmatrix} \widetilde{g}_0. \quad (A38)$$

In uniform media, the Fourier transform of the scalar Green's function is

$$\widetilde{g}_0 = -\frac{1}{2w} \exp\left(-w\,|x - x'|\right) \quad (A39)$$

in terms of

$$w = \sqrt{u^2 + v^2 + \varepsilon\mu\kappa^2}, \quad (A40)$$

because this $\widetilde{g}_0$ solves the Fourier-transformed scalar wave equation (A24),

$$\left(\partial_x^2 - w^2\right) \widetilde{g}_0 = \delta(x - x'). \quad (A41)$$

One can also verify that $\widetilde{g}_0$ is the scalar Green's function (A28) Fourier-transformed with respect to $y - y'$ and $z - z'$. For $x \neq x'$ we replace

$(\partial_x^2 - \varepsilon\mu\kappa^2)\tilde{g}_0$ in the Green's function (A38) by $(u^2 + v^2)\tilde{g}_0$ and $\partial_x\tilde{g}_0$ by $-\partial_x'\tilde{g}_0$. We get

$$\tilde{G}_0 = \frac{1}{\varepsilon\kappa^2}\begin{pmatrix} u^2 + v^2 & -iu\,\partial_x' & -iv\,\partial_x' \\ iu\,\partial_x & -u^2 - \varepsilon\mu\kappa^2 & -uv \\ iv\,\partial_x & -uv & -v^2 - \varepsilon\mu\kappa^2 \end{pmatrix}\tilde{g}_0 \quad \text{for } x \neq x'.$$

(A42)

Now imagine that $\varepsilon(x)$ and $\mu(x)$ vary in an arbitrary way. We introduce the electric and magnetic scalar Green's functions g_E and g_M defined by

$$\nabla \cdot \frac{1}{\mu}\nabla\tilde{g}_E - \varepsilon\kappa^2\tilde{g}_E = \nabla \cdot \frac{1}{\varepsilon}\nabla\tilde{g}_M - \mu\kappa^2\tilde{g}_M = \delta(x - x'). \quad (A43)$$

In the case of uniform media we have

$$\tilde{g}_E = \mu\tilde{g}_0, \quad \tilde{g}_M = \varepsilon\tilde{g}_0. \quad (A44)$$

The scalar Green's functions also obey reciprocity relations,

$$\begin{aligned} g_E\left(\mathbf{r}, \mathbf{r}', \kappa\right) &= g_E\left(\mathbf{r}', \mathbf{r}, -\kappa\right), \\ g_M\left(\mathbf{r}, \mathbf{r}', \kappa\right) &= g_M\left(\mathbf{r}', \mathbf{r}, -\kappa\right), \end{aligned} \quad (A45)$$

as one finds along similar lines as in our derivation of the reciprocity relation (A17). We are going to use the scalar Green's functions as vital ingredients in G. Other building blocks are the unity vectors

$$\mathbf{n}_E = \frac{1}{\sqrt{u^2 + v^2}}\begin{pmatrix} 1 \\ 0 \\ 0 \end{pmatrix} \times \begin{pmatrix} 0 \\ u \\ v \end{pmatrix} = \frac{1}{\sqrt{u^2 + v^2}}\begin{pmatrix} 0 \\ -v \\ u \end{pmatrix} \quad (A46)$$

$$\mathbf{n}_E' = -\mathbf{n}_E.$$

They project onto the transversal–electric (TE) polarization of electromagnetic waves (Landau and Lifshitz, vol. VIII, 1984). We obtain

$$\mathbf{n}_E \otimes \mathbf{n}_E' = \frac{1}{u^2 + v^2}\begin{pmatrix} 0 & 0 & 0 \\ 0 & -v^2 & uv \\ 0 & uv & -u^2 \end{pmatrix}, \quad (A47)$$

$$\nabla \times \mathbf{n}_E \otimes \mathbf{n}_E' \times \overleftarrow{\nabla'} = (\nabla \times \mathbf{n}_E) \otimes (\nabla' \times \mathbf{n}_E')$$

$$= \begin{pmatrix} -u^2 - v^2 & iu\,\partial_x' & iv\,\partial_x' \\ -iu\,\partial_x & -\dfrac{u^2\,\partial_x\partial_x'}{u^2 + v^2} & -\dfrac{uv\,\partial_x\partial_x'}{u^2 + v^2} \\ -iv\,\partial_x & -\dfrac{uv\,\partial_x\partial_x'}{u^2 + v^2} & -\dfrac{v^2\,\partial_x\partial_x'}{u^2 + v^2} \end{pmatrix}. \quad (A48)$$

The Green's function turns out to be given by the expressions

$$\tilde{G} = \tilde{G}_E + \tilde{G}_M \quad \text{for} \quad x \neq x', \tag{A49}$$

$$\tilde{G}_E = \mathbf{n}_E \tilde{g}_E \otimes \mathbf{n}_E', \quad \tilde{G}_M = -\frac{\nabla \times \mathbf{n}_E \tilde{g}_M \otimes \mathbf{n}_E' \times \overleftarrow{\nabla'}}{\varepsilon\varepsilon'\kappa^2}. \tag{A50}$$

To prove the validity of formulae (A49)–(A50) consider $\tilde{G}$ in uniform media first. In this case we replace the expression $\partial_x \partial_x' \tilde{g}_M$ that, according to Eq. (A48), occurs in $\nabla \times \mathbf{n}_E \tilde{g}_M \otimes \mathbf{n}_E' \times \nabla'$ by $-w^2 \varepsilon \tilde{g}_0$. We see from the relations

$$\frac{u^2 w^2 + \kappa'^2 v^2}{u^2 + v^2} = u^2 + \kappa'^2, \qquad \frac{v^2 w^2 + \kappa'^2 u^2}{u^2 + v^2} = v^2 + \kappa'^2,$$

$$\frac{uvw^2 - \kappa'^2 uv}{u^2 + v^2} = uv \tag{A51}$$

that $\tilde{G}$ indeed agrees with $\tilde{G}_0$ of Eq. (A42). Now turn to non-uniform media. For planar dielectrics where ε and μ vary only in x, we obtain from the definition (A46)

$$\nabla \cdot \frac{1}{\mu} \mathbf{n}_E \tilde{g}_E = 0, \tag{A52}$$

which implies for $x \neq x'$

$$\nabla \times \frac{1}{\mu} \nabla \times \mathbf{n}_E \tilde{g}_E = -\mathbf{n}_E \left(\nabla \cdot \frac{1}{\mu} \nabla \tilde{g}_E \right) = -\varepsilon \kappa^2 \mathbf{n}_E \tilde{g}_E \tag{A53}$$

from the representation (A6) of the double vector product and the definition (A43) of the scalar Green's functions. Consequently, the term $\mathbf{n}_E \tilde{g}_E \otimes \mathbf{n}_E'$ in the expression (A50) of the Green's function satisfies the wave equation (A37) for $x \neq x'$. Similarly, from

$$\nabla \cdot \frac{1}{\varepsilon} \mathbf{n}_E \tilde{g}_M = 0 \tag{A54}$$

follows

$$\left(\frac{1}{\mu} \nabla \times \frac{1}{\varepsilon} \nabla \times + \kappa^2 \right) \mathbf{n}_E \tilde{g}_M \otimes \mathbf{n}_E' = 0 \tag{A55}$$

and hence

$$\left(\frac{1}{\varepsilon} \nabla \times \frac{1}{\mu} \nabla \times + \kappa^2 \right) \frac{\nabla \times \mathbf{n}_E \tilde{g}_M \otimes \mathbf{n}_E' \times \overleftarrow{\nabla'}}{\varepsilon\varepsilon'} = 0. \tag{A56}$$

Therefore, also the second term in the expression (A50) obeys the wave equation (A37) for $x \neq x'$. When x approaches x' the behaviour of the Green's function is dominated by the delta function in the inhomogeneous wave equation (A37). Close to x' we can regard the medium as uniform. But in this case we have already proved that $\widetilde{G}$ agrees with $\widetilde{G}_0$ for $x \neq x'$. Consequently, $\widetilde{G}$ indeed is the Fourier-transformed Green's function for $x \neq x'$. Note that our results (A49)–(A50) are not valid at $x = x'$, but we never need $\widetilde{G}$ there, because we approach $x' \to x$ as a limit with $x' \neq x$ in the Green's function.

Having established the Green's function we calculate the stress of the quantum vacuum according to our general procedure. We obtain for the correlation function (A10) in the limit $\mathbf{r} \to \mathbf{r}'$ the expression

$$\tau = \frac{\hbar c}{\pi} \int_0^\infty \left(-\kappa^2 \varepsilon \, G + \frac{1}{\mu} \nabla \times G \times \overleftarrow{\nabla'} \right) d\kappa \tag{A57}$$

$$= \frac{\hbar c}{4\pi^3} \int_0^\infty \int_{-\infty}^{+\infty} \int_{-\infty}^{+\infty} \left(-\kappa^2 \varepsilon \, \widetilde{G} + \frac{1}{\mu} \nabla \times \widetilde{G} \times \overleftarrow{\nabla'} \right) du \, dv \, d\kappa.$$

Consider the G_E contribution to the Green's function (A50). We obtain from Eqs. (A47)–(A48) that the corresponding off-diagonal elements of τ are odd functions of u and v. Such functions vanish in the integral (A57). For the G_M contribution we obtain, using the wave equation (A55) and the reciprocity relations (A45),

$$-\kappa^2 \varepsilon \, \widetilde{G}_M + \frac{1}{\mu} \nabla \times \widetilde{G}_M \times \overleftarrow{\nabla'} = \frac{1}{\varepsilon} \nabla \times \mathbf{n}_E \widetilde{g}_M \otimes \mathbf{n}'_E \times \overleftarrow{\nabla'} - \mu \kappa^2 \mathbf{n}_E \widetilde{g}_M \otimes \mathbf{n}'_E \tag{A58}$$

in the limit $x' \to x$. Therefore we can apply the same argument as for the G_E term, thus proving that τ and consequently σ are diagonal matrices,

$$\sigma = \mathrm{diag}\left(\sigma_{xx}, \sigma_{yy}, \sigma_{zz} \right). \tag{A59}$$

In this way we have shown that, in planar dielectrics, the vacuum force has no sideways component – no shear, which is not surprising in view of the lateral symmetry. The only component of Maxwell's stress tensor that generates a vacuum force is σ_{xx}, because σ in the force density (A8) does not depend on y and z either. Note that σ_{xx} contains the trace of τ according to the definition (A9). We get from Eqs. (A47)–(A48)

$$\mathrm{Tr}\left\{ \mathbf{n}_E \widetilde{g} \otimes \mathbf{n}'_E \right\} = -\widetilde{g},$$

$$\mathrm{Tr}\left\{ \nabla \times \mathbf{n}_E \widetilde{g} \otimes \mathbf{n}'_E \times \overleftarrow{\nabla'} \right\} = -\left(u^2 + v^2 + \partial_x \partial'_x \right) \widetilde{g} \tag{A60}$$

and obtain the G_E contribution

$$\mathrm{Tr}\,\tau_E = \frac{\hbar c}{4\pi^3} \int_0^\infty \int_{-\infty}^{+\infty} \int_{-\infty}^{+\infty} \left(\kappa^2 \varepsilon - \frac{1}{\mu}(u^2 + v^2 + \partial_x \partial'_x) \right) \widetilde{g}_E \, du \, dv \, d\kappa \tag{A61}$$

and the equivalent expression for $\operatorname{Tr} \tau_M$ with ε and μ interchanged and g_M used instead of g_E. Finally, we obtain for the σ_{xx} component of the vacuum stress (A9)

$$\sigma_{xx} = \sigma_{xx}^E + \sigma_{xx}^M \qquad (A62)$$

where for the G_E contribution we apply the explicit expressions (A47)–(A48) and the abbreviation (A40),

$$\begin{aligned}
\sigma_{xx}^E &= -\frac{\hbar c}{8\pi^3} \int_0^\infty \int_{-\infty}^{+\infty} \int_{-\infty}^{+\infty} \frac{1}{\mu} \left(w^2 - \partial_x \partial_x' \right) \tilde{g}_{ES} \, du \, dv \, d\kappa \\
&= -\frac{\hbar c}{4\pi^2} \int_0^\infty \int_0^\infty \frac{1}{\mu} \left(w^2 - \partial_x \partial_x' \right) \tilde{g}_{ES} \, u \, du \, d\kappa, \qquad (A63)
\end{aligned}$$

putting $v = 0$ in the last line, as the scalar Green's functions depend only on $u^2 + v^2$ but not on the angle of (u, v). The integration over this angle gives 2π. Similarly, we get for the G_M contribution

$$\sigma_{xx}^M = -\frac{\hbar c}{4\pi^2} \int_0^\infty \int_0^\infty \frac{1}{\varepsilon} \left(w^2 - \partial_x \partial_x' \right) \tilde{g}_{MS} \, u \, du \, d\kappa. \qquad (A64)$$

In our result (A62) with the expressions (A63) and (A64) the subscript S indicates that we subtracted the (infinite) contribution from the bare Green's function G_0, because, as we argued before, this contribution would come from the self-interaction of the dielectric and is unphysical. Here G_0 is taken as the Green's function in a uniform medium with ε and μ approaching their local values in a manner consistent with the reciprocity relations (A45):

$$\tilde{g}_{ES} = \tilde{g}_E - \sqrt{\mu\mu'}\,\tilde{g}_0, \quad \tilde{g}_{MS} = \tilde{g}_M - \sqrt{\varepsilon\varepsilon'}\,\tilde{g}_0 \qquad (A65)$$

where we employ the scalar Green's function (A39) symmetrized as

$$\tilde{g}_0 = -\frac{1}{2\sqrt{ww'}} \exp\left(-\left| \int_{x'}^x w \, dx \right| \right) \qquad (A66)$$

in agreement with the reciprocity relations (A45). Our procedure simplifies Lifshitz's classic result (Lifshitz, 1955) and generalizes it to planar dielectrics with arbitrary $\varepsilon(x)$ and $\mu(x)$ profiles. Our result saves considerable labour in calculating the vacuum stress, because it only depends on the scalar Green's functions.

To see how this theory works we consider the dielectric sandwich illustrated in Fig. A2 that consists of three uniform regions with constants $\varepsilon_l, \varepsilon, \varepsilon_r$ and μ_l, μ, μ_r. The outer dielectrics are idealized to be infinitely thick, the left outer plate extends from $-\infty$ to 0, the inner region ranges from 0 to a and the right outer plate from a to ∞. We could solve the scalar wave equations (A43) with the corresponding boundary conditions (continuity of $\widetilde{g}_E$ and $\mu^{-1}\partial_x\widetilde{g}_E$ or of $\widetilde{g}_M$ and $\varepsilon^{-1}\partial_x\widetilde{g}_M$). It is more instructive, however, to solve this problem by thinking about the structure of the scalar waves, about multiple propagations and reflections that will occur, such that we can immediately write down the solution. The scalar Green's functions describe waves emitted at x' that are multiple reflected at the dielectric interfaces. We suppress the E and M indices of the scalar Green's functions, because our argument runs simultaneously for both cases. We represent $\widetilde{g}_E$ as some $\mu\widetilde{g}$ and $\widetilde{g}_M$ as, another, $\varepsilon\widetilde{g}$. Between the left and right plates, waves propagate from x' to x forward and backward, depending on whether $x' < x$ or $x < x'$. The waves are reflected at the interfaces with reflectivities ϱ_l and ϱ_r. Multiple reflections and propagations between the outer layers add up to

$$\sum_{m=0}^{\infty} \left(e^{-2aw}\varrho_l\varrho_r\right)^m = \left(1 - e^{-2aw}\varrho_l\varrho_r\right)^{-1}. \tag{A67}$$

The expression

$$\widetilde{g}_1 = -\frac{e^{-w(x-x')} + e^{-w(x+x')}\varrho_l + e^{w(x-x')} + e^{w(x+x'-2a)}\varrho_r}{2w\left(1 - e^{-2aw}\ \varrho_l\varrho_r\right)} \tag{A68}$$

contains all possible multiple reflections and propagations from x' with the strength $-(2w)^{-1}$ of the bare Green's function (A39). However, if $x' < x$ the direct backward propagation $-(2w)^{-1}\exp[w(x - x')]$ from x' to x is not possible, only indirect ones that involve multiple reflections. Hence we need to subtract $-(2w)^{-1}\exp[w(x - x')]$ in this case. Similarly, if $x < x'$ we need to subtract $-(2w)^{-1}\exp[-w(x - x')]$. The required subtractions are conveniently described by putting

$$\widetilde{g} = \widetilde{g}_0 + \widetilde{g}_1 + \widetilde{g}_2 \quad \text{and hence} \quad \widetilde{g}_S = \widetilde{g}_1 + \widetilde{g}_2 \tag{A69}$$

in terms of the bare Green's function (A39) and

$$\widetilde{g}_2 = \frac{1}{2w}\left(e^{w(x'-x)} + e^{w(x-x')}\right). \tag{A70}$$

For the dielectric sandwich we are considering, ε and μ are constant between the outer plates. Consequently, the Casimir stress σ_{xx} must be constant.[2] Therefore, the terms of the two scalar Green's functions that remain x-dependent in the limit $x' \to x$ must vanish in σ_{xx} – they turn out to disappear due to the derivatives in expressions (A63) and (A64). We only need to pay attention to the terms that are constant in

---

[2] The components σ_{yy} and σ_{zz} that do not generate a force may vary in x and usually do.

this limit, the expressions containing $\exp[\pm w(x - x')]$. These are, in the limit $x' \to x$,

$$\tilde{g}_S = -\frac{1}{w\left(1 - e^{-2aw}\,\varrho_l\varrho_r\right)} + \frac{1}{w} = -\frac{1}{w\left(\varrho_l^{-1}\varrho_r^{-1}\,e^{2aw} - 1\right)}. \tag{A71}$$

In this way we obtain Lifshitz's result (Lifshitz, 1955)

$$\begin{aligned}
\sigma_{xx} &= \frac{\hbar c}{4\pi^3} \int_0^\infty \int_{-\infty}^{+\infty} \int_{-\infty}^{+\infty} \left(A_{EE}^{-1} + A_{MM}^{-1}\right) w \, du \, dv \, d\kappa \\
&= \frac{\hbar c}{2\pi^2} \int_0^\infty \int_0^\infty \left(A_{EE}^{-1} + A_{MM}^{-1}\right) w \, u \, du \, d\kappa
\end{aligned} \tag{A72}$$

with

$$A_{EE} = \varrho_{El}^{-1}\,\varrho_{Er}^{-1}\,e^{2aw} - 1, \quad A_{MM} = \varrho_{Ml}^{-1}\,\varrho_{Mr}^{-1}\,e^{2aw} - 1. \tag{A73}$$

The A_{EE} and A_{MM} result from multiple reflections and appear as typical resonance denominators. They would produce poles where the reflection at the two dielectric boundaries exactly matches the $\exp(2aw)$ propagation back and forth in the cavity. The integral (A72) for the vacuum stress rapidly converges, the main contribution coming from wave numbers w smaller than the cavity size a, but note these are purely imaginary wave numbers!

Now turn to the Green's functions in the outer plates themselves. There the only terms that depend on $\exp[\pm w(x - x')]$ are the ones describing direct propagation, but in $\tilde{g}_S$ they are subtracted from $\tilde{g}$ by the bare Green's function (A39). Consequently, the Casimir stress vanishes in the outer plates; σ_{xx} jumps from (A72) to zero. The force density (A8) is concentrated in the interface (as a delta function). The force points inward if σ_{xx} is positive and outward if σ_{xx} is negative; the Casimir force is attractive for $\sigma_{xx} > 0$ and repulsive for $\sigma_{xx} < 0$. Formula (A72) remains valid even when the outer dielectrics have non-uniform ε and μ profiles, because it only depends on the reflectivities and not on the details of the dielectric structures. But, in such cases σ_{xx} is not strictly zero in the outer layers. For uniform plates we use the electromagnetic reflection coefficients (Jackson, 1998)

$$\begin{aligned}
\varrho_{El} &= \frac{\mu_l w - \mu w_l}{\mu_l w + \mu w_l}, & \varrho_{Ml} &= -\frac{\varepsilon_l w - \varepsilon w_l}{\varepsilon_l w + \varepsilon w_l}, \\
\varrho_{Er} &= \frac{\mu_r w - \mu w_r}{\mu_r w + \mu w_r}, & \varrho_{Mr} &= -\frac{\varepsilon_r w - \varepsilon w_r}{\varepsilon_r w + \varepsilon w_r},
\end{aligned} \tag{A74}$$

where the w_l, w and w_r are defined by Eq. (A40) with the ε and μ in the corresponding regions. In the case of perfect reflection – Casimir's original case – we take the limit $\varepsilon_l, \varepsilon_r \to \infty$ with $\mu_l = \mu_r = 1$

and we assume vacuum between the plates, $\varepsilon = \mu = 1$. In this case we get $\varrho_E = \varrho_M = -1$, the phase jump of π at a perfect mirror. We obtain from the three-dimensional Lifshitz integral (A72) expressed in spherical coordinates with w as radius (A40)

$$
\begin{aligned}
\sigma_{xx} &= \frac{\hbar c}{2\pi^3} \int_0^\infty \int_0^{2\pi} \int_0^{\pi/2} \frac{w^3 \sin \vartheta \, d\vartheta \, d\varphi \, dw}{e^{2aw} - 1} \\
&= \frac{\hbar c}{\pi^2} \int_0^\infty \frac{w^3 \, dw}{e^{2aw} - 1}.
\end{aligned}
\tag{A75}
$$

We use Eq. 2.3.14.4 of Prudnikov et al. (1992)

$$
\int_0^\infty \frac{x^{2m-1}}{e^x - 1} \, dx = (-1)^{m+1} \frac{(2\pi)^{2m}}{4m} B_{2m}
\tag{A76}
$$

for positive integers m with the Bernoulli numbers B_{2m}. In this way we deduce Casimir's classic result from Lifshitz's theory, expressed in terms of Maxwell's stress

$$
\sigma_{xx} = \frac{\hbar c \pi^2}{240 a^4} = \sigma_{Casimir}.
\tag{A77}
$$

As an extra bonus, we easily obtain the vacuum stress between a perfect electric mirror with $\varepsilon_l \to \infty$, $\mu_l = 1$ and a perfect magnetic mirror with $\varepsilon_r = 1$, $\mu_r \to \infty$ (Boyer, 1974). In this case $\varrho_{Er} = \varrho_{Mr} = 1$ and hence

$$
\sigma_{xx} = -\frac{\hbar c}{\pi^2} \int_0^\infty \frac{w^3 \, dw}{e^{2aw} + 1}.
\tag{A78}
$$

Applying Eq. 2.3.14.4 of Prudnikov et al. (1992) we get the relationship

$$
\int_0^\infty \frac{x^{2m-1}}{e^x + 1} \, dx = (1 - 2^{1-2m}) \int_0^\infty \frac{x^{2m-1}}{e^x - 1} \, dx.
\tag{A79}
$$

In this way we derive Boyer's (1974) result

$$
\sigma_{xx} = -\frac{7}{8} \sigma_{Casimir}.
\tag{A80}
$$

The vacuum force is repulsive: an electric and a magnetic mirror repel each other. The vacuum force is also repulsive if the three sandwiched materials obey the relationship (Dzyaloshinskii et al., 1961)

$$
\varepsilon_l > \varepsilon > \varepsilon_r, \quad \mu_l = \mu = \mu_r = 1
\tag{A81}
$$

over a sufficiently wide range of imaginary frequencies, because in this case $\varrho_{El}, \varrho_{Ml} < 0$ and $\varrho_{Er}, \varrho_{Mr} > 0$. (One could of course swap ε_l and ε_r to the same effect.) Such a repulsive Casimir–Lifshitz force has been observed between gold, bromobenzene and silica (Munday et al., 2009).

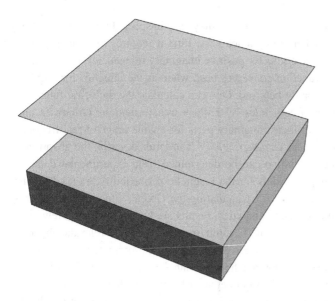

Fig. A3. Quantum levitation. The repulsive Casimir force of a negatively refracting material levitates a mirror on zero-point energy.

One would also provoke strongly repulsive Casimir forces by inserting a negatively refracting material between two perfect mirrors, as Fig. A3 shows. In such a medium, electromagnetic waves attain negative phase delays – a negatively refracting material of thickness b seems to reduce the cavity size a to $a' = a - 2b$ (Leonhardt and Philbin, 2006). For $a < 2b$ the effective size a' is negative. In this case, the integral in Lifshitz's formula (A72) is proportional to

$$\frac{1}{e^{2a'w} - 1} = -\frac{1}{e^{-2a'w} - 1} - 1 = -\frac{1}{e^{2|a'|w} - 1} - 1. \quad \text{(A82)}$$

The -1 constitutes an infinite contribution to σ_{xx} that, however, does not generate a force (A8), because it does not depend on x. In the idealized case of a non-dispersive negatively refracting material we obtain the effective vacuum stress (Leonhardt and Philbin, 2007)

$$\sigma_{xx} = -\frac{\hbar c \pi^2}{240 a'^4}, \quad a' = a - 2b < 0. \quad \text{(A83)}$$

In practice, such materials are dispersive and optical gain would be required to sustain a repulsive Casimir force (Leonhardt and Philbin, 2007). But for $a \lesssim 2b$ the force of the quantum vacuum may become relatively strong, strong enough to lift light mirrors, levitating them on, literally, nothing.

We see that Lifshitz theory reproduces Casimir's result, but the theory is also significantly more flexible. In principle, Lifshitz theory is applicable to calculating the stress of the quantum vacuum in dielectric bodies of arbitrary geometric shapes and with frequency-dependent dielectric functions ε and μ that may vary in space. In practice, Lifshitz theory is a rather heavy machinery. First it requires the calculation of the dielectric functions for positive imaginary frequencies, because here the Green's function converges best, whereas the integrals for real frequencies are badly behaved. One can calculate the dielectric functions for imaginary frequencies by Hilbert transformations (Ablowitz and Fokas, 1997) of their imaginary parts for a wide range of real frequencies (Munday and Capasso, 2007). The latter describe the absorption in the material and so can be determined from experimental data. In addition, simple models for the magnetic permeability are used (Munday and Capasso, 2007). Calculating the Green's function usually is a major challenge. Here analytical results are known only for a few simple geometries. Even the Casimir force in the most relevant physical situation, the sphere above a plate, is only approximately known for a plate–sphere separation much larger than the sphere radius, in Proximity Force Approximation (Blocki et al., 1977). For difficult geometries one could apply brute-force numerical techniques (Rodriguez et al., 2007) but they are not straightforward either. The next step of the procedure is to subtract the diverging contribution from the Green's function and to take the limit $\mathbf{r} \to \mathbf{r}'$ in the expressions (A22) and (A23) that serve to calculate the require correlation functions (A11). Finally, one obtains the vacuum stress tensor from Eqs. (A9) and (A10). Remarkably, this laborious procedure may produce some relatively simple and elegant results.

It took more than half a century before an exact expression for the next obvious generalization of Lifshitz's classic result (A72) was derived (Philbin and Leonhardt, 2009): the vacuum stress between moving plates. Imagine the right plate is moving relative to the left plate in the y direction. For the velocity of the moving plate we use the common relativistic notation $c\beta$, in order to avoid confusing it with the integration variables u and v. One might expect that the moving plate drags electromagnetic modes, even when they are in their vacuum states, and so causes friction. But after considerable algebra came a surprise (Philbin and Leonhardt, 2009): the friction is exactly zero. The vacuum stress tensor depends on the velocity of the moving plate, but it is diagonal and so it does not generate a shear force that would result in friction. Given the complexity of the calculation (Philbin and Leonhardt, 2009) the result for the vacuum stress is relatively simple:

$$\sigma_{xx} = \frac{\hbar c}{4\pi^3} \int_0^\infty \int_{-\infty}^{+\infty} \int_{-\infty}^{+\infty} \Omega\, w\, du\, dv\, d\kappa \tag{A84}$$

$$\Omega = \frac{(A_{EE} + A_{MM})(u^2 + v^2 - i\kappa\beta)^2 - (A_{EM} + A_{ME})w^2 v^2 \beta^2}{A_{EE}A_{MM}(u^2 + v^2 - i\kappa\beta)^2 - A_{EM}A_{ME}w^2 v^2 \beta^2}$$

with the expressions (A40) and (A73) as before and

$$A_{EM} = \varrho_{El}^{-1}\varrho_{Mr}^{-1}e^{2aw} - 1, \quad A_{ME} = \varrho_{Ml}^{-1}\varrho_{Er}^{-1}e^{2aw} - 1. \qquad (A85)$$

The reflection on the moving boundary is described in a co-moving frame. As the Lorentz transformation from the rest frame to the moving frame mixes electric and magnetic fields, the moving boundary mixes the polarizations of light, leading to the mixed terms A_{EM} and A_{ME}. In the co-moving frame we get the reflection coefficients

$$\varrho_{Er} = \frac{\mu_r(ic\kappa')w - w_r}{\mu_r(ic\kappa')w + w_r}, \quad \varrho_{Mr} = -\frac{\varepsilon_r(ic\kappa')w - w_r}{\varepsilon_r(ic\kappa')w + w_r}, \qquad (A86)$$

$$w_r = \sqrt{u'^2 + v^2 + \varepsilon_r(ic\kappa')\mu_r(ic\kappa')\kappa'^2} \qquad (A87)$$

in terms of the Doppler-shifted frequencies and wave numbers

$$\kappa' = \gamma(\kappa + i\beta u), \quad u' = \gamma(u - i\beta\kappa), \quad \gamma = \frac{1}{\sqrt{1 - \beta^2}}. \qquad (A88)$$

Note that the numerator in Eq. (A84) becomes the denominator if the sums of pairs of A's are replaced by products. Despite the presence of imaginary terms in the integrand, the expression (A84) is a real number. This is easily seen by considering a Taylor expansion in u of the integrand. Since the permittivities and permeabilities are real on the positive imaginary frequency axis (Landau and Lifshitz, Vol. VIII, 1984), this expansion consists of real terms even in u and imaginary terms odd in u; the latter vanish after the integration with respect to u. When $\beta = 0$, when the plates do not move relative to each other, Eq. (A84) reduces to Lifshitz's result (A72). Although formula (A84) is significantly more complicated than Lifshitz's, it has a certain elegance of its own.

Lifshitz theory shows that the Casimir force appears as the electromagnetic force of surface charges at dielectric interfaces where the vacuum stimulates dipoles to align such that they attract each other. The Casimir force is not always attractive, but can also become repulsive, depending on the geometry and the electromagnetic properties of the material used. For example, the Casimir forces between electric and magnetic dielectrics can be repulsive (Boyer, 1974; Buks and Roukes, 2002; Henkel and Joulain, 2005). Repulsive Casimir forces have been observed using three different dielectric media: gold, bromobenzene and silica (Munday et al., 2009). In an extreme and simple case (Leonhardt and Philbin, 2007) the Casimir repulsion of a negatively refracting material (Veselago, 1968; Pendry, 2000) may levitate mirrors on, literally, nothing: on zero-point energy.

Appendix B
State reconstruction in quantum mechanics

Most laws of physics make conditional statements; they predict what happens when some idealized conditions are met. For example, dynamical laws describe how physical systems evolve in time if the initial conditions are specified at a given time. The initial conditions are not fixed: on the contrary, they specify the degrees of freedom, whereupon the evolution is governed by the dynamical laws usually formulated as differential equations. The initial conditions uniquely determine the physical properties of the system at any time and hence we may regard them as describing the state of the system. In non-relativistic quantum mechanics the dynamical law is the Schrödinger equation (1.28) or the von Neumann equation (6.5) and the quantum state is encoded in the density matrix. We argued in Section 5.2.3 that the quantum state is not observable – we cannot see quantum objects as they are – but the state determines the evolution of observable quantities. Can we reconstruct the state from the way an ensemble of quantum objects evolves?

In Section 5.2.3 we have already seen an example of state reconstruction: optical homodyne tomography. Here the Wigner function is reconstructed from the measured quadrature distributions $\mathrm{pr}(q, \theta)$ of an ensemble of equally prepared single modes of light. As we discussed in Section 3.1.1, the quadratures $\hat{q}$ and $\hat{p}$ behave like the position and momentum of a harmonic oscillator, the electromagnetic oscillator of the single mode. Moreover, the phase-shifted quadrature $\hat{q}_\theta$ is the position $\hat{x}$ of the harmonic oscillator evolved in time $t = \theta$ (with dimensionless frequency set to unity). So the quadrature probability distribution $\mathrm{pr}(q, \theta)$ is the position probability distribution $\mathrm{pr}(x, t)$ of a harmonic-oscillator wave packet evolving in time. As we have seen, the Wigner function and hence the quantum state can be reconstructed from $\mathrm{pr}(x, t)$. In this appendix we show how to reconstruct the density matrix from the evolution of one-dimensional wave packets in general, not only for harmonic oscillators. Our procedure was inspired by, and leads to, efficient numerical recipes for sampling the density matrix in optical homodyne tomography (Leonhardt et al., 1996; Leonhardt, 1997a). The state reconstruction depends on a general property of the one-dimensional Schrödinger equation (Leonhardt and Raymer, 1996) that was not known until 1996, 70 years after the Schrödinger equation was discovered. The Schrödinger equation turns out to be remarkably well-designed for state reconstruction such that one may wonder whether there is a deeper reason behind it. Let us present the result and then prove the underlying theorem on the Schrödinger equation.

We describe the quantum state in terms of the density matrix ρ_{mn} in energy representation (as $\langle m | \hat{\rho} | n \rangle$ with the energy eigenstates $\{|n\rangle\}$). For simplicity, we focus on bound states with discrete energy levels, but the continuous part of the spectrum can be included in a similar formalism as well (Leonhardt and Schneider, 1997). The probability distribution $\mathrm{pr}(x, t)$ for the position x evolves in time t as

$$\mathrm{pr}(x, t) = \sum_{mn} \rho_{mn} \, \psi_m(x, t) \, \psi_n^*(x, t). \tag{B1}$$

We show that the quantum state can be reconstructed from the observed positions x of the quantum particle at the time t as

$$\rho_{mn} = \left\langle\!\!\left\langle \frac{\partial}{\partial x} \Big(\psi_m^*(x, t) \, \varphi_n(x, t) \Big) \right\rangle\!\!\right\rangle_{x, t} \tag{B2}$$

where the double brackets denote the average over x and t,

$$\langle\!\langle F(x, t) \rangle\!\rangle_{x, t} \equiv \lim_{T \to \infty} \frac{1}{T} \int_{-T/2}^{+T/2} \int_{-\infty}^{+\infty} F(x, t) \, \mathrm{pr}(x, t) \, \mathrm{d}x \, \mathrm{d}t. \tag{B3}$$

For the harmonic oscillator, we do not need to observe the position over an infinite time, because the motion is periodic, but otherwise the observation time should be infinite in principle or sufficiently long in practice (Leonhardt, 1997b). The function $\varphi_n(x, t)$ in the reconstruction formula (B2) denotes an irregular wave function evolving in time. What are irregular wave functions?

Both the regular and the irregular wave functions oscillate in time t with the frequencies ω_n that correspond to the energy levels,

$$\phi_n(x, t) = \phi_n(x) \exp(-\mathrm{i}\omega_n t), \tag{B4}$$

and they obey the stationary Schrödinger equation,

$$\left(-\frac{1}{2}\frac{d^2}{dx^2} + V(x)\right)\phi_n = \omega_n\phi_n. \tag{B5}$$

Here we scaled the units of position and momentum such that $\hbar = 1$ and the potential $V(x)$ and the energies ω_n are dimensionless. As the Schrödinger equation (B5) is real, the time-independent wave functions $\phi_n(x)$ can be assumed to be real. The regular wave functions are normalized,

$$\int_{-\infty}^{+\infty} \psi_n^2(x)\,dx = 1. \tag{B6}$$

Consider the Wronskian of the two solutions ψ_n and φ_n of the Schrödinger equation[1]

$$W_n = \psi_n(x)\frac{d\varphi_n(x)}{dx} - \frac{d\psi_n(x)}{dx}\varphi_n(x). \tag{B7}$$

We obtain by differentiation with respect to x and application of the Schrödinger equation that the Wronskian is a constant

$$\frac{dW_n}{dx} = 0. \tag{B8}$$

It turns out that, for state reconstruction, the Wronskian of the regular and irregular wave functions is required to be

$$W_n = 2. \tag{B9}$$

The regular wave function must tend to zero for $x \to \pm\infty$, because they are normalizable. So the Wronskian (B7) can only remain 2 if the irregular wave functions grow to infinity for $x \to \pm\infty$. Consequently, the $\varphi_n(x)$ are not normalizable. Only one of the two fundamental solutions of the stationary Schrödinger equation describes a physical state for a given ω_n: in one dimension the energy of a bound state never is degenerate. But, although the irregular wave functions do not describe physical states, they play a vital role in state reconstruction.

To give examples of irregular wave functions, consider the case of the harmonic oscillator with potential

$$V(x) = \frac{x^2}{2}. \tag{B10}$$

In Section 3.2.2 we calculated the energies

$$\omega_n = n + \frac{1}{2} \tag{B11}$$

---

[1] The Wronskian resembles the scalar product (2.17) of a zero-dimensional field theory evolving in x.

with integer n. In the same section we encountered the irregular wave function (3.34) for the vacuum state. The Wronskian condition (B9) determines the prefactor of the wave function (3.34) such that we obtain

$$\varphi_0(x) = \pi^{3/4} \exp\left(-\frac{x^2}{2}\right) \text{erfi}(x) \qquad (B12)$$

with $\text{erfi}(x)$ defined in Eq. (3.35). We construct the irregular wave functions for the excited states by the appropriate application of creation operators, similar to the mathematical creation (3.28) of the Fock states,

$$\varphi_n(x) = \frac{\hat{a}^{\dagger n}}{\sqrt{n!}} \varphi_0(x), \quad \hat{a}^{\dagger} = \frac{1}{\sqrt{2}}\left(x - \frac{d}{dx}\right). \qquad (B13)$$

This procedure guarantees that the φ_n obey the Schrödinger equation of the harmonic oscillator, because we can adopt the arguments of Section 3.2.2 for any wave functions, not only the regular ones. We only need to show in addition that the Wronskian remains 2. To see this, we note that for both regular and irregular wave functions

$$\phi_{n+1} = \frac{\hat{a}^{\dagger}}{\sqrt{n+1}} \phi_n = \frac{1}{\sqrt{2n+2}}\left(x\,\phi_n - \frac{d\phi_n}{dx}\right) \qquad (B14)$$

and hence

$$\frac{d\phi_{n+1}}{dx} = \frac{1}{\sqrt{2n+2}}\left((2n+2)\,\phi_n + x\,\frac{d\phi_n}{dx} - x^2\phi_n\right), \qquad (B15)$$

using the Schrödinger equation (B5) with the potential (B10) and the energies (B11). From the relationships (B14) and (B15) follows

$$W_{n+1} = W_n. \qquad (B16)$$

Consequently, the "excited irregular states" (B13) qualify as irregular wave functions for state reconstruction. Figure B1 shows some plots of regular and irregular wave functions. Figure B2 displays the derivative of products of regular and irregular wave functions. The figure illustrates how these functions sense the characteristic patterns of products of wave functions, of ψ_n^2 in the specific case shown in the figure. Hence they are called *pattern functions* (Leonhardt et al., 1996). In this way the density matrix elements ρ_{nn} are inferred from the probability distribution (B1).

Before we prove our theorem (B2) it is instructive to discuss general regular and irregular wave functions in the *semiclassical approximation* that we derive without assuming prior knowledge. (Readers familiar with semiclassical quantum mechanics (Fröman and Fröman, 1965; Landau and Lifshitz, Vol. III, 1981; Fröman and Fröman, 2005) may skim through this paragraph.) Suppose we represent a wave function in terms of a slowly varying amplitude and a rapidly oscillating phase as

$$\phi_n(x) = \mathcal{A}(x) \exp\left(i \int k(x)\,dx\right). \qquad (B17)$$

Fig. B1. Regular and irregular wave functions. Top: regular wave functions for the first three energy levels of the harmonic oscillator. Bottom: the corresponding irregular wave functions. The irregular wave functions grow exponentially in the classically forbidden region where the regular wave functions exponentially decrease. We also see that the irregular wave functions are out of phase relative to the regular ones.

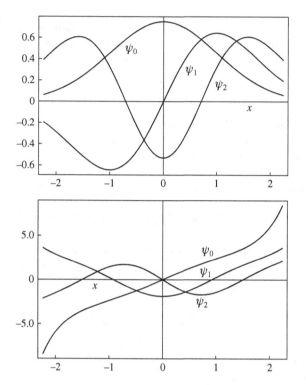

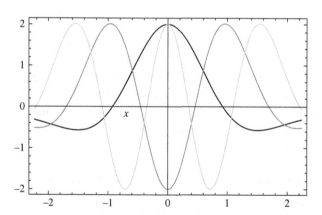

Fig. B2. Pattern functions for state reconstruction, $d(\psi_n \varphi_n)/dx$. Black line: pattern function for the vacuum state $n = 0$. The function senses the characteristic shape of the quadrature probability distribution of the vacuum. We also see how the regular wave function ψ_0 compensates for the exponential growth of the irregular φ_0. Grey line: pattern function for $n = 1$; light-grey line: $n = 2$. We see that the functions oscillate between ± 2, a feature that can be proved in the semiclassical approximation (Leonhardt, 1997a).

Here $k(x)$ denotes the position-dependent wave number that corresponds to the local wavelength

$$\lambda(x) = \frac{2\pi}{k(x)} \tag{B18}$$

and $\mathcal{A}(x)$ describes the amplitude of the wave function $\phi_n(x)$. We assume for the time being that k and $\mathcal{A}$ are real. We insert the representation (B17) into the Schrödinger equation (B5) and obtain

$$k^2 \mathcal{A} - \frac{d^2 \mathcal{A}}{dx^2} - i \left(2 \frac{d\mathcal{A}}{dx} k + \frac{dk}{dx} \mathcal{A} \right) = 2 \left(\omega_n - V \right) \mathcal{A}. \tag{B19}$$

As the right-hand side of this equation is real, the imaginary part of the left-hand side must vanish, which gives a differential equation for $\mathcal{A}$ with the general solution

$$\mathcal{A} = \frac{C}{\sqrt{k(x)}}. \tag{B20}$$

The amplitude $\mathcal{A}(x)$ varies less than the phasor $\exp(i \int k \, dx)$ in the wave function (B17) if

$$\left| \frac{d\lambda}{dx} \right| \ll 1, \tag{B21}$$

as we see by differentiating (B18) and comparing the result with the ratio between the derivatives of the amplitude (B20) and $\mathcal{A}$ times the derivative of $\exp(i \int k \, dx)$. The regime (B21) defines the validity range of the semiclassical approximation, also known as the *WKB or WKBJ approximation* (Wentzel–Kramers–Brillouin–Jeffreys). In this regime we can neglect the second derivative of the amplitude in the Schrödinger equation (B19), because

$$\frac{16\pi^2}{k^2 \mathcal{A}} \frac{d^2 \mathcal{A}}{dx^2} = 2\lambda \frac{d^2\lambda}{dx^2} - \left(\frac{d\lambda}{dx} \right)^2. \tag{B22}$$

Thus neglecting $d^2 \mathcal{A}/dx^2$ we obtain the dispersion relation

$$k(x) = \sqrt{2 \left(\omega_n - V(x) \right)}. \tag{B23}$$

The wave number (B23) has two roots that differ by a sign and so the Schrödinger wave is, in general, a superposition of the type

$$\phi_n(x) = \frac{A}{\sqrt{k}} \exp \left(i \int k \, dx \right) + \frac{B}{\sqrt{k}} \exp \left(-i \int k \, dx \right). \tag{B24}$$

In the classically allowed region where the potential $V(x)$ lies below the energy ω_n the wave number is real and the wave function consists

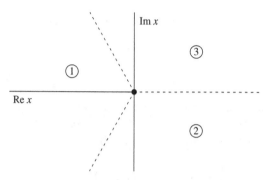

Fig. B3. Stokes lines. The dotted lines show the Stokes lines in the complex plane around a turning point of the classical motion. At a Stokes line, the phase of the stationary wave function is purely imaginary. In the semiclassical approximation, one of the two fundamental waves in the superposition (B24) becomes exponentially larger than the other. The superposition is not completely maintained: a contribution from the dominant term turns out to spill over onto the exponentially small term. The numbers label the regions where the superposition (B24) is valid.

of a superposition of right- and left-moving waves. The semiclassical approximation is no longer valid near the classical turning points x_0 with

$$V(x_0) = \omega_n \tag{B25}$$

where $k(x)$ becomes proportional to $(x_0 - x)^{1/2}$ and the wavelength (B18) diverges. Nevertheless, we can apply the semiclassical approximation to work out the reflection at the turning point (Furry, 1947). For this, we analytically continue both the representation (B24) and the exact wave function in the complex plane (Ablowitz and Fokas, 1997). Consider the vicinity of one turning point as shown in Fig. B3. We assume that the classically allowed region lies to the left of the turning point and the classically forbidden region to the right. By analytic continuation we connect the two regions. In the classically forbidden region, where $\omega_n < V(x)$, the wave number (B23) is purely imaginary and a regular wave function should exponentially decay. There are two more lines in the complex plane around the turning point where the phase $\int k \, dx$ is purely imaginary, for the following reason. Near the turning point at x_0 the wave number k, given by Eq. (B23), is proportional to $(x_0 - x)^{1/2}$. Hence the integral of k is proportional to $(x_0 - x)^{3/2}$ that is purely imaginary in the three directions shown in Fig. B3. The lines of purely imaginary phase, called the *Stokes lines*, then continue beyond the immediate vicinity of the turning point. At a Stokes line one of the two fundamental waves in the superposition (B24) becomes exponentially larger than the other.

So the superposition (B24) may not be completely maintained within the accuracy of the semiclassical approximation – some contribution from the exponentially dominant wave may spill over onto the exponentially suppressed wave. Any such additional contribution must be linear in the amplitude of the dominant wave in order to comply with the superposition principle, the linearity of the Schrödinger equation. This mathematical feature, known as the *Stokes phenomenon*,[2] turns out to be necessary to ensure the Schrödinger wave function remains single valued, whereas the semiclassical waves (B17) by themselves are multi-valued functions. Let us use indices ranging from 1 to 3 to distinguish the coefficients A_l and B_l in the regions separated by Stokes lines shown in Fig. B3. If we choose the root (B23) with positive real part in the classically allowed region, the wave number is positively imaginary on Stokes line 1, negatively imaginary on Stokes line 2, the classically forbidden region, and positively imaginary on Stokes line 3. Consequently, we obtain the connection formulae (Furry, 1947)

$$
\begin{aligned}
A_2 &= A_1 + \alpha B_1, & B_2 &= B_1, \\
A_3 &= A_2, & B_3 &= B_2 + \beta A_2, \\
A_{1'} &= A_3 + \gamma B_3, & B_{1'} &= B_3.
\end{aligned}
\tag{B26}
$$

After completing one loop in the complex plane around the turning point we obtain the coefficients $A_{1'}$ and $B_{1'}$. But after the loop the phase $(x_0 - x)^{3/2}$ has switched sign and therefore the A and B terms are interchanged. Furthermore, the amplitude $(x_0 - x)^{-1/4}$ has gained a factor of $-i$ after the loop. Consequently (Furry, 1947),

$$
A_1 = -iB_{1'}, \quad B_1 = -iA_{1'}. \tag{B27}
$$

We obtain $iA_1 = B_2 + \beta A_2 = \beta A_1 + (1 + \alpha\beta)B_1$, which implies that $\alpha = \beta = i$. For the B_1 component we get along similar lines $iB_1 = (\gamma + \alpha + \alpha\beta\gamma)B_1 + (1 + \beta\gamma)A_1$, which then gives $\gamma = i$. Consequently, the connection rules (B26) are uniquely determined with (Furry, 1947)

$$
\alpha = \beta = \gamma = i. \tag{B28}
$$

We deduce from the connection rules that a regular wave function, which is exponentially decaying in the classically forbidden region,

$$
\psi_n \sim \frac{A_n}{\sqrt{|k|}} \exp\left(-\left|\int k \, dx\right|\right), \tag{B29}
$$

---

[2] George G. Stokes noticed this feature in connection with the wave theory of the rainbow (Nussenzveig, 1992).

corresponds, in the classically allowed region, to

$$\psi_n \sim \frac{2A_n}{\sqrt{k}} \cos\left(\int k\,dx - \frac{\pi}{4}\right). \tag{B30}$$

We see that a stationary wave attains the phase shift $\pi/4$ upon reflection at a turning point. A bound state is enclosed by two turning points, one on the left and one on the right, the one considered here in detail. The phase integral from the right turning point to the left one and back to the right turning point, including the phase shifts, must give an integer multiple of 2π, because the wave function is single-valued. In this way we obtain the *Bohr–Sommerfeld quantization*

$$\oint k\,dx = 2\pi\left(n + \frac{1}{2}\right). \tag{B31}$$

The irregular wave function

$$\varphi_n \sim \frac{1}{A_n\sqrt{k}} \sin\left(\int k\,dx - \frac{\pi}{4}\right) \tag{B32}$$

satisfies the Wronskian condition (B9) for state reconstruction and grows in the classically forbidden region like

$$\varphi_n \sim \frac{1}{A_n\sqrt{|k|}} \exp\left(+\left|\int k\,dx\right|\right) \tag{B33}$$

where we neglect the exponentially suppressed contribution. We could add any multiple of the regular wave function ψ_n to φ_n without changing the Wronskian (B7), but the wave function (B32) is sufficiently irregular with minimal amplitude. Using the semiclassical approximation, we have seen how the irregular wave functions grow for $x \to \pm\infty$.

After our brief excursion into semiclassical wave mechanics we return to the proof of the reconstruction theorem (B2) on the Schrödinger equation (Leonhardt and Raymer, 1996; Richter and Wünsche, 1996). We assume that $\omega_m \geq \omega_n$ such that $\psi_m\varphi_n$ decays for $x \to \pm\infty$, which does not restrict state reconstruction, because for $\omega_m < \omega_n$ we can use the Hermiticity of the density matrix $\rho_{mn} = \rho_{nm}^*$. For the spatial derivatives of the $\psi_m^*\varphi_n$ to project the density matrix elements out of the probability distribution (B1), they must be orthonormal to products of the wave functions; the quantity

$$D_{m'n'}^{mn} \equiv \lim_{T\to\infty} \frac{1}{T} \int_{-T/2}^{+T/2} \int_{-\infty}^{+\infty} \psi_{m'}\psi_{n'}^* \frac{\partial(\psi_m^*\varphi_n)}{\partial x}\,dx\,dt \tag{B34}$$

must give

$$D_{m'n'}^{mn} = \delta_{mm'}\,\delta_{nn'}. \tag{B35}$$

In quantum mechanics we are used to orthogonal sets for wave functions, but in quantum state reconstruction we are concerned with products of wave functions, because such products appear in observable quantities. The time average with the stationary waves (B4) selects some frequencies,

$$\omega_m - \omega_n = \omega_{m'} - \omega_{n'},\tag{B36}$$

but, in general, it does not determine the matrix element (B34) completely. We need to prove that the quantity

$$G_{m'n'}^{mn} \equiv \int_{-\infty}^{+\infty} \psi_{m'}(x)\psi_{n'}(x)\frac{\mathrm{d}}{\mathrm{d}x}\Big(\psi_m(x)\varphi_n(x)\Big)\,\mathrm{d}x\tag{B37}$$

gives, if the frequency constraint (B36) is satisfied,

$$G_{m'n'}^{mn} = \delta_{mm'}\,\delta_{nn'}.\tag{B38}$$

From expression (B37) we already obtain the Wronskian condition (B9) of the irregular wave functions: for $m = m'$ and $n = n'$ we get

$$\begin{aligned}
G_{mn}^{mn} &= \int_{-\infty}^{+\infty}\left(\psi_m^2\psi_n\frac{\mathrm{d}\varphi_n}{\mathrm{d}x} + \psi_m\varphi_n\psi_n\frac{\mathrm{d}\psi_m}{\mathrm{d}x}\right)\mathrm{d}x\\
&= W_n + \int_{-\infty}^{+\infty}\psi_m\varphi_n\frac{\mathrm{d}(\psi_m\psi_n)}{\mathrm{d}x}\,\mathrm{d}x
\end{aligned}\tag{B39}$$

using the conservation (B8) of the Wronskian (B7) and the normalization (B6) of the regular wave functions. On the other hand, partial integration in the definition (B37) gives

$$G_{mn}^{mn} = -\int_{-\infty}^{+\infty}\psi_m\varphi_n\frac{\mathrm{d}(\psi_m\psi_n)}{\mathrm{d}x}\,\mathrm{d}x.\tag{B40}$$

Adding Eqs. (B39) and (B40) leads to the condition (B9) for $G_{mn}^{mn} = 1$.

An elegant proof (Richter and Wünsche, 1996) of the orthonormality of $\mathrm{d}(\psi_m\varphi_n)/\mathrm{d}x$ on products of wave functions with constraint (B36) involves a fourth-order differential equation for products of wave functions. We obtain from the Schrödinger equation (B5) by differentiation

$$\left(\hat{D}_4 - 4(\omega_m - \omega_n)^2 - 4(\omega_m + \omega_n)\frac{\mathrm{d}^2}{\mathrm{d}x^2}\right)\phi_m\phi_n = 0\tag{B41}$$

with the fourth-order differential operator

$$\hat{D}_4 = -\frac{\mathrm{d}^4}{\mathrm{d}x^4} + 8V\frac{\mathrm{d}^2}{\mathrm{d}x^2} + 12\frac{\mathrm{d}V}{\mathrm{d}x}\frac{\mathrm{d}}{\mathrm{d}x} + 4\frac{\mathrm{d}^2V}{\mathrm{d}x^2}.\tag{B42}$$

Consider the Hermitian conjugate of this operator,

$$
\begin{aligned}
\hat{D}_4^\dagger &= -\frac{\mathrm{d}^4}{\mathrm{d}x^4} + \frac{\mathrm{d}^2}{\mathrm{d}x^2}8V - 12\frac{\mathrm{d}}{\mathrm{d}x}\frac{\mathrm{d}V}{\mathrm{d}x} + 4\frac{\mathrm{d}^2V}{\mathrm{d}x^2} \\
&= -\frac{\mathrm{d}^4}{\mathrm{d}x^4} + 8V\frac{\mathrm{d}^2}{\mathrm{d}x^2} + 4\frac{\mathrm{d}V}{\mathrm{d}x}\frac{\mathrm{d}}{\mathrm{d}x}.
\end{aligned}
\tag{B43}
$$

The differential equation (B41) is not Hermitian, so products of wave functions are not orthonormal to themselves, but we obtain by differentiation

$$
\frac{\mathrm{d}}{\mathrm{d}x}\hat{D}_4^\dagger = \hat{D}_4\frac{\mathrm{d}}{\mathrm{d}x}.
\tag{B44}
$$

As we will see, this relationship implies that the *derivatives* of $\psi_m\varphi_n$ are orthonormal on $\psi_{m'}\psi_{n'}$. Let us write

$$
\psi_m\varphi_n = \frac{\mathrm{d}h_{mn}}{\mathrm{d}x}.
\tag{B45}
$$

We obtain from the fourth order differential equation (B41) with relationship (B44)

$$
\frac{\mathrm{d}}{\mathrm{d}x}\cdot\left(\hat{D}_4^\dagger - 4(\omega_m - \omega_n)^2 - 4(\omega_m + \omega_n)\frac{\mathrm{d}^2}{\mathrm{d}x^2}\right)h_{mn} = 0.
\tag{B46}
$$

We integrate this equation and put the integration constant to zero, as h_{mn} is only determined in formula (B45) up to a constant anyway. Consequently,

$$
\left(\hat{D}_4^\dagger - 4(\omega_m - \omega_n)^2 - 4(\omega_m + \omega_n)\frac{\mathrm{d}^2}{\mathrm{d}x^2}\right)h_{mn} = 0.
\tag{B47}
$$

The rest is a brief calculation,

$$
\begin{aligned}
\delta &\equiv 4(\omega_m + \omega_n - \omega_{m'} - \omega_{n'})\,G_{m'n'}^{mn} \\
&= 4(\omega_m + \omega_n)\int_{-\infty}^{+\infty}\psi_{m'}\psi_{n'}\frac{\mathrm{d}^2 h_{mn}}{\mathrm{d}x^2}\,\mathrm{d}x \\
&\quad - 4(\omega_{m'} + \omega_{n'})\int_{-\infty}^{+\infty}h_{mn}\frac{\mathrm{d}^2\psi_{m'}\psi_{n'}}{\mathrm{d}x^2}\,\mathrm{d}x \\
&= \int_{-\infty}^{+\infty}\psi_{m'}\psi_{n'}\hat{D}_4^\dagger h_{mn}\,\mathrm{d}x - \int_{-\infty}^{+\infty}h_{mn}\hat{D}_4\,\psi_{m'}\psi_{n'}\,\mathrm{d}x \\
&= 0,
\end{aligned}
\tag{B48}
$$

where we used the frequency constraint (B36) in order to arrive at the third line. As δ vanishes but $\omega_m + \omega_n - \omega_{m'} - \omega_{n'} \neq 0$ for $m \neq m'$ and $n \neq n'$ and the constraint (B36), G_{mn}^{mn} must vanish in this case.

We have thus proved that $\partial(\psi_m^*(x,t)\varphi_n(x,t))/\partial x$ is orthonormal to products of wave functions. This orthonormality theorem critically depends on the structure of the Schrödinger equation. We may as well turn the tables and ask whether this result was a coincidence or a principle. Has Nature chosen the Schrödinger equation so that the states of physical objects can be inferred from measurement? What are the conditions imposed on dynamical laws such that quantum states can be reconstructed?

References

Abbas, G. L., Chan, V. W. S., and Yee, T. K. (1983), *Opt. Lett.* **8**, 419.

Abbott, L. F., and Pi, S. Y. (1986), *Inflationary Cosmology* (World Scientific, Singapore).

Ablowitz, M. J., and Fokas, A. S. (1997), *Complex Variables* (Cambridge University Press, Cambridge).

Abramowitz, M., and Stegun, I. A. (1970), *Handbook of Mathematical Functions* (Dover, New York).

Achilles, D., Silberhorn, C., Śliwa, C., Banaszek, K., Walmsley, I. A., Fitch, M. J., Jacobs, B. C., Pittman, T. B., and Franson, J. D. (2004), *J. Mod. Opt.* **51**, 1499.

Agarwal, G. S. (1981), *Phys. Rev. A* **24**, 2889.

Agrawal, G. (2001), *Nonlinear Fiber Optics* (Academic Press, San Diego).

Aharonov, Y., Albert, D. Z., and Au, C. K. (1981), *Phys. Rev. Lett.* **47**, 1029.

Almeida, L. B. (1994), *IEEE Trans. Signal Proc.* **42**, 3084.

Altland, A., and Simon, B. (2006), *Condensed Matter Field Theory* (Cambridge University Press, Cambridge).

Arthurs, E., and Kelly, J. L., Jr. (1965), *Bell Syst. Tech. J.* **44**, 725.

Aspect, A., Grangier, P., and Roger, G. (1981), *Phys. Rev. Lett.* **47**, 460.

Aspect, A., Grangier, P., and Roger, G. (1982a), *Phys. Rev. Lett.* **49**, 91.

Aspect, A., Dalibard, J., and Roger, G. (1982b), *Phys. Rev. Lett.* **49**, 1804.

Autumn, K., Sitti, M., Liang, Y. A., Peattle, A. M., Hansen, W. R., Sponberg, S., Kenny, T. W., Fearing, R., Israelachvili, J. N., and Full, R. J. (2002), *Proc. Nat. Acad. Sci.* **99**, 12252.

Bachor, H., and Ralph, T. C. (2004), *A Guide to Experiments in Quantum Optics* (Wiley-VCH, Berlin).

Bacry, H., Grossmann, A., and Zak, J. (1975), *Phys. Rev. B* **12**, 1118.

Badurek, G., Hradil, Z., Lvovsky, A. I., Molina Terriza, G., Rauch, H., Řeháček, J., Vaziri, A., and Zawisky, M. (2004), *Maximum Likelihood Estimation in Experimental Quantum Physics*, in *Quantum State Estimation*, ed. Paris, M. G. A., and Řeháček, J. (Springer, Berlin).

Ballentine, L. E. (1990), *Quantum Mechanics* (Prentice-Hall, Englewood Cliffs, NJ).

Banaszek, K., and Wódkiewicz, K. (1997), *Phys. Rev. A* **55**, 3117.

Barnett, S. M., and Phoenix, S. J. D. (1989), *Phys. Rev. A* **40**, 2404.

Barnett, S. M., and Phoenix, S. J. D. (1991), *Phys. Rev. A* **44**, 535.

Barnett, S. M., and Radmore, P. M. (2002), *Methods in Theoretical Quantum Optics* (Clarendon Press, Oxford).

Bekenstein, J. D. (1973), *Phys. Rev. D* **7**, 2333.

Bekenstein, J. D. (2001), *Stud. Hist. Philos. Mod. Phys.* **32**, 511.

Bekenstein, J. D., and Mayo, A. E. (2001), *Gen. Rel. Grav.* **33**, 2095.

Bell, J. S. (1964), *Physics* **1**, 195.

Bell, J. S. (1987), *Speakable and Unspeakable in Quantum Mechanics* (Cambridge University Press, Cambridge).

Bennett, C. H. (1992), *Phys. Rev. Lett.* **68**, 3121.

Bennett, C. H., and Brassard, G. (1984), Quantum cryptography: public key distribution and coin tossing, in *Proceedings of the IEEE International Conference on Computers, Systems, and Signal Processing*, Bangalore, 175.

Bennett, C. H., Brassard, G., Josza, R., Peres, A., and Wooters, W. K. (1993), *Phys. Rev. Lett.* **70**, 1895.

Bertrand, J., and Bertrand, P. (1987), *Found. Phys.* **17**, 397.

Birrell, N. D., and Davies, P. C. W. (1982), *Quantum Fields in Curved Space* (Cambridge University Press, Cambridge).

Bjorken, J. D., and Drell, S. D. (1965), *Relativistic Quantum Fields* (McGraw-Hill, New York).

Blocki, J., Randrup, J., Swiatecki, W. J., and Tsang, C. F. (1977), *Ann. Phys.* **105**, 427.

Böhmer, B., and Leonhardt, U. (1995), *Opt. Commun.* **118**, 181.

Bohr, N. (1935), *Phys. Rev.* **48**, 696.

Bordag, M., Mohideen, U., and Mostepanenko, V. M. (2001), *Phys. Rep.* **353**, 1.

Born, M. (1956), *Physics in My Generation* (Pergamon, London).

Born, M., and Wolf, E. (1999), *Principles of Optics* (Cambridge University Press, Cambridge).

Boschi, D., Branca, S., De Martini, F., Hardy, L., and Popescu, S. (1998), *Phys. Rev. Lett.* **80**, 1121.

Boulware, D. G. (1980), *Ann. Phys.* **124**, 169.

Bouwmeester, D., Ekert, A., and Zeilinger, A. (2001), *The Physics of Quantum Information* (Springer, Berlin).

Bouwmeester, D., Pan, J. W., Mattle, K., Eibl, M., Weinfurter, H., and Zeilinger, A. (1997), *Nature*, **390**, 575.

Boyd, R. W. (1992), *Nonlinear Optics* (Academic Press, San Diego).

Boyer, T. H. (1974), *Phys. Rev. A* **9**, 2078.

Braginsky, V. B., Khalili, F. Y., and Thorne, K. S. (1995), *Quantum Measurement* (Cambridge University Press, Cambridge).

Brandão, F. G. S. L., and Plenio, M. (2008), *Nature Phys.* **4**, 873.

Braunstein, S. L. (1990), *Phys. Rev. A* **42**, 474.

Braunstein, S. L., and Kimble, H. J. (1998), *Phys. Rev. Lett.* **80**, 869.

Breitenbach, G., Müller, T., Pereira, S. F., Poizat, J. P., Schiller, S., and Mlynek, J. (1995), *J. Opt. Soc. Am. B* **12**, 2304.

Brendel, J., Schütrumpf, S., Lange, R., Martienssen W., and Scully, M. O. (1988), *Europhys. Lett.* **5**, 223.

Brout, R., Massar, S., Parentani, R., and Spindel, P. (1995a), *Phys. Rep.* **260**, 329.

Brout, R., Massar, S., Parentani, R., and Spindel, P. (1995b), *Phys. Rev. D* **52**, 4559.

Brumfiel, G. (2007), *Nature* **448**, 245.

Brunner, W. H., Paul, H., and Richter, G. (1965), *Ann. Phys.* (Leipzig) **15**, 17.

Bruss, D. (2002), *J. Math. Phys.* **43**, 4237.

Buhmann, S. Y., and Welsch, D. G. (2007), *Progr. Quant. Electron.* **31**, 51.

Buks, E., and Roukes, M. L. (2002), *Nature* **419**, 119.

Bužek, V., and Knight, P. L. (1995), *Prog. Opt.* **34**, 1.

Cahill, K. E. (1965), *Phys. Rev.* **138**, 1566.

Cahill, K. E., and Glauber, R. J. (1969a), *Phys. Rev.* **177**, 1857.

Cahill, K. E., and Glauber, R. J. (1969b), *Phys. Rev.* **177**, 1882.

Callen, B. C., and Welton, T. A. (1951), *Phys. Rev.* **83**, 34.

Campos, R. A., Saleh, B. E. A., and Teich, M. C. (1989), *Phys. Rev. A* **40**, 1371.

Carmichael, H. J. (1987), *J. Opt. Soc. Am. B* **4**, 1588.

Carmichael, H. J. (1993), *An Open Systems Approach to Quantum Optics* (Springer, Berlin).

Carmichael, H. J. (2003), *Statistical Methods in Quantum Optics 1: Master Equations and Fokker–Planck Equations* (Springer, Berlin).

Carmichael, H. J. (2007), *Statistical Methods in Quantum Optics 2: Non-Classical Fields* (Springer, Berlin).

Casimir, H. B. G. (1948), *Proc. Kon. Nederland. Akad. Wetensch.* **B51**, 793.

Caves, C. M. (1982), *Phys. Rev. D* **26**, 1817.

Chan, H. B., Aksyuk, V. A., Kleiman, R. N., Bishop, D. J., and Capasso, F. (2001), *Science* **291**, 1941.

Clauser, J. F., Horne, M. A., Shimony, A., and Holt, R. A. (1969), *Phys. Rev. Lett.* **23**, 880.

Cohen Tannoudji, C., Diu, B., and Laloë, F. (1977), *Quantum Mechanics* (Wiley, New York).

Cohen Tannoudji, C., Dupont Roc, J., and Grynberg, G. (1989), *Photons and Atoms* (Wiley, New York).

Cohen Tannoudji, C., Dupont Roc, J., and Grynberg, G. (1992), *Atom–Photon Interactions* (Wiley, New York).

Cohendet, O., Combe, P., Sirugue, M., and Sirugue Collin, M. (1988), *J. Phys. A* **21**, 2875.

Cohendet, O., Combe, P., and Sirugue Collin, M. (1990), *J. Phys. A* **23**, 2001.

Cornwell, J. F. (1984), *Group Theory in Physics* (Academic, London).

Dakna, M., Anhut, T., Opatrný, T., Knöll, L., and Welsch, D. G. (1997), *Phys. Rev. A* **55**, 3184.

Damour, T., and Ruffini, R. (1976), *Phys. Rev. D* **14**, 332.

Davydov, A. S. (1991), *Quantum Mechanics* (Pergamon, Oxford).

Dicke, R. H. (1946), *Rev. Sci. Instrum.* **17**, 268.

Dieks, D. (1982), *Phys. Lett.* **92A**, 271.

Dirac, P. A. M. (1984), *The Principles of Quantum Mechanics* (Clarendon, Oxford).

Dowling, J. P., Schleich, W. P., and Wheeler, J. A. (1991), *Ann. Phys.* (Leipzig) **48**, 423.

Dowling, J. P., Agarwal, G. S., and Schleich, W. P. (1994), *Phys. Rev. A* **49**, 4101.

Dragoman, D. (1997), *Prog. Opt.* **37**, 1.

Dragoman, D. (2002), *Prog. Opt.* **43**, 433.

Dusek, M., Lütkenhaus, N., and Hendrych, M. (2006), *Prog. Opt.* **49**, 381.

Dzyaloshinskii, I. E., Lifshitz, E. M., and Pitaevskii, L. P. (1961), *Adv. Phys.* **10**, 165.

Einstein, A. (1905a), *Ann. Phys.* **17**, 132.

Einstein, A. (1905b), *Ann. Phys.* **17**, 891.

Einstein, A., Rosen, N., and Podolsky, B. (1935), *Phys. Rev.* **47**, 777.

Eisert, J., and Wolf, M. M. (2007), *Gaussian Quantum Channels*, in *Quantum Information with Continuous Variables of Atoms and Light*, ed. Cerf, N., Leuchs, G., and Polzik, E. S. (Imperial College Press, London).

Ekert, A. (1991), *Phys. Rev. Lett.* **67**, 661.

Engen, G. F., and Hoer, C. A. (1972), *IEEE Trans. Instrum. Meas.* **21**, 470.

Erdélyi, A., Magnus, W., Oberhettinger, F., and Tricomi, F. G. (1953), *Higher Transcendental Functions* (McGraw-Hill, New York).

Fetter, A. L., and Walecka, J. D. (1971), *Quantum Theory of Many-Particle Systems* (McGraw-Hill, New York).

Fizeau, H. (1851), *C. R. Acad. Sci.* (Paris) **33**, 349.

Fox, M. (2006), *Quantum Optics: An Introduction* (Oxford University Press, Oxford).

Fradkin, E. (1991), *Field Theories of Condensed Matter Systems* (Westview Press, Boulder).

Freire Jr, O. (2006), *Stud. Hist. Phil. Mod. Phys.* **37**, 577.

Fresnel, A. J. (1818), *Ann. Chim. Phys.* **9**, 57.

Freyberger, M., Vogel, K., and Schleich, W. P. (1993), *Phys. Lett. A* **176**, 41.

Friedman, J. R., Patel, V., Chen, W., Tolpygo, S. K., and Lukens, J. E. (2000), *Nature* **406**, 43.

Fröman, N., and Fröman, P. O. (1965), *JWKB Approximation* (North-Holland, Amsterdam).

Fröman, N., and Fröman, P. O. (2005), *Physical Problems Solved by the Phase-Integral Method* (Cambridge University Press, Cambridge).

Furry, W. H. (1947), *Phys. Rev.* **71**, 360.

Furusawa, A., Sorensen, J. L., Braunstein, S. L., Fuchs, C. A., Kimble, H. J., and Polzik, E. S. (1998), *Science* **282**, 706.

Garay, L. J., Anglin, J. R., Cirac, J. I., and Zoller, P. (2000), *Phys. Rev. Lett.* **85**, 4643.

Gardiner, C. W. (1991), *Quantum Noise* (Springer, Berlin).

Gardiner, C. W., and Zoller, P. (2004), *Quantum Noise: A Handbook of Markovian and Non-Markovian Quantum Stochastic Methods with Applications to Quantum Optics* (Springer, Berlin).

Garrison, J. C., and Chiao, R. Y. (2008), *Quantum Optics* (Oxford University Press, Oxford).

Gel'fand, I. M., and Shilov, G. E. (1965), *Generalized Functions* (Academic, San Diego).

Gerry, C., and Knight, P. L. (2004), *Introductory Quantum Optics* (Cambridge University Press, Cambridge).

Ghirardi, G. C., Rimini, A., and Weber, T. (1986), *Phys. Rev. D* **34**, 470.

Gisin, N., Ribordy, G., Tittel, W., and Zbinden, H. (2002), *Rev. Mod. Phys.* **74**, 145.

Glauber, R. J. (1963), *Phys. Rev. Lett.* **10**, 84.

Glauber, R. J., and Lewenstein, M. (1991), *Phys. Rev. A* **43**, 467.

Gordon, W. (1923), *Ann. Phys.* (Leipzig) **72**, 421.

Green, M. B., Schwarz, J. H., and Witten, E. (1987), *Superstring Theory* (Cambridge University Press, Cambridge).

Haken, H. (1981), *Light* (Elsevier, Amsterdam).

Hanamura, E., Kawabe, Y., and Yamanaka, A. (2007), *Quantum Nonlinear Optics* (Springer, Berlin).

Hawking, S. W. (1974), *Nature* **248**, 30.

Heisenberg, W. (1969), *Der Teil und das Ganze* (Piper, Munich). English translation: Heisenberg, W. (1971), *Physics and Beyond; Encounters and Conversations* (Harper and Row, New York).

Henkel, C., and Joulain, K. (2005), *Europhys. Lett.* **72**, 929.

Herman, G. T. (1980), *Image Reconstruction from Projections: The Fundamentals of Computerized Tomography* (Academic, New York).

Hertz, H. (1887), *Ann. Phys.* **31**, 982.

Hogan, J. (2007), *Nature* **448**, 240.

Hong, C. K., Ou, Z. Y., and Mandel, L. (1987), *Phys. Rev. Lett.* **59**, 2044.

Hradil, Z., Řeháček, J., Fiurášek, J., and Ježek, M. (2004), *Maximum Likelihood Methods in Quantum Mechanics*, in *Quantum State Estimation*, ed. Paris, M. G. A., and Řeháček, J. (Springer, Berlin).

Hradil, Z., Mogilevtsev, D., and Řeháček, J. (2006), *Phys. Rev. Lett.* **96**, 230401.

Hudson, R. L. (1974), *Rep. Math. Phys.* **6**, 249.

Husimi, K. (1940), *Proc. Phys. Math. Soc. Japan* **22**, 264.

Itzykson, C., and Zuber, J. B. (2006), *Quantum Field Theory* (Dover, Mineola).

Jackson, J. D. (1998), *Classical Electrodynamics* (Wiley, New York).

Jacobson, T. (1999), *Prog. Theor. Phys. Suppl.* **136**, 1.

Jammer, M. (1989), *The Conceptual Development of Quantum Mechanics* (McGraw-Hill, New York).

Jordan, P. (1935), *Z. Phys.* **94**, 531.

Kenyon, I. R. (2008), *The Light Fantastic: A Modern Introduction to Classical and Quantum Optics* (Oxford University Press, Oxford).

Kerr, R. (1963), *Phys. Rev. Lett.* **11**, 237.

Kiesel, T., Vogel, W., Parigi, V., Zavatta A., and Bellini, M. (2008), *Phys. Rev. A* **78**, 021804.

Kim, Y. H., Yu, R., Kulik, S. P., Shih, Y., and Scully, M. O. (2000), *Phys. Rev. Lett.* **84**, 1.

Kim, Y. H., Yu, R., Kulik, S. P., and Shih, Y. (2001), *Phys. Rev. Lett.* **86**, 1370.

Klauder, J. R. (1966), *Phys. Rev. Lett.* **16**, 534.

Klauder, J. R., and Sudarshan, E. C. G. (2006), *Fundamentals of Quantum Optics* (Dover, Mineola).

Klimov, A. B., Sanchez Soto, L. L., Yustas, E. C., Söderholm, J., and Björk, G. (2005), *Phys. Rev. A* **72**, 033813.

Knöll, L., Scheel, S., and Welsch, D. G. (2001), QED in dispersing and absorbing media, in *Coherence and Statistics of Photons and Atoms*, ed. Peřina, J. (Wiley, New York).

Korolkova, N. (2007), Polarization squeezing and entanglement, in *Quantum Information with Continuous Variables of Atoms and Light*, ed. Cerf, N., Leuchs, G., and Polzik, E. S. (Imperial College Press, London).

Korolkova, N., and Loudon, R. (2005), *Phys. Rev. A* **71**, 032343.

Lambropoulos, P., and Petrosyan, D. (2006), *Fundamentals of Quantum Optics and Quantum Information* (Springer, Berlin).

Lamoreaux, S. K. (1997), *Phys. Rev. Lett.* **78**, 5.

Lamoreaux, S. K. (1999), *Am. J. Phys.* **67**, 850.

Lamoreaux, S. K. (2005), *Rep. Prog. Phys.* **68**, 201.

Lanczos, C. (1998), *Linear Differential Operators* (Dover, New York).

Landau, L. D., and Lifshitz, E. M. (1982), *Mechanics*, Vol. I (Butterworth Heinemann, Oxford).

Landau, L. D., and Lifshitz, E. M. (1987), *The Classical Theory of Fields*, Vol. II (Butterworth Heinemann, Oxford).

Landau, L. D., and Lifshitz, E. M. (1981), *Quantum Mechanics: Non-Relativistic Theory*, Vol. III (Butterworth Heinemann, Oxford).

Landau, L. D., and Lifshitz, E. M. (1996), *Statistical Physics*, Vol. V (Butterworth Heinemann, Oxford).

Landau, L. D., and Lifshitz, E. M. (1987), *Fluid Mechanics*, Vol. VI (Butterworth Heinemann, Oxford).

Landau, L. D., and Lifshitz, E. M. (1984), *Electrodynamics of Continuous Media*, Vol. VIII (Butterworth Heinemann, Oxford).

Landau, L. D., and Lifshitz, E. M. (1980), *Statistical Physics, Part 2*, Vol. IX (Butterworth Heinemann, Oxford).

Leggett, A. J. (2003), *Found. Phys.* **10**, 1469.

Lehner, J., Leonhardt, U., and Paul, H. (1996), *Phys. Rev. A* **53**, 2727.

Leonhardt, U. (1993), Quantum theory of simple optical instruments, *Humboldt University PhD Thesis*.

Leonhardt, U. (1993), *Phys. Rev. A* **48**, 3265.

Leonhardt, U. (1994), *Phys. Rev. A* **49**, 1231.

Leonhardt, U. (1995), *Phys. Rev. Lett.* **74**, 4101.

Leonhardt, U. (1996), *Phys. Rev. A* **53**, 2998.

Leonhardt, U. (1997a), *Measuring the Quantum State of Light* (Cambridge University Press, Cambridge).

Leonhardt, U. (1997b), *Phys. Rev. A* **55**, 3164.

Leonhardt, U. (2002), *Nature* **415**, 406.

Leonhardt, U. (2003), *Rep. Prog. Phys.* **66**, 1207.

Leonhardt, U. (2006a), *Science* **312**, 1777.

Leonhardt, U. (2006b), *New J. Phys.* **8**, 118.

Leonhardt, U., and Paul, H. (1993a), *Phys. Rev. A* **47**, R2460.

Leonhardt, U., and Paul, H. (1993b), *Phys. Scr.* **T48**, 45.

Leonhardt, U., and Paul, H. (1993c), *Phys. Rev. A* **48**, 4598.

Leonhardt, U., and Paul, H. (1994), *Phys. Lett. A* **193**, 117.

Leonhardt, U., and Philbin, T. G. (2006), *New J. Phys.* **8**, 247.

Leonhardt, U., and Philbin, T. G. (2007), *New J. Phys.* **9**, 254.

Leonhardt, U., and Philbin, T. G. (2009), *Prog. Opt.*, **53**, 69.

Leonhardt, U., and Piwnicki, P. (1999), *Phys. Rev. A* **60**, 4301.

Leonhardt, U., and Piwnicki, P. (2000), *Phys. Rev. Lett.* **84**, 822.

Leonhardt, U., and Raymer, M. G. (1996), *Phys. Rev. Lett.* **76**, 1985.

Leonhardt, U., and Schneider, S. (1997), *Phys. Rev. A* **56**, 2549.

Leonhardt, U., and Tyc, T. (2009), *Science* **323**, 110.

Leonhardt, U., Böhmer, B., and Paul, H. (1995a), *Opt. Commun.* **119**, 296.

Leonhardt, U., Vaccaro, J. A., Böhmer, B., and Paul, H. (1995b), *Phys. Rev. A* **51**, 84.

Leonhardt, U., Munroe, M., Kiss, T., Richter, T., and Raymer, M. G. (1996), *Opt. Commun.* **127**, 144.

Lieb, E. H. (1990), *J. Math. Phys.* **31**, 594.

Lifshitz, E. M. (1955), *Zh. Eksp. Teor. Fiz.* **29**, 94 [(1956), *Soviet Physics JETP* **2**, 73].

Lindblad, G. (1976), *Commun. Math. Phys.* **48**, 119.

Lo, H. K., Popescu, S., and Spiller, T. (2000), *Introduction to Quantum Computation and Information* (World Scientific, Singapore).

Loudon, R. (2000), *The Quantum Theory of Light* (Clarendon Press, Oxford).

Loudon, R., and Knight, P. L. (1987), *J. Mod. Opt.* **34**, 709.

Louisell, W. H. (1973), *Quantum Statistical Properties of Radiation* (Wiley, New York).

Luis, A. (2002), *Phys. Rev. A* **66**, 013806.

Luis, A.. and Korolkova, N. V. (2006), *Phys. Rev. A* **74**, 043817.

Luks, A., and Peřinova, V. (2002), *Prog. Opt.* **43**, 295

Lütkenhaus, N., and Barnett, S. M. (1995), *Phys. Rev. A* **51**, 3340.

Lvovsky, A. I. (2004), *J. Opt. B* **6**, S556.

Lvovsky, A. I., and Raymer, M. G. (2009), *Rev. Mod. Phys.* **81**, 299.

Lvovsky, A. I., Hansen, H., Aichele, T., Benson, O., Mlynek, J., and Schiller, S. (2001), *Phys. Rev. Lett.* **87**, 050402.

Lynch, R. (1995), *Phys. Rep.* **256**, 367.

Mandel, L., and Wolf, E. (1995), *Optical Coherence and Quantum Optics* (Cambridge University Press, Cambridge).

Massar, S., and Spindel, P. (2006), *Phys. Rev. D* **74**, 085031.

Mattle, K., Michler, M., Weinfurter, H., Zeilinger, A., and Zukowski, M. (1995), *Appl. Phys. B* **60**, S111.

Mensky, M. B. (1993), *Continuous Quantum Measurements and Path Integrals* (IOP Publishing, Bristol).

Meystre, P., and Sargent, M. III (2007), *Elements of Quantum Optics* (Springer, Berlin).

Milonni, P. W. (1994), *The Quantum Vacuum* (Academic, London).

Milton, K. A. (2001), *The Casimir Effect* (World Scientific, Singapore).

Misner, C., Thorne, K. S., and Wheeler, J. A. (1999), *Gravitation* (Freeman, New York).

Moncrief, V. (1980), *Astrophys. J.* **235**, 1038.

Moyal, J. E. (1949), *Proc. Cambridge Philos. Soc.* **45**, 99.

Munday, J. N., and Capasso, F. (2007), *Phys. Rev. A* **75**, 060102.

Munday, J. N., Capasso, F., and Parsegian, V. A. (2009), *Nature* **457**, 170.

Natterer, F. (1986), *The Mathematics of Computerized Tomography* (Wiley, Chichester).

Neergaard Nielsen, J. S., Nielsen, B. M., Hettich, C., Molmer, K., and Polzik, E. S. (2006), *Phys. Rev. Lett.* **97**, 083604.

Negele, J. W., and Orland, H. (1998), *Quantum Many Particle Systems* (Westview Press, Boulder).

Nielsen, M. A., and Chuang, I. L. (2000), *Quantum Computation and Quantum Information* (Cambridge University Press, Cambridge).

Noh, J. W., Fougères, A., and Mandel, L. (1991), *Phys. Rev. Lett.* **67**, 1426.

Noh, J. W., Fougères, A., and Mandel, L. (1992a), *Phys. Rev. A* **45**, 424.

Noh, J. W., Fougères, A., and Mandel, L. (1992b), *Phys. Rev. A* **46**, 2840.

Noh, J. W., Fougères, A., and Mandel, L. (1993a), *Phys. Scr.* **T48**, 29.

Noh, J. W., Fougères, A., and Mandel, L. (1993b), *Phys. Rev. Lett.* **71**, 2579.

Nussenzveig, H. M. (1974), *Introduction to Quantum Optics* (Gordon and Breach, New York).

Nussenzveig, H. M. (1992), *Diffraction Effects in Semiclassical Scattering* (Cambridge University Press, Cambridge).

Oppenheimer, J. R., and Snyder, H. (1939), *Phys. Rev.* **56**, 455.

Orzag, M. (2007), *Quantum Optics: Including Noise Reduction, Trapped Ions, Quantum Trajectories, and Decoherence* (Springer, Berlin).

Ou, Z. Y., and Kimble, H. J. (1995), *Phys. Rev. A* **52**, 3126.

Ou, Z. Y., Pereira, S. F., Kimble, H. J., and Peng, K. C. (1992), *Phys. Rev. Lett.* **68**, 3663.

Ourjoumtsev, A., Tualle Brouri, R., Laurat, J., and Grangier, P. (2006), *Science* **312**, 83.

Parentani, R. (1995), *Nucl. Phys. B* **454**, 227.

Parentani, R. (2004), *preprint* arxiv:/astro-ph/0404022.

Paul, H. (1982), *Rev. Mod. Phys.* **54**, 1061.

Paul, H. (1995), *Photonen: eine Einführung in die Quantenoptik* (Teubner, Stuttgart).

Paul, H., and Jex, I. (2004), *Introduction to Quantum Optics: From Light Quanta to Quantum Teleportation* (Cambridge University Press, Cambridge).

Paul, H., Törmä, P., Kiss, T., and Jex, I. (1996), *Phys. Rev. Lett.* **76**, 2464.

Pauli, W. (1933), *Die allgemeinen Prinzipien der Wellenmechanik*, in *Handbuch der Physik*, ed. Geiger, H., and Scheel, K. (Springer, Berlin). English translation: Pauli, W. (1980), *General Principles of Quantum Mechanics* (Springer, Berlin).

Paye, J. (1992), *IEEE J. Quant. Electron.* **28**, 2262.

Peierls, R. (1979), *Surprises in Theoretical Physics* (Princeton University Press, Princeton, NJ).

Pendry, J. B. (2000), *Phys. Rev. Lett.* **85**, 3966.

Pendry, J. B., Schurig, D., and Smith, D. R. (2006), *Science* **312**, 1780.

Peng, J. S., and Li, G. X. (1998), *Introduction to Modern Quantum Optics* (World Scientific, Singapore).

Perelomov, A. M. (1986), *Generalized Coherent States and Their Applications* (Springer, Berlin).

Perez, A. (1995), *Quantum Theory: Concepts and Methods* (Kluwer, Dordrecht).

Peřina, J. (1991), *Quantum Statistics of Linear and Nonlinear Optical Phenomena* (Kluwer, Dordrecht).

Peřina, J., Hradil, Z., and Jurco, B. (1994), *Quantum Optics and Fundamentals of Physics* (Kluwer, Dordrecht).

Philbin, T. G., and Leonhardt, U. (2009), *New J. Phys.* **11**, 033035.

Philbin, T. G., Kuklewicz, C., Robertson, S., Hill, S., König, F., and Leonhardt, U. (2008a), *Science* **319**, 1367.

Philbin, T. G., Kuklewicz, C., Robertson, S., Hill, S., König, F., and Leonhardt, U. (2008b), *Supporting Online Material* of the paper above.

Pitaevskii, L. P. (2006), *Phys. Rev. A* **73**, 047801.

Pitaevskii, L. P., and Stringari, S. (2003), *Bose–Einstein Condensation* (Clarendon Press, Oxford).

Πλάτωνος Πολιτεία. English translation: Plato (1935), *Republic*, Book VII, p. 514, in *The Loeb Classical Library* **L276**, Vol. VI (Harvard University Press, Cambridge, MA).

Plenio, M., and Knight, P. L. (1998), *Rev. Mod. Phys,* **70**, 101.

Plenio, M., and Virmani, S. (2007), *Quant. Inf. Comp.* **7**, 1.

Polzik, E. S., Carry, J., and Kimble, H. J. (1992), *Phys. Rev. Lett.* **68**, 3020.

Popper, K. R. (1982), *Quantum Theory and the Schism in Physics* (Hutchinson, London).

Prudnikov, A. P., Brychkov, Yu.A., and Marichev, O. I. (1992), *Integrals and Series* (Gordon and Breach, New York).

Quan, P. M. (1957), *C. R. Acad. Sci.* (Paris) **242**, 465.

Quan, P. M. (1957/58), *Archive Rat. Mech. Analysis* **1**, 54.

Raabe, C., and Welsch, D.-G. (2005), *Phys. Rev. A* **71**, 013814.

Radon, J. (1917), *Berichte über die Verhandlungen der Königlich–Sächsischen Gesellschaft der Wissenschaften zu Leipzig, Mathematisch–Physische Klasse* **69**, 262.

Raymer, M. G., Cooper, J., Carmichael, H. J., Beck, M., and Smithey, D. T. (1995), *J. Opt. Soc. Am. B* **12**, 1801.

Reck, M., Zeilinger, A., Bernstein, H. J., and Bertani, P. (1994), *Phys. Rev. Lett.* **73**, 58.

Řeháček, J., Hradil, Z., Mogilevtsev, D., and Hradil, Z. (2008), *New J. Phys.* **10**, 043022.

Reid, M. D. (1988), *Phys. Rev. A* **40**, 913.

Reid, M. D., and Drummond, P. D. (1988), *Phys. Rev. Lett.* **60**, 2731.

Reznik, B. (2003), *Found. Phys.* **33**, 167.

Reznik, B., Retzker, A., and Silman, J. (2005), *Phys. Rev. A* **71**, 054301.

Richter, T., and Wünsche, A. (1996), *Acta Phys. Slov.* **46**, 487.

Risken, H. (1996), *The Fokker–Planck Equation: Methods of Solutions and Applications* (Springer, Berlin).

Robertson, H. P. (1929), *Phys. Rev.* **34**, 163.

Rodriguez, A., Ibanescu, M., Iannuzzi, D., Joannopoulos, J. D., and Johnson, S. G. (2007), *Phys. Rev. A* **76**, 032106.

Rousseaux, G., Mathis, C., Maissa, P., Philbin, T. G., and U. Leonhardt (2008), *New J. Phys.* **10**, 053015 (2008).

Rovelli, C. (1998), *Living Rev. Rel.* **1**, 1.

Sakharov, A. N. (1968), *Sov. Phys. Dok.* **10**, 1040.

Schaffer, S. (1979), *J. Hist. Astron.* **10**, 42.

Schleich, W. P. (2001), *Quantum Optics in Phase Space* (Wiley-VCH, Berlin).

Schleich, W., and Scully, M. O. (1984), *General Relativity and Modern Optics*, in *New Trends in Atomic Physics, Proceedings of the Les Houches Summer School, Session XXXVIII*, ed. Stora, R., and Grynberg, G. (North-Holland, Amsterdam).

Schleich, W. P., Walther, H., and Wheeler, J. A. (1988), *Found. Phys.* **18**, 953.

Schleich, W. P., Pernigo, M., and Fam Le Kien (1991), *Phys. Rev. A* **44**, 2172.

Schleich, W. P., Bandilla, A., and Paul, H. (1992), *Phys. Rev. A* **45**, 6652.

Schrödinger, E. (1926), *Naturwissenschaften* **14**, 664.

Schrödinger, E. (1935a), *Proc. Cam. Phil. Soc.* **31**, 555.

Schrödinger, E. (1935b), *Naturwissenschaften* **23**, 807.

Schurig, D., Mock, J. J., Justice, B. J., Cummer, S. A., Pendry, J. B., Starr, A. F., and Smith, D. R. (2006), *Science* **314**, 977.

Schwarzschild, K. (1916), *Sitzber. Deut. Akad. Wiss. Berlin, Kl. Math.-Phys. Tech.*, 189.

Schwinger, J. (1952), *U. S. Atomic Energy Commission Report No. NYO 3071* (U. S. GPO, Washington, DC); reprinted in Biederharn, L. C., and van Dam, H. (1965), *Quantum Theory of Angular Momenta* (Academic, New York).

Scully, M. O., and Zubairy, M. S. (1997), *Quantum Optics* (Cambridge University Press, Cambridge).

Shapiro, J. H., and Wagner, S. S. (1984), *IEEE J. Quantum Electron.* **QE-20**, 803.

Shapiro, J. H., Yuen, H. P., and Machado Mata, J. A. (1979), *IEEE Trans. Inf. Theory* **IT-25**, 179.

Shen, Y. R. (1984), *The Principles of Nonlinear Optics* (Wiley, New York).

Shih, Y. (2009), *An Introduction to Quantum Optics: Photon and Bi-Photon Physics* (Taylor and Francis, London).

Silberhorn, C. (2007), *Contemporary Phys.* **48**, 143.

Smithey, D. T., Beck, M., Raymer, M. G., and Faridani, A. (1993), *Phys. Rev. Lett.* **70**, 1244.

Stenholm, S. (1992), *Ann. Phys.* (New York) **218**, 233.

Strang, G. (2005), *Linear Algebra and Its Applications* (Brooks Cole, Boston, MA).

Sudarshan, E. C. G. (1963), *Phys. Rev. Lett.* **10**, 277.

Takei, N., Lee, N., Moriyama, D., Neergaard Nielsen, J. S., and Furusawa, A. (2006), *Phys. Rev. A* **74**, 060101.

Tatarskii, V. I. (1983), *Sov. Phys. Usp.* **26**, 311.

Tijms, H. (2007), *Understanding Probability* (Cambridge University Press, Cambridge).

Törmä, P. (1998), *Phys. Rev. Lett.* **81**, 2185.

Törmä, P., and Jex, I. (1999), *J. Opt. B* **1**, 8.

Törmä, P., Jex, I., and Stenholm, S. (1995), *Phys. Rev. A* **52**, 4853.

Törmä, P., Jex, I., and Schleich, W. P. (2002), *Phys. Rev. A* **65**, 052110.

Tsvelik, A. M. (2007), *Quantum Field Theory in Condensed Matter Physics* (Cambridge University Press, Cambridge).

Tyc, T., and Sanders, B. (2004), *J. Phys. A: Math. Gen.* **37**, 7341.

Unruh, W. G. (1976), *Phys. Rev. D* **14**, 870.

Unruh, W. G. (1981), *Phys. Rev. Lett.* **46**, 1351.

Unruh, W. G. (2008), *Phil. Trans. Roy. Soc. A* **366**, 2905.

Unruh, W. G., and Schützhold, R. (2007), *Quantum Analogues: From Phase Transitions to Black Holes and Cosmology* (Springer, Berlin).

Varcoe, B. T. H., Brattke, S., Weidinger, M., and Walther, H. (2000), *Nature* **403**, 743.

Vedral, V. (2005), *Modern Foundations of Quantum Optics* (World Scientific, Singapore).

Veselago, V. G. (1968), *Sov. Phys. Usp.* **10**, 509.

Visser, M. (1998), *Class. Quant. Grav.* **15**, 1767.

Vogel, W., and Grabow, J. (1993), *Phys. Rev. A* **47**, 4227.

Vogel, K., and Risken, H. (1989), *Phys. Rev. A* **40**, 2847.

Vogel, K., and Schleich, W. P. (1992), More on interference in phase space, in *Lectures delivered at Les Houches, Session LIII, Systèmes Fondamentaux en Optique Quantique* (Elsevier, Amsterdam).

Vogel, W., and Welsch, D. G. (2006), *Quantum Optics: An Introduction* (Akademie Verlag, Berlin).

Volovik, G. E. (2003), *The Universe in a Helium Droplet* (Clarendon Press, Oxford).

Vourdas, A. (2004), *Rep. Prog. Phys.* **67**, 267.

Walker, N. G. (1987), *J. Mod. Opt.* **34**, 15.

Walker, N. G., and Carroll, J. E. (1984), *Electron. Lett.* **20**, 981.

Walls, D. F., and Milburn, G. J. (2008), *Quantum Optics* (Springer, Berlin).

Weinberg, S. (2000a), *The Quantum Theory of Fields: Foundations* (Cambridge University Press, Cambridge).

Weinberg, S. (2000b), *The Quantum Theory of Fields: Modern Applications* (Cambridge University Press, Cambridge).

Weinberg, S. (2000c), *The Quantum Theory of Fields: Supersymmetry* (Cambridge University Press, Cambridge).

Weyl, H. (1950), *The Theory of Groups and Quantum Mechanics* (Dover, New York).

Wigner, E. P. (1932), *Phys. Rev.* **40**, 749.

Wigner, E. P. (1960), *Comm. Pure Appl. Math.* **13**, 1.

White, R. W. (1973), *J. Acoust. Soc. Am.* **53**, 1700.

Wódkiewicz, K. (1984a), *Phys. Rev. Lett.* **52**, 1064.

Wódkiewicz, K. (1984b), *Phys. Lett. A* **115**, 304.

Wootters, W. K. (1987), *Ann. Phys. (New York)* **176**, 1.

Wootters, W. K., and Zurek, W. H. (1982), *Nature* **299**, 802.

Wünsche, A. (1996), *Quantum Semiclass. Opt.* **8**, 343.

Yamomoto, Y., and Imamoglu, A. (1999), *Mesoscopic Quantum Optics* (Wiley, New York).

Yuen, H. P., and Chan, V. W. S. (1983), *Opt. Lett.* **8**, 177.

Yuen, H. P., and Shapiro, J. H. (1978a), Quantum statistics of homodyne and heterodyne detection, in *Coherence and Quantum Optics IV*, ed. Mandel, L., and Wolf, E. (Plenum, New York).

Yuen, H. P., and Shapiro, J. H. (1978b), *IEEE Trans. Inf. Theory* **IT-24**, 657.

Yuen, H. P., and Shapiro, J. H. (1980), *IEEE Trans. Inf. Theory* **IT-26**, 78.

Zavatta, A., Viciani, S., and Bellini, M. (2004), *Science* **306**, 660.

Zou, X. Y., Wang, L. J., and Mandel, L. (1991), *Phys. Rev. Lett* **67**, 318.

Index

Printed in the United States
By Bookmasters